Plants of Commercial Values

Plants of Commercial Values

Editor

Bikarma Singh

(B.Sc. Honours, M.Sc. Gold Medalist, Ph.D., NABET)
CSIR-Indian Institute of Integrative Medicine
(Council of Scientific & Industrial Research)
Ministry of Science and Technology, Govt. of India
Canal Road, Jammu-180001, Jammu & Kashmir State

CRC Press
Taylor & Francis Group
Boca Raton London New York

CRC Press is an imprint of the
Taylor & Francis Group, an **informa** business

NEW INDIA PUBLISHING AGENCY
New Delhi – 110 034

CRC Press
Taylor & Francis Group
6000 Broken Sound Parkway NW, Suite 300
Boca Raton, FL 33487-2742

First issued in paperback 2023

© 2020 New India Publishing Agency
CRC Press is an imprint of Taylor & Francis Group, an Informa business

No claim to original U.S. Government works

International Standard Book Number-13: 978-1-03-265446-1 (pbk)
International Standard Book Number-13: 978-0-3678-1936-1 (hbk)

Print edition not for sale in South Asia (India, Sri Lanka, Nepal, Bangladesh, Pakistan or Bhutan)

Publisher's Note
The publisher has gone to great lengths to ensure the quality of this reprint but points out that some imperfections in the original copies may be apparent.

Library of Congress Cataloging-in-Publication Data
A catalog record has been requested

Visit the Taylor & Francis Web site at
http://www.taylorandfrancis.com

and the CRC Press Web site at
http://www.crcpress.com

This book entitled 'Plants of Commercial Values' is devoted to Fifty nine topmost Authors of Fourteen different research organizations, and family members of Editor (Jammu Singh, Sam Devi, Manish Singh, Bishander Singh, Aryan Singh, Aditi Singh, Simla Singh and Ankush Singh) who allowed untimely to carry out research work late night in office to complete the assigned task and research.

Preface

Plant systems are the primary source of food and medicine directly contribute to the development of new drugs in addition to agrochemicals, fragrance, flavour, cosmetics, fine chemicals and nutraceuticals. Earth is a repository of 2,97,326 plant species distributed on every continent except Antarctica, and of these 21,000 species listed by World Health Organization (WHO) is medicinal plants. It is of record that only one-third of medicinal plants discovered across the globe are studied for their chemical composition and medicinal application for product development. According to WHO 25% prescribed drugs considered essential for human survival and animal health care is obtained from plants.

As we enter the new decades of twenty-first century, research on plants and their application for human health care is once again assuming a prominent position. This book deals with the commercial valuable plant species and ongoing R&D to highlight their importance in current scenerio. To understand the biodiversity across the earth requires understanding the importance of usages and environmental role of plants. It is the objective of the present book to provide this understanding to industries and pharma companies for discovery of new drugs and medicines. Renewed emphasis on developing medicinal and aromatic products from native plants has encouraged new botanical endeavors. Efforts to feed the growing populations in the most developing countries have positioned biological scientists at the cutting edge of genetic engineering with the creation of transgenic plants and chemical researchers for development of botanical medicine by using herbs for therapeutic values. Major pharmaceutical companies are currently conducting extensive research on plant materials collected from high altitude forests of himalayan regions and other geographic locations for their potential medicinal, aromatic and nutraceutical values. Substances obtained from the plants and microbes remain the basis for a large proportion of the commercial medications used today for the treatment of chronic illness such as heart disease, high blood pressure, pain, asthma and other associated human problems.

As plants and fermented microbes are rich source of novel drugs that form the ingredients in traditional system of medicine, and approximately 90% raw botanicals used in the manufacture of Ayurveda, Amchi (Tibetan medicine),

Homoeopathy, Siddha and Unani systems of medicine is largely prepared by using plants from wild and captive cultivated source from agriculture at farms and gardens. The usages of plants for human food and medicine is an age-old tradition of civilization, and there is a strong well proved belief that plants keep the mind in tune with nature and maintains proper balance. The use and the search for medicine and nutrients supplements derived from plants have accelerated the discovery in recent years, and well established example is nobel prize winner, Professor Youyou Tu in 2015 for key contribution to the discovery of artimisinin from a plant called *Artemisia annua* for treating malaria. Discovery of artimisinin has saved millions of lives and represent one of the significant contribution of China to global health.

Botanists, natural product chemists, pharmacologists and microbiologists were from plants combing their research investigation for phytochemicals that could be developed for treatment of chronic illness of human kind. Interest in various natural products obtained from plants for diverse useful is attributed to their different bioactivities, low toxicity and environmental sustainability. Today's, folklore herbals, drugs, food supplements, nutraceuticals, pharmaceutical intermediates, bioactive natural products and lead compounds from synthetic drugs are of high demand. Biotech plants have emerged amazingly fast as a boon for science and society. Genetic engineerings are playing a significant role in modern agriculture, pharmaceutical and environmental sectors, to meet the increasing demands of food, fuels, fibers, perfumes, cosmetics, minerals, vitamins, antibiotics, narcotics and other health-related drugs and fine chemicals. It is mentioned in literatures that the people of earlier civilization distinguishes plants suitable for nutritional purpose from others with a definitive pharmacological action observed by applying plants in their daily use. This relationship has grown between plants and humans, and many new plants have come to be used as drugs and nutraceutical.

Himalayas, Indo-Myanmar (former Indo-Burma) and Western Ghats of India is a repository of unique medicinal and aromatic plant species. Interactions between the advance science and understading of natural ecosystem helped lots in maintaining the richness of species and genetic materials for discovery and for sustainance to man-kind. Different human societies use plants according to their beliefs, knowledge, and earlier experiences gathered from their ancestor. Their knowledge about the usages of the plants is not known to scientists, unless and until, such records dessiminated through publications or any other permanent records. These hidden sacreds need to be explored in future for planning and new discoveries in science. Biotechnological intervention, molecular investigation and new science analysis on the usefulness of plants in recent decades have resulted in portrayals of relationship that have impacted understanding and interpretations of origin and diversification of commercial valuable plants. Therefore, it has been challenging issues to keep track of all new developments especially that deals with value addition and product formulations. It has been relevance for the scientists and the researchers for asking and being able to answer the significant scientific questions related to discovery and future strategies on commercial values and societal benefits. Content of this book attempts to fill this need of the hours.

I, on behalf of all authors, confident that this particular book will be a useful tool for academia and industries. The present issue, therefore, builds upon the excellent research articles given by fifty nine scientists of fourteen topmost research organizations and institutes of India. This book, therefore, is a coherent statement of the current status and title '**Plants of Commercial Values**', and goes beyond the papers presented here in different chapters for globalization. This book starts with first chapter on medicinal-cum-nutraceutical mushroom, *Morchella esculenta*, which is one of the richest sources of proteins, fibers, vitamins, amino acids and calories, because of its nutrient composition, mycologist called it as the costliest and the superior mushroom. This chapter is followed by medicinal usages of *Woodfordia fructicosa* in curing peptic ulcer of human-being and *Tinospora cordifolia* as multifaceted elixer plant in Arurvedic system of medicine. Modelling and conservation status of one of the rare plant of Himalaya, *Magnolia campbellii* is discussed in the fourth chapter of this book. *Bunium persicum* is yet an important plant of temperate regions and its seeds used as carminative substances in various food recipes, is presented in chapter fifth of this publication. It was then followed by *Dysoxylum binectariferum*, one of the threatened and important lead molecule plants of Northeastern India and Western Ghats of South India, and, biochemical analysis of hemiparasitic taxa at mitigator of pollutants in chapter six and chapter eight, respectively. Recently a tissue culture raised variety of banana was first time introduced by CSIR-IIIM Jammu for commercial cultivation with aim to double the income of the farmers and chapter nine discussed the importance of tissue culture banana as commercial scale in this book. Chapter nine discussed about the commercial value of the husk derived from seeds of *Plantago ovata*, an important medicine herb, used as emollient and laxative in the treatment of dysentery and diarrhoea, and dried seeds contain over 30% mucilage.

Nobel prized plant *Artemisia annua* is known for the content of anti-malarial drug called as artemisinin is presented in chapter ten. This plant species is widely distributed in the subtropical and temperate zones worldwide and also known for its usages in traditional system of medicine for the treatment of various ailments associated with mankind and animal health care. Threatened subsistence of *Spinifex littoreus,* value of plant growth promoting bacteria, systematics of starch grains, techniques for improvement of commercially valuable plants, sustainability of *Mucuna pruriens* and phytochemical screening of *Eulaliopsis binata* are also explained by different authors in chapter eleven to sixteen of this book. Rhododendrons are one of the most important multiferous flowering plants of Himalayas, and commonly known as Burans. Flowers and leaves of *Rhododendron* exhibits many nutritional, medicinal, and aromatic properties and has a number of uses in folklore medicine. Chapter seventeen presented in this book provide information on the checklist of Indian rhododendrons, their traditional usages, phytochemistry and potentials for value addition in near future. Chapter eighteen and nineteen deals with *Cyperus pangorei* and biosynthesis of nanoparticles using leaf extacts of three medicinal plants (*Spondias mombin, Stachytarpheta jamaicensis, Syzygium samarangensis*) as commercial plants for rural prosperity and pollution indicator. An important technique required for biotechnological intervention is plant tissue culture techniques, and is presented in chapter twenty with special reference to commercial crop improvement. Aromatic plants are valued for their aromas, tastes and their applications in treatment of various illness, and mostly prefered in cosmatics, perfumes, confectionery foods and medicines. Chapter twenty-one deals with aromatic wealth of Himalaya, potential of value addition and product development from essential oils. Before, all these twenty one chapters, this book starts with introduction section where importance of plants, their value addition, and future perspectives are discussed as editor choice.

We are sure that this book will serve as a stimulus for continued research in biological sciences, chemical sciences, pharmacological and clinical studies on plants and microbes which will add and contribute to value addition in the form of medicine discovery and products development. In addition to research suggestions contained within each of the chapters, an introduction section emphasizes particular research avenues for attention. One of the most challenging is how to deal effectively with endangered and commercial viable plants presented in this book such as Morels, Giloy, *Woodfordia, Panax, Bunium, Magnolia, Plantago, Cyperus, Dysoxylum, Mucuna, Rhododendron*, high yielding variety Banana, *Artemisia, Spondius, Stachytarpheta, Syzygium* and others. Detail studies presented in different chapters on biology, chemistry, pharmacolgy and commercial aspects will help more precisely test biogeographic theory and chemical hypothesis. Examining chemical constituents of particular species will also be more feasible by studying this book.

We have tried to convey a maximum of knowledge through this book regarding potential plants for medicine and product values in a minimum of words, and believe that there is always scope for improvement. Readers are the best panel of judges to evaluate the content of this particular book. I, on behalf of all team members, hoping and believe that the readers have a moral obligation to convey suggestions on this book entitled "**Plants of Commercial Values**' in near future for improvement. It would be of greatest pleasure for me if this book could attract students of botany, ecology, chemistry, pharmacology, zoology and strategies planners like forest departments, tourists and industries, who have something in their mind in relation to drug discovery, value addition and product development. This book will have a way of providing a new level of future perspectives in understanding different areas of sciences. Research should be continually encouraged to successfully achieve such objectives so that we can take maximum advantage of what could be offered in a sustainable way that will be beneficial in and of itself. Without contributions from fifty nine authorities, it would have been hard for me to imagine a more thorough explanation on the content and prosperity of this renouned publication.

Bikarma Singh
Scientist & Editor
CSIR-Indian Institute of Integrative Medicine Jammu
(Council of Scientific & Industrial Research)
Ministry of Science and Technology
Govt. of India, Canal Road, Jammu-180001, J&K

Acknowledgements

Book entitled '**Plants of Commercial Values**' must inevitably be deeply indebted to authors whose precious scientific contributions are presented in the form of high quality research in twenty one different chapters. Credit of this book goes to fifty nine scientists, professors and research scholars of India's topmost fourteen scientific organizations and institutes such as CSIR-Indian Institute of Integrative Medicine (IIIM Jammu, J&K), CSIR-Central Drug Research Institute (CDRI Lucknow, UP), CSIR-Central Institute of Medicinal and Aromatic Plants (CIMAP Lucknow, UP), CSIR-National Botanical Research Institute (NBRI Lucknow, UP), CSIR-North East Institute of Science and Technology (Jorhat, Assam), Midnapore City College (Paschim Medinipure, West Bengal), North-Eastern Hill University (Shillong, Meghalaya), University of Jammu (Jammu, J&K), Veer Kunwar Singh University (Ara, Bihar), Vidyasagar University (Midnapore, West Bengal), Lady Brabourne College (Kolkata, West Bengal), Himachal Pradesh University (Shimla, Himachal Pradesh), University of Delhi (Delhi), and Jawaharlal Nehru University (New Delhi). All these institutes/organizations are well known for their quality research in the field of drug discovery, value addition and product development. These organizations are also responsible for grooming young minds and encouraging new generation people for pursuing career in sciences and allied areas.

Thanks goes to renowned research workers of twenty four various divisions and sections such as plant sciences, cGMP, clinical microbial division, department of biological sciences, department of botany, department of molecular bioprospection, ecological engineering laboratory, genetic resource and agrotechnology division, instrumentation section, biological sciences and technology division, medicinal chemistry division, natural product chemistry division, pharmacology division, plant biotechnology department, plant breeding department, biodiversity and applied botany section, value addition centre, department of zoology, PKPD toxicology and formulation division, school of life sciences, soil science department, sophisticated analytical instrument facility (SAIF) division, department of biosciences, plant taxonomy and biosystematics, and molecular taxonomy laboratory, whose top-class research articles included in this book for readers and strategic planners.

Presented invaluable contributions comes from various themes on which research articles were called for includes scope of medicinal plants, commercial applications, product development and value addition, techniques and trends leads to improvement in plant functioning, roles of microbes, ecological niche modelling, societal benefits, elicitators and plant tissue cultures. I am sure, this publication will greatly help different researchers in the task of new research activities and will cover different thematic areas in which various research organizations are currently working and their potential value in new era of society and science.

My heartiest thanks and gratitude goes to worthy Director IIIM Jammu, Dr Ram A. Vishwakarma, and Director NBRI Lucknow, Professor Saroj Kanta Barik, as they continuously encouraged me for my good works and provide guidance. I owe a special debt of gratitude to fifty eight authors for giving their best research articles for this publication. I thank Er. Rajneesh Anand, Head Chatha and Deputy Director, IIIM Jammu, and my Head of Department and my other Departmental Collegues who continuously encouraged me for my hardwork and supported me whenever required. Heartliest thanks goes to my Ph.D. Supervisors, Dr. S.K. Borthakur, Department of Botany, Gauhati University, and Dr. (Mrs) S.J. Phukan, Botanical Survey of India, for making me as a true researcher, hardworker and mould my career. I also extend thank to Dr Sougata Sarkar, who encouraged me to write book chapters. I also thank my staff members, Kiran Koul, Sudhir Nanda, Rajesh, Sumit and Gara associated with me, and my lovable family members who over the years have always assisted me with unfailing courtesy. I also dedicate this book to my wife (Manisha), Son (Aryan) and daughter (Aditi) for always making me smile. Thanks to Kamlesh Singh, IIIM, IT Section for help in formating and page setting of the book.

Finally, I wish to thank all researchers, government officers, collectors and farmers who had helped directly or indirectly in this publication. We give abundant thanks to our synantherological colleagues for making this book possible.

I, on behalf of all authors, would like to extend my gratitude to NIPA, New Delhi Publisher who agreed to publish these research outcomes and timely brought for public dessimination. I also appreciate all the excellent help that I received during the production of this book from Mr. Sumit Jain, Mr. Raj Kumar and Miss Rita Verma. Thanks for their valuable time in correction of few grammer mistakes and advised on formate designs for different chapters and parts of this book.

I am sure this book will attract academicians, industrialist and pharmaceutical companies for their research, product development and value addition from commercial valuable plants of Indian Himalayas and elsewhere across the globe. Those who know about the importance of plants will be able to appreciate all authors and value relates to how much these research papers have contributed to growth of Indian economy and other countries for future development.

Bikarma Singh
Scientist & Editor
CSIR-Indian Institute of Integrative Medicine Jammu
(Council of Scientific & Industrial Research)
Ministry of Science of Technology
Govt. of India, Canal Road, Jammu-180001

Contents

Editor, Authors and Addresses
Number in parentheses indicate chapter number

Plants of Commercial Values

Editor

Dr. Bikarma Singh (B.Sc. Honours, M.Sc. Gold Medalist, Ph.D., NABET)
Scientist, Plant Sciences (Biodiversity and Applied Botany Division), Section Incharge-Value Addition Centre, CSIR-Indian Institute of Integrative Medicine, Jammu 180001, Jammu & Kashmir, INDIA. Assistant Professor Academy of Scientific and Innovative Research Anusandhan Bhawan, New Delhi -110001 INDIA

Authors and Addresses

Bikarma Singh *(1,2 ,8,17,21)*
Plant Sciences (Biodiversity and Applied Botany Division & Value Addition Centre) CSIR-Indian Institute of Integrative Medicine, Jammu 180001, Jammu & Kashmir, INDIA Academy of Scientific and Innovative Research, Anusandhan Bhawan, New Delhi -110001 INDIA; drbikarma@iiim.ac.in, drbikarma@iiim.res.in

Ram A. Vishwakarma *(6)*
Medicinal Chemistry Division, CSIR - Indian Institute of Integrative Medicine, Canal Road Jammu - 180001, Jammu & Kashmir, INDIA; ram@iiim.ac.in

Saroj Kanta Barik *(4)*
CSIR-National Botanical Research Institute, Lucknow 226001, Uttar Pradesh, INDIA sarojkbarik@gmail.com

Inshad Ali Khan *(2)*
Clinical Microbiology Division, CSIR-Indian Institute of Integrative Medicine, Jammu 180001 Jammu & Kashmir, INDIA; iakhan@iiim.ac.in

Yash Pal Sharma *(1)*
Department of Botany, University of Jammu, Jammu-180006, Jammu & Kashmir, INDIA yashdbm3@yahoo.co.in

Brijesh Kumar *(1)*
Sophisticated Analytical Instrument Facility Division, CSIR-Central Drug Research Institute Lucknow 226001, India; brijesh_kumar@cdri.res.in

Bishander Singh *(1)*
Department of Botany, Veer Kunwar Singh University, Ara 802301, Bihar

TN Lakhanpal *(1)*
Former Professor, Department of Biosciences, Himachal Pradesh University, Simla 171005 Himachal Pradesh

Anil Katare *(2)*
cGMP Division, CSIR_Indian Institute of Integrative Medicine, Jammu 180001, Jammu & Kashmir, INDIA; akkatare@iiim.ac.in

Durga Prasad Mindala *(2)*
Clinical Microbiology Division, CSIR-Indian Institute of Integrative Medicine, Jammu 180001 Jammu & Kashmir, INDIA

Prasoon Kumar Gupta *(2)*
Natural Product Chemistry Division, CSIR-Indian Institute of Integrative Medicine, Jammu 180001, Jammu & Kashmir, INDIA; guptap@iiim.ac.in

Surjeet Singh *(2)*
Pharmacology Division, CSIR-Indian Institute of Integrative Medicine, Jammu 180001, Jammu & Kashmir, INDIA

Neha Sharma *(3)*
Natural Product Chemistry Division, CSIR-Indian Institute of Integrative Medicine, Jammu 180001, Jammu & Kashmir, INDIA; nehasharma.ns27@gmail.com

Naresh Kumar Satti *(3)*
Natural Product Chemistry Division CSIR-Indian Institute of Integrative Medicine, Jammu 180001, Jammu & Kashmir, INDIA

Dibyendu Adhikari *(4)*
Department of Botany, North-Eastern Hill University, Shillong-793022, East Khasi Hills Meghalaya, INDIA; dibyenduadhikari@gmail.com

Prem Prakash Singh *(4)*
Department of Botany, North-Eastern Hill University, Shillong-793022, East Khasi Hills Meghalaya, INDIA

Raghuvar Tiwary *(4)*
Department of Botany, North-Eastern Hill University, Shillong-793022, East Khasi Hills Meghalaya, INDIA

Sajan Thakur *(5)*
Ecological Engineering Laboratory, Department of Botany, University of Jammu, Jammu-180006 Jammu & Kashmir, INDIA

Harish Chander Dutt *(5)*
Ecological Engineering Laboratory, Department of Botany, University of Jammu, Jammu-180006 Jammu & Kashmir, INDIA; hcdutt@rediffmail.com

Vikas Kumar *(6)*
Preformulation Laboratory, PKPD Toxicology & Formulation Division, CSIR-Indian Institute of Integrative Medicine, Canal Road, Jammu - 180001, Jammu & Kashmir, INDIA

Sonali S. Bharate *(6)*
Preformulation Laboratory, PKPD, Toxicology & Formulation Division, CSIR - Indian Institute of Integrative Medicine, Canal Road, Jammu - 180001, Jammu & Kashmir, INDIA

Sandip B. Bharate *(6)*
Medicinal Chemistry Division, CSIR - Indian Institute of Integrative Medicine, Canal Road Jammu - 180001, Jammu & Kashmir, INDIA; sbharate@iiim.ac.in

Shibdas Maity *(7)*
Plant Taxonomy, Biosystematics and Molecular Taxonomy Laboratory, Department of Botany and Forestry, Vidyasagar University, Midnapore-721102, West Bengal, INDIA

Souradut Ray *(7)*
Plant Taxonomy, Biosystematics and Molecular Taxonomy Laboratory, Department of Botany and Forestry, Vidyasagar University, Midnapore-721102, West Bengal, INDIA

Amal Kumar Mondal *(7,11,13,16,18,19)*
Plant Taxonomy, Biosystematics and Molecular Taxonomy Laboratory, Department of Botany and Forestry, Vidyasagar University, Midnapore-721102, West Bengal, INDIA amalcaebotvu@gmail.com

Sabha Jeet *(8)*
Genetic Resources and Agrotechnology Division, CSIR-Indian Institute of Integrative Medicine Jammu 180001, Jammu & Kashmir, INDIA; sabhajeet@iiim.ac.in

VP Rahul *(8)*
Genetic Resources and Agrotechnology Division, CSIR-Indian Institute of Integrative Medicine Jammu 180001, Jammu & Kashmir, INDIA

Rajendra Bhanwaria *(8)*
Genetic Resources and Agrotechnology Division, CSIR-Indian Institute of Integrative Medicine Jammu 180001, Jammu & Kashmir, INDIA

Chandra Pal Singh *(8)*
Chatha Centre, Genetic Resources and Agrotechnology Division, CSIR-Indian Institute of Integrative Medicine, Jammu 180001, Jammu & Kashmir, INDIA

Rajendra Gochar *(8)*
Chatha Centre, Genetic Resources and Agrotechnology Division, CSIR-Indian Institute of Integrative Medicine, Jammu 180001, Jammu & Kashmir, INDIA

Amit Kumar *(8)*
Chatha Centre, Genetic Resources and Agrotechnology Division, CSIR-Indian Institute of Integrative Medicine, Jammu 180001, Jammu & Kashmir, INDIA

Kaushal Kumar *(8)*
Chatha Centre, Genetic Resources and Agrotechnology Division, CSIR-Indian Institute of Integrative Medicine, Jammu 180001, Jammu & Kashmir, INDIA

Jagannath Pal *(8)*
Chatha Centre, Genetic Resources and Agrotechnology Division, CSIR-Indian Institute of Integrative Medicine, Jammu 180001, Jammu & Kashmir, INDIA

Sougata Sarkar *(9)*
Genetic Resources and Agrotechnology Division, CSIR-Indian Institute of Integrative Medicine Jammu 180001, Jammu & Kashmir, INDIA; recalling.de.parted@gmail.com

Sana Khan *(9)*
Plant Biotechnology Department, Department, CSIR-Central Institute of Medicinal and Aromatic Plants (CIMAP), Near Kukrail Picnic Spot Road, Lucknow-226015,Uttar Pradesh INDIA

Janhvi Pandey *(9)*
Soil Science Department, CSIR-Central Institute of Medicinal and Aromatic Plants (CIMAP) Near Kukrail Picnic Spot Road, Lucknow-226015, Uttar Pradesh, INDIA

Raj Kishori Lal *(9)*
Plant Breeding Department, CSIR-Central Institute of Medicinal and Aromatic Plants (CIMAP) Near Kukrail Picnic Spot Road, Lucknow-226015,Uttar Pradesh, INDIA; rajkishorilal @gmail.com

Praveen Kumar Verma *(10)*
Medicinal Chemistry Division, CSIR-Indian Institute of Integrative Medicine, Canal Road Jammu - 180001, Jammu & Kashmir, INDIA; praveen.ihbt@gmail.com

Aliya Tabassum *(10)*
Medicinal Chemistry Division, CSIR-Indian Institute of Integrative Medicine, Canal Road Jammu - 180001, Jammu & Kashmir, INDIA

Sanghapal D. Sawant *(10)*
Medicinal Chemistry Division, CSIR-Indian Institute of Integrative Medicine, Canal Road Jammu - 180001, Jammu & Kashmir, INDIA; sdsawant@iiim.ac.in

Tamal Chakraborty *(11)*
Plant Taxonomy, Biosystematics and Molecular Taxonomy, UGC-DRS-SAP & DBT-BOOST-WB Funded Department, Department of Botany & Forestry, Vidyasagar University Midnapur- 721102, West Bengal, INDIA

Sanjukta Mondal Parui *(11)*
Post Graduate Department of Zoology, Lady Brabourne College, P1/2, Suhrawardy Avenue Kolkata – 700 017, West Bengal, INDIA

Prachi Sharma *(12)*
Department of Botany, University of Delhi, Delhi-110007, INDIA

Ratul Baishya *(12)*
Department of Botany, University of Delhi, Delhi-110007, INDIA; rbaishyadu@gmail.com

Shilpa Dinda *(13,16)*
Department of Biological Sciences, Midnapore City College, Kuturia, Bhadutata, Midnapore Paschim Medinipure-721129, West Bengal, INDIA

Ujjal Jyoti Phukan *(14)*
School of Life Sciences, Jawaharlal Nehru University, New Delhi-110067, INDIA ujjwal.phukan@gmail.com

Vigyasa Singh *(14)*
Department of Molecular Bioprospection, CSIR-CIMAP, Lucknow-26015, INDIA vigyasa105@gmail.com

Susheel Kumar Singh *(15)*
Biotechnology Division, CSIR-Central Institute of Medicinal and Aromatic Plants P.O. CIMAP Lucknow-226015, INDIA; susheelbt05@gmail.com

Sunita Singh Dhawan *(15)*
Biotechnology Division, CSIR-Central Institute of Medicinal and Aromatic Plants P.O. CIMAP Lucknow-226015, INDIA.

Debashree Ghosh *(16)*
Plant Taxonomy, Biosystematics and Molecular Taxonomy Laboratory, UGS-DRS-SAP Department of Botany and Forestry, Vidyasagar University, Midnapore-721102, West Bengal INDIA.

Souradut Ray (16)
Plant Taxonomy, Biosystematics and Molecular Taxonomy Laboratory, UGS-DRS-SAP Department of Botany and Forestry, Vidyasagar University, Midnapore-721102, West Bengal INDIA.

Sumit Singh *(17)*
Plant Sciences (Biodiversity and Applied Botany Division & Value Addition Centre), CSIR-Indian Institute of Integrative Medicine, Jammu 180001, Jammu & Kashmir, India; Academy of Scientific and Innovative Research, Anusandhan Bhawan, New Delhi, INDIA.

Anup Kumar Bhunia *(18)*
Plant Taxonomy, Biosystematics and Molecular Taxonomy Laboratory, UGC-DRS-SAP and DBT-BOOST-WB Funded Department, Department of Botany & Forestry, Vidyasagar University, Midnapore-721102, West Bengal, INDIA.

Sk. Md. Abu Imam Saadi *(19)*
Plant Taxonomy, Biosystematics and Molecular Taxonomy Laboratory, UGS-DRS-SAP and DBT-BOOST-WB Funded Department, Department of Botany and Forestry, Vidyasagar University, Midnapore-721102, West Bengal, INDIA

Gitasree Borah *(20)*
Medicinal Aromatic and Economic Plants Group, Biological Science and Technology Division CSIR-North East Institute of Science and Technology, Jorhat-785006, Assam, INDIA

Manabi Paw *(20)*
Medicinal Aromatic and Economic Plants Group, Biological Science and Technology Division
CSIR-North East Institute of Science and Technology, Jorhat-785006, Assam, INDIA

Mohan Lal *(20)*
Medicinal Aromatic and Economic Plants Group, Biological Science and Technology Division
CSIR-North East Institute of Science and Technology, Jorhat-785006, Assam, INDIA
drmohanlal80@gmail.com

Sneha *(21)*
CSIR-Central Institute of Medicinal and Aromatic Plant, P.O. CIMAP, Lucknow-226015, INDIA

Rajneesh Anand *(21)*
Instrumentation Division, CSIR-Indian Institute of Integrative Medicine, Jammu 180001
Jammu −180001, Jammu & Kashmir, INDIA; ranand@gmail.com

Introduction

Editor's Choice: Bikarma Singh

Plants are the basic requirement upon which all other living organisms depend. In biology, the species is a basic unit of classification to any taxonomic rank, and compared in all fields of biology, from to systematic to anatomy, ecology, genetics engineering, chemistry, pharmacology, microbiology, paleontology and physiology. Phenomenon of evolution suggest that species of plant originated from algae (Chlophyta) as they colonized and invaded the vacant land-niches, and therefore, evolution of this group offers unique challenges for evolutionary researchers. The process of succession leading to adaptations causes this group of organisms to migrate from water towards landmasses. Forests gets dominated by giant pteridophytes, and primitive group of gymnosperms such as conifers and cycads, and todays they are available on the earth as fossil fuels. In prehistoric time, these groups of plants were replaced by seed bearing vascular plants with passage of times, and there is evidence that these groups were reported to occur at 360 million years ago. Recent inventions of molecular and fossil evidence have complicated the evolutionary sequence that explains how plants get evolved from simple to complex forms. Vascular angiosperms alleged to have an advantage over gymnosperms, and now most of them dominated the earth. Flowering plants were the most recent groups to appear since flowers and leaves attract pollinator and promotes reproductive functions. They appeared in 130 million years ago, and now there is enormous diversity. Living fossil plants species *Gingko biloba* proves that the evolution is a plastic theory that can accommodate speedy changes and provide stability for millions of years, howevers, nowaday many conifers (*Pinus*, *Cupressus*), cycads and gnetoplaytes (*Gnetum*, *Ephedra Welwitschia*) dominanted the high altitude regions of Himalayas and elsewhere in the world.

Plants and their Importance

As plants are basis for survival of living organisms, usages of plants to humans and other life is unsteadly and life would not have been possible without this group of community. They provide us with all the food we eat as nutrition. Animals and others living microbes live and thrive on plants. Ecologically, plants sustain atmosphere by producing oxygen, and absorbing carbondioxide by a natural phenomenon called photosynthesis, and in this way plants maintain the oxygen content of the air for survival of living organisms on the earth. Without plants we would starve to death, and die of suffocation. There is a enough evidence that plants maintain ozone layer that protects life on earth from sun's ultra-violet radiation. Biogeochemical cycles responsible for the movement of water from soil to air by a process called transpiration. Nutrient, oxygen and calories that human and animals intake are based on two parameters, firstly, plants has the ability to store energy of the sun and convert them to carbohydrates, starches, essential oils, fats and other fine secondary metabolite constituents of the plant body, and, secondly, they have the ability to synthesize simple types of chemical compounds to complex necessary for the existence of life.

Plants community create ecosystem for living organisms, and there is a full proven record that earth is a repository of 297,326 plant species distributed across the globe except Iceland of North Atlantic and Southernmost continent Antarctica. Out of these 21,000 species listed by World Health Organization (WHO) are medicinal plants. Addition to this evidence, WHO estimated that 25% prescribed drugs considered essential for human survival and animal health care is obtained from plants. Still today 70% population of the world depends on wild plants for their daily uses. It is of the record that 100,000 natural compounds isolated from plants, and many of these have similar active molecules are yet to be explored from economic and commercial point of view. As we enter the new decades of twenty-first century, the research on plants and their application for human health care is once again assuming a prominent place.

Some of the most powerful and useful compounds that are discovered from plants, as for example, anti-malarial artimisinin from *Artemisia annua* and cannabidiol (CBD) are isolated and characterized from plant *Cannabis sativa* or local hemp for cancer treatment. Published evidence say the Nabiximols drug was approved by Canadian authorities in 2005 to alleviate pain associated with multiple sclerosis. Unfortunately, only one-third of medicinal plants discovered across the world have been studied for their chemical composition,

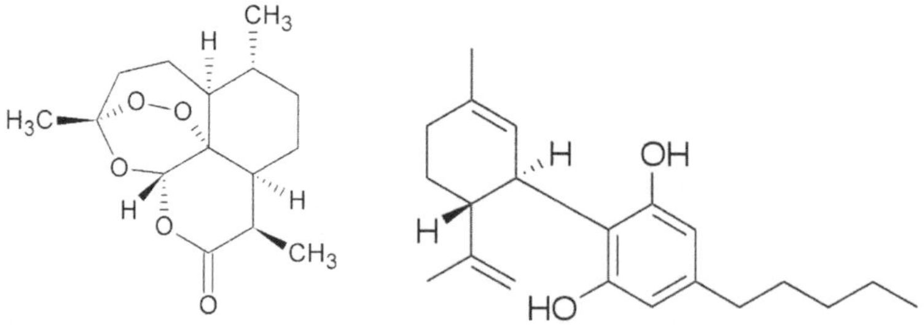

Fig. 1: Structure of life saving drug Artemisinin (left) and Cannabidiol (right) from plants

and still medicinal properties of several bioactive constituents for commercial product development and drug discovery are unknown.

Plants provide many products for human use, and there is a strong evidence that single plant would be responsible for giving food and shelter to thousands of living organisms in the form of microbes, worms, insects, reptiles, birds and mammals. Now folklore herbal nutraceuticals, pharmaceutical intermediates, bioactive natural products and isolated and characterized lead compounds from synthetic drugs are of high demand in market and all these are obtained directly or indirectly from plants and microbial communities.

Hotspot and Plant Diversity

British Biologist Norman Myers invented the term *Biodiversity Hotspot* in 1988 as a geographic region characterized by levels of species endemism. There are several factor causing serious levels of habitat degradation imposing threats to extinction of species and loss of geographic biodiversity. Such globally recognized regions have the maximum diversity and the most threatened reservoirs of plant and animal species. As far as India is concerned, four biodiversity hotspots regions, *viz.*, Himalayas (share with Bhutan and eastern Nepal), Indo-Myanmar (share with Myanmar, China and eastern Bangladesh), Sundaland (Nicobar island) and Western Ghats (including Sri Lanka) were globally placed rich regions of biodiversity. Indo-Myanmar hotspot includes parts of north-eastern India, and extends to over an area of about 2 million sq. km in tropical Asia. This region is represented by wide diversity of climates and vegetation habitat patterns. Published reports suggested that 1,300 birds, and 13,500 plant species abode to these regions, but now, this hotspot is deteriorating at faster rate, and many mammal, examples, primates langurs and gibbons are under threats and moving a step forward towards extinction.

Another Indian hotspot, the Himalayas includes western, central (Sikkim) and eastern Himalayas and their altitudinal zonation varies between 500 m to 8000 m above sea level which results in diversity of ecosystems that range from alluvial grasslands, subtropical broad-leaved forests, temperate forests, mixed conifers to alpine meadows above tree-line zone. This hotspot has 163 threatened species such as one-horned rhinoceros, and wild asian-water buffaloes. According to one study, 8,000 species of plants are flourishing in Himalayas, and out of these, one-third categorized as endemic and threatened. Himalayan quail, cheer pheasant, tragopan, vulture, golden langur, tahr, sambar, snow leopard, bear and blue sheep were categorized as endemic and threatened due to human interferance in virgin forests for their greedy needs.

Western Ghats mountaineous forests of southwestern India are called 'Sahyadri Hills'. Wide variation of rainfall coupled with geographical location produces a great variety of forests types and vegetation richness in this hotspot. It's a part of Malabar Botanical Province, and 5,800 species of flowering plants and 2,100 species reported are endemic. The state of Karnataka alone harbours 3,900 species, while Nilgiris Hills have 2,611 species of flowering plants. Climate and altitudinal gradient has resulted vegetation differences and includes dry scrub, moist deciduous, semi-evergreen, evergreen forests and sholas coupled with the high altitude grasslands.

The Sundland hotspot lies in South-east Asia and includes Nicobar-island of India, Thailand, Singapore, Indonesia Brunei and Malaysia. These belt of rich in marine ecosystems including Mangroves, seagrassbeds and coral reefs, however, in many places terrestrial ecosystem is also dominant. Dolphins, Whales, Crocodiles and lobsters are threatened marine biodiversity.

It can be concluded that Himalayas, Indo-Myanmar, Sundaland and Western Ghats represents India unique repository of medicinal and aromatic plant and animal species. There is urgent need of interaction between the advance science and understanding of natural ecosystem that would help in maintaining species richness and conservation of genetic materials for discovery and for survival of man-kind in coming years to come.

Application of Commercial Plants

Medicinal, aromatic and nutraceutical plants are a matter of special concern to people due to their therapeutic use. The reason being they posses unique and valuable properties due to presence of bioactive chemical constituents which helps in building tissue and growth of cells. Natural essential oils derived from plants are used as cosmetics, perfumes and flavors in confectionary foods and beverages. Since at the time immorial, India, China, Egypt and Iran societies are coming using essential oils in various ways for herbal medicine. Research contribution reveals the value on use of aromatic plants. Integrating essential oils and justifying their therapeutic applications, agricultural scientist and botanist have been motivated to expand their cultivation parameters and now it has created a huge competitive global market. It is estimated that 2,000 species of herbs distributed in 60 families, as for examples, apiaceae, lamiaceae and asteraceae as the dominant ones are cultivated for the best quality essential oils. In current scenerio, 3000 types of essential oils available for use in different product development, and out of these, 300 species are of commercial importance to international markets. Essential oils interface with microbial activities are causing destruction of germs without causing any adverse effects on human health and animal health care.

On the other hand, herbal botanicals are drug substance that could be used by humans for curing illness, and such botanicals has no or very less side effect. They have the ability to fight against a wide range of germs, monocytes and insects, and this is possible only due to their unique chemical compositions, which produces unique pharmacological activity with reference to some validated models. Hence, we can say that the economic importance of plants are many, but it need proper documentation and research for new discovery.

It can be concluded that plants has many uses in pharmaceutical and food industries, and it may be due to their biological, pharmacological and therapeutic potentials. Natural products, essential oils and fine chemicals derived from plants has diverse usage as drug compounds and can also be used as synthetic anti-biotics. Carcinogen of many compounds used as preservatives in a variety of health products, cosmetics and food to increase storage time, but again question of safety for consumers and animal health arises.

Value Addition from Plants

Herbal botanicals and synthesized products prepared from plants and plant parts considered as natural antidote for human ailments and disorders. Herbal business offers unlimited opportunity to add value for new era for development of nutraceuticals and medicines, and several value-added products are used in animal and human care. Value addition of plants can be done directly through minor processing or indirectly through maintaining quality standard of different formulations prepared from plants. *Lavender, Geranium, Lemongrass, Mints, Ocimums,* and other aroma bearing plants were regarded as the primary ingredient for many new and existing herbal products. In fact, lavender oil derived from an important aroma bearing plant, *Lavendula angustifolia*, a varietal crop released by CSIR-IIIM Jammu for high altitude farmers, is known for its herbal fragrance worldwide. Oils of this plant species is very popular in development of human used value-added products such as soaps, sachets and dried bouquets. Sale of lavender seedlings can boost profits to farmers and rural communities in earning their livelihood. Finally, it is very important to explore, manage and conserve available natural resources at national and international levels in a sustainable way and this will guarantee the safety and healthy life for the new generation to come.

Synopsis and Current Issues

Editor's Choice: Bikarma Singh

New value added products prepared from natural herbal botanicals are replacing the existing market products on a regular basis due to best quality improvement from safety point of view. Since human civilization, natural resources such as plants and microbes are the basic raw material for development of products. Wild plants provide food and medicine to people and prefered by many tribal communities over cultivated crops due to their chemical constituents rich in nutrients and free from adultration. Increasing global demand for chemicals and herbal products derived from medicinal and aromatic plants have opened up a new entrepreneurial opportunity to process these plant resources for value addition, thereby, generating employment avenues for people. Civilizations that has came up from different parts of the world made use of plant species growing in their areas, and extended its knowledge to different community. Historical evidence indicate that different species were used by different cultural people at different decades for treating the same illness or the others.

India's vascular flora comprised of 18,664 plant species, making this country as one of the world's tenth-richest country in the globe. Himalayan belts support approximately 50% India's plant species, and more than 4,000 species recorded from these regions are threatened and endemics. These regions are rich in indigenous knowledge on wild food and herbal medicine. Market triggered due to globalization, as well as pursuit of commercial herbal products that satisfy human wants is intrinsic to flavour and fragrance, pharmaceutical, nutraceuticals and confectionery food industries. The rate at which new plant products are appearing in market is growing at faster rate and the reasons for this progress are many. Therefore, we can truly say that the degree of newness is an indicator of the difference between the new product and the existing one.

Plants of medicinal and therapeutic values can be defined as plants whose parts such as roots, barks, stems, leaves, flowers, fruits and seeds or chemical substances derived from these parts are used in different systems of medicine such as allopathy, aamchi, ayurveda, siddha, unani, homeopathy and herbo-mineral globally known for their curative properties. In the context of globalization, all plants that have been alleged to have drug botanicals related to human health or which have been proven to be useful as drugs by western standards for their chemical constituents can be used in the formulation of drugs are nutraceutical products.

Indian medicinal and aromatic plants have tremendous demand in world market. The export data says that parts of medicinal plants, herbal extracts and natural compound isolated from plant has increased from INR 450 crores in 1999-2000 to INR 3600 crores in 2007-2008. The exports of essential oils have increased from INR 30 crores in 1980 to INR 250 crores in 2010. Many countries are importing value added essential oil from India for preparation of different flavour and fragrance products for human use. Hence, this type of world scenerio has placed India as one of the fast growing country in the world.

Ministry of Environment, Forests and Climate Change sought to understand the nature and the diversity of genetically engineered crops that may move to product commercialization. New product development and value addition to various commodities can be achieved through processing the raw samples to produce a commercial marketable products varies from the raw products. Aromatic oils, aroma chemicals, absolutes from flowers, oleoresins from spices, resins, resinoids, gums, water hydrosols and similar plant products are used in fragrance, flavour, aromatherapy and pharmaceutical industries.

With reference to contents of the book "Plants of Commercial Values", a breif chapter-wise synopsis with reference to key points and results are summarized below in different paragraphs.

Chapter 1 described by Singh et al. presented the importance of medicinal morel mushrooms and their nutrient compositions. According to them, morels are Himalayan endemic plants used in traditional system of medicines, and recorded to have importance in Chinese and Western Pharmacopeia. *Morchella esculenta* Dill.ex Pers. is the most prized edible fungi known for unique culinary preparations. Due to various pharmacological activities such as anti-oxidant, anti-inflammatory, anti-cancerous and immuno-stimulatory properties, this group of morel had wide application for discovering new drugs and medicines. Phytochemicals such as polysaccharide group, phenolic compound, tocopherol, ascorbic acid and vitamin D are the most active compunds present in morels. Future investigation is required to study the life cycle and ecological association of different morels in high altitude regions, and their association with coniferous and broad-leaved trees such as *Cedrus deodara, Pinus gerardiana, Picea smithiana* and *Quercus* species.

Katare et al. describes *Woodfordia fructicosa* (L.) Kurz. as an important potent medicinal plant in curing peptide ulcer in chapter 2 of this book. According to research team, *W. fructicosa* has tremendous phamacological applications.

This chapter provides a composition for treating ulcers such as stress induced ulcer, peptic ulcer, cold restraint, drug and acid induced ulcer. Herbal formulation in the form of capsules were conducted on marker based stability studies as per ICH Q1 guidelines at accelerated and real time stability studies (40±2°C, 75% ± 5% RH & 30±2°C, 65%±5% RH respectively), and also stability testing was done as per stability specifications at intervals of 0, 3, 6 months for accelerated stability studies and 0, 3, 6, 9, 12, 18, 24, 36 months for real time. This botanical can be considered as safe for human use.

Neha and Satti research deals on *Tinospora cordifolia* (Willd.) Miers ex Hook.f. & Thomson presented in chapter 3. This species is one of the most versatile liana rich in various active compounds having diverse chemical structures. According to this investigation, limited research works have been done on the biological properties and the probable medicinal applications of the individual isolated molecules actually responsible for the medicinal nature of the plant. This research studies make *T. cordifolia* as indigenous drug and a novel candidate for bio-prospection and drug development for the treatment of diseases such as cancer, ulcers, liver disorders, heart diseases, diabetes and post-menopausal syndrome. Thus, a detailed chemical investigation with the aim of isolating novel molecules, their structure elucidation, quantification, validation and pharmacological evaluation in terms of its anti-cancer potential is described and presented in this particular chapter.

Chapter 4 was presented by Adhikari et al. on *Magnolia campbelii* Hook.f. & Thomson, one of the deciduous tree species placed under family Magnoliaceae, and valued for medicinal and ornamental uses. The species is threatened by habitat degradation as well as its utility for timber and construction work. The authors presented and modelled the environmental niche and delineated the potential habitats and distributional areas of *Magnolia campbelii*. The results of the study has implications for *in situ* conservation of the species as well as identifying suitable areas for reintroduction.

Thakur and Dutt discussed *Bunium persicum* (Boiss.) Fedtsch., as an important economic plant of temperate region of Himalayas. The seeds of this species used as carminative substance in various food recipes and described in chapter 5. As the species is restricted in its habitat, therefore, requires mass cultivation practices. This species has typical Palaeo-Mediterranean distribution. In India, this species is distributed wild only in North-West Himalaya (J&K, HP & UK) which makes its occurrence rare in the country. The species is specifically available between 1,850-3,100 meters above mean sea level. Phylogenetically the species is near to *Carum carvi* L., therefore *C. carvi* is the potential adulterant of *B. persicum* as suggested by author in this chapter.

Further, Kumar et al. describes chapter 6 and presented their research on *Dysoxylum binectariferum* (Roxb.) Hook., an evergreen tree species native to India. Apart from its utility in timber market, this species has tremendous medicinal properties because of the presence of its unique secondary metabolites. According to them, the hydroalcoholic extracts of *D. binectariferum* has wide range of pharmacological activities such as immunomodulatory, anti-inflammatory, anti-cancer, and anti-leishmanial. As per this chapter, the chromone alkaloid rohitukine is the major constituent of the plant, possess anti-fertility, anti-leishmanial, anti-adipogenic, anti-ulcerogenic, anti-cancer and immunomodulatory activities. The most extensive and advanced therapeutic property of this chromone alkaloid is its cyclin-dependent kinase inhibition activity, which has a direct link with its anti-cancer effect. It is also an important preclinical lead coded IIIM-290, which is directly derived from rohitukine *via* synthetic modification, and this has demonstrated promising oral efficacy in animal models of various cancer types. The high-value medicinal applications of this plant have necessitated the need of bringing this tree species under captive cultivation as described in this communication.

Chapter 7 were presented by Maity et al., and in their investigation, three plant species *Macrosolen cochinchinensis*, *Loranthus parasiticus* and *Viscum album* control pollution which is indicated by some biochemical analysis undertaken by authors. Total chlorophyll content, leaf extract pH, relative water content and ascorbic acid test were also carried out and presented the results. Authors claim that *Macrosolen cochinchinensis* is highly pollutant tolerant species among the selected three species. Thus, in future this plant species may play an important role as pollution indicator. Besides, these three hemiparasite plant species some valuable ethnomedicinal important have been recorded our countries, tribal people used leaves and stem barks in treatment of abortion, miscarriage, vaginal bleeding pregnancy, treat circulatory and respiratory system problems as discussed in chapter 7.

Jeet et al. describes chapter 8 where agrotechnology of banana (variety BHIM) has been enumerated and presented for the first time by CSIR-Indian Institute of Integrative Medicine Jammu for the commercial cultivation in Himalayan Shivalik range. The aim of study was to investigate the potential of tissue culture (TC) banana production in a diverse environment of Shivalik range, and to make the state self-sufficient in banana production, employment generation and generate revenue for the farmers and also do studies on the post harvest handling and marketing of TC banana. Chapter 9 deals with *Plantago* species as described by Sarkar et al. According to this chapter, plant tissue culture is an important aspect for production of new varietal crops.

Similarly, Chapter 10 deals with *Artemisia annua* L., one of the important medicinal herbs of Himalayas known for artemisinin which is used as antimalarial drug. According to Verma et al., this species widely distributed in the subtropical, temperate and sub-tropical zones worldwide and traditionally used for the treatment of various ailments. Essential oil of this species contain ketone, camphor, caryophyllene oxide, 1,8-cineole and α-pinene, and are known for their medicinal properties. Artemisinin was discovered in 1971 by a Chinese medicinal chemistry scientist Youyou Tu (Nobel Prize in 2015). As the demand for artemisinin remains high worldwide, exploring the chemical and the genetic variation especially knowledge about its mechanisms for high-yielding would be more important. Plant extract of this species are also known for other biological activities such as anti-hypertensive, anti-microbial, immuniosuppresive, and anti-parasitic activities. Therefore, an overview of present status of medicinal applications, phytochemistry, and future perspectives for the research possibilities on the active ingredient artemisinin from *A. annua* is broadly covered as indicated by the authors.

Chapter 11 described by Mondal et al. deals with coastal grass, *Spinifex littoreus* (Burm.f.) Merr. This forms strong and stable dunes. In this chapter, salient morphological features along with different crucial micro-morphological peculiarities have been discussed. Chapter 12 described by Sharma and Baishya focused on soil salinity as one of the most serious environmental issues that limit agricultural productivity, especially for commercial scale cultivated plants. According to their research, the most of food crops belong to glycophytes and hence, they are not able to withstand salinity stress. Salt stress significantly affects the various morphological, biochemical and physiological processes of the plant which leads to poor plant growth. Therefore, there is an emergent need to find out mitigation strategies to cope with the deleterious impacts of salt stress. Plant breeding techniques, efficient resource management and genetic engineering have been employed to cope up with salt stress but these were of limited success as salt tolerant trait is complex both genetically and physiologically. Therefore, using plant growth promoting bacteria seems to be a cost effective and sustainable approach for salinity stress management. This chapter endeavor to provide an overview of the various important mechanisms like facilitation of nutrients uptake, production of phytohormones, ACC deaminase production, and siderophore production employed by the plant growth promoting bacteria to enhance their growth under salinity stress as indicated and highlighted by the authors.

Dinda and Mondal describes chapter 13, and highlighted that people meet their need by taking pulses as their daily meal. Pulses are also called 'Nutrient Powerhouse' as they contain high proteins, fibers, vitamins and amino acids.

Besides, the starch grains are useful taxonomic tool in plant systematic, and therefore, the authors investigated studies on morphological variations of starch grains in seeds, and presented in this book. Besides, chapter 14 deals with techniques suitable for improvement of commercial valueable plants. According to Phukan and Singh, commercially valuable plants are one of the main sources of economy, growth, development and yield of many commercial crops are severely affected by various environmental factors. In recent era, new approaches have been developed combining recombinant DNA technology with tissue culture techniques to obtain improved varieties with specific desired traits. This method had an edge over other technologies but it still carries lots of complications. In this communication, authors discussed transcription factors (TFs) act as master regulators or central players of a particular pathway. Chapter focused on TF families such as ERF, WRKY, NAC and bHLH. Further, chapter 15 deals with *Mucuna pruriens* (L.) DC., which ensures sustainability and societal enabling through agricultural practices. As per Singh and Dhawan, green revolution provided a sufficient amount of foods, but the way resources were utilized causes the degradation of natural resources in various aspects. Authors aim to develop and improve the agri-economic growth of farmers by introducing elite varieties of *Mucuna pruriens* because this species are known for its tonifying, strengthening and all around beneficial properties. It also acts like aphrodisiac, anti-snake venom, anti-depressant, anti-diabetic and anti-microbial as presented by authors.

Eulaliopsis binata (Retz.) C.E.Hubb., is a valuable grass grows only in Midnapore district for commercial purposes. Livelihood of many people depends on the production of these grasses. Chapter 16 described by Dinda et al. deals with this particular species. Phytochemical investigation reveals alkaloids, flavonoids and phenolic compounds are its major chemical constituents. Authors investigated morpho-taxonomy, anatomical characterization and phytochemicals of this species as indicated in the book chapter.

Rhododendrons placed under the family Ericaceae is one of the most important flowering and multiferous keystone plant of Asia. Wide distribution of majority of this species concentrated in Himalayas. Singh and Singh describes chapter 17 with special reference to Indian rhododendrons. As per authors, rhododendron exhibits many nutritional, medicinal and aromatic properties as evident from tribal knowledge and application of different parts such as flowers and leaves in folk medicine and for preparation of local herbal wines. Further, rhododendrons are repository of several bioactive chemicals such as taraxerol, hyperoside, betulinic acid, quercitin, arbutin, rutin, coumaric acid, and several other compounds present in minor quantities. This chapter provide up-dated checklist of Indian rhododendrons, chemistry, associated traditional knowledge, value added products, threat and possible ecological requirements of rhododendrons for conservation and future direction for research is highlighted in this particular chapter.

Chapter 18 deals with mat grass (*Cyperus pangorei* Rottb.). Bhunia and Mandal presented this chapter and says cultivation of its valuable and expensive products can play an important role in rural areas such as Sabong and Pingla in Paschim Medinipur district of West Bengal in India. Mat grasses play the role of an act to fill the empty employment field to the resource of poor farming community to support and secured their livelihoods and maintain the economical balance. This grass usually cultivated by poor and marginal farmers. According to them, major focus of this investigation is to establish the scientific principles behind processing, separation of clum strand, texture and strength properties of the mat fibers and increase mat production automatically. Economical prospects will be increased which will be helpful for entire rural people, as indicated in this communication. Again, Saadi and Mondal describes chapter 19 on the use of engineered nanomaterials. According to them, Silver nanoparticles (AgNPs) are now used to enhance seed germination, plant growth and as anti-microbial agents to control plant diseases. The presented chapter investigated the synthesis of silver nanoparticles using the aqueous solution of three medicinally important plants (such as- *Spondius mombin* L., *Stachytarpheta jamaicensis* (L.) Vahl and *Syzygium samarangense* (Blume) Merr & Perry and leaf extract at room temperature (35°C) used as sample. Findings suggested that the seed germination percentage, relative seed germination rate, relative shoot and root growth and germination index of the tested plant depends upon concentrate gradient of AgNPs as indicated by author.

Borah et al. describes chapter 20, and suggested that plant tissue culture is an important and efficient technique for the production of desired plant products with different improved variety in aseptic conditions. This technique has wide scope for creation, conservation, and utilization of genetic variability for genetic improvement of field, fruits, vegetables, forest crops and medicinal and aromatic plants. The main objective of this chapter is to collects the information and techniques of tissue culture for the genetic improvement of the medicinal plants. As per authors, tissue culture technique in combination with molecular and biotechnological intervention techniques has been successively used to incorporate specific traits through gene transfer as mentioned in this chapter.

The last chapter 21 described by Singh et al. deals with focus on aromatic wealth of Himalayas and potential of value addition from essential oils. According to the authors, developments of processsing industries are directly linked to specific needs, available resources and intervention of technological capabilities. Plant resources have been used over the decades for human welfare in promotion of health as medicine, flavour and fragrance. Resurgence of public interest in herbal products has created a huge market for plant based products which not only satisfies human needs, but also provides quality and safety insurance. Aromatic plants and their essential oils has high demand in international market for development of value added products, for isolation of commercial important compounds used for drug discovery programmes, preparation and formulation for development of new nutraceutical products and semi-synthetic derivatives for medicines. Therefore, authors presented an account of different aromatic wealth of Himalaya, extraction techniques and their major chemical constituents, which will be helpful for pharmaceutical industries in development of value added products for use in near future.

Inventory of life supporting and life saving species in different zones of the country would be an important contribution for man-kind. Global consumer preference for natural products to synthetic ingredients in perfumery, flavouring, pharmaceutical and many other industries has created tremendous potential for natural products from medicinal and aromatic plants. There is an urgent need to initiate collaborative programmes involving taxonomists, biotechnologists, chemists, pharmacologists and commercial expert engineers for bioprospection of natural resources, and search for new value-added products from natural resources for human and animal health care. The book 'Plant of Commercial Values' will be an important contribution of science to man-kind.

1

Morchella esculenta Dill. ex Pers., An Important Medicinal Mushroom of Himalaya: Traditional Usages, Phytochemistry, Pharmacology and Need for Scientific Intervention

Bikarma Singh[1*], Yash Pal Sharma[2*], Brijesh Kumar[3], Bishander Singh[4] and TN Lakhanpal[5]

[1]Plant Sciences (Biodiversity and Applied Botany Division), CSIR-Indian Institute of Integrative Medicine, Jammu-180001, Jammu &Kashmir, INDIA
[2]Department of Botany, University of Jammu, Jammu-180006, Jammu &Kashmir, INDIA
[3]Sophisticated Analytical Instrument Facility Division
CSIR-Central Drug Research Institute, Lucknow-226001, INDIA
[4]Department of Botany, Veer Kunwar Singh University, Ara-802301, Bihar, INDIA
[5]Department of Biosciences, Himachal Pradesh University, Simla-171005
Himachal Pradesh, INDIA
*Email: *drbikarma@iiim.res.in*
yashdbm3@yahoo.co.in,brijesh_kumar@cdri.res.in
bishander85@gmail.com tezlakhanpal@rediffmail.com

ABSTRACT

Mushrooms are in use for centuries in traditional system of medicines, and have importance in greater Chinese and western pharmacopeia. *Morchella* species are well known across the globe as a popular and prized edible fungi and unique culinary saprophytes. They have a long history of usage in the formulation of different traditional medicine for centuries due to their anti-oxidant, anti-inflammatory, anti-cancerous and immuno-stimulatory properties. Inspite of high demand and day-by-day increasing importance, cultivation of *M. esculenta* Dill.ex Pers and other morels are limited, however, there is a report that some species of *Morchella* are successfully cultivated in China. Phytochemical studies reveals that polysaccharide groups, phenolic compounds, tocopherols, ascorbic acid and vitamin D are the most active molecules present in morels. In addition to these compounds, morels are also rich in carbohydrates, amino acids, fatty acids, organic acids and several kinds of macro-and micronutrients. Literature revealed that no data on reproductive biology of *M. esculenta* are available till date. Future investigation is required to study the life cycle and the ecological association of different morels to high altitude regions, and causes of their association with coniferous and broad leaved trees such as *Cedrus deodara, Pinus gerardiana, Picea smithiana, Quercus* and *Castanopsis* species. There is an urgent need to explore all traditional knowledge associated with morels, phytochemistrys and associated pharmological studies of major compounds, reproductive cycle and their basic biology for future conservation of this precious mushrooms for man-kind.

Keywords: Precious Mushroom, *Morchella esculenta* traditional uses, Phytochemicals, Nutrients, Pharmacology, Conservation, Product development.

INTRODUCTION

The genus *Morchella* Dill. ex Pers., commonly referred to as morels (Family: Morchellaceae; Order: Pezizales; Class: Pezizomycetes; Division Ascomycota), comprises of a variety of edible species, highly appreciated among the gastronomists for their desirable taste, quality and culinary value in many cuisines for their unique aroma and delicate flavour (Tietel and Masaphy 2018). It has been regarded as group of 'booster' mushrooms as they have inspired many researchers to keep their research and development on them so as to achieve their artificial cultivation (Lakhanpal et al. 2010). The species under the genus are consumed as food across the globe, while in Himalayan belts of India, tribal populations cook M*orchella* especially, *Morchella esculenta* Dill. ex Pers. with other vegetables. It is reported that for food-flavoring purposes, morels are fermented *in-vitro* as a mycelia in a submerged culture, and these cultures have been characterized and optimized (Litchfteld et al. 1963, Mau et al. 2004, Tietel and Masaphy 2018).

Morels are rich in protein, fibre, vitamins and calories (Robinson and Davidson 1959), and because of their nutrient composition regarded as the superior mushrooms. Samajpati (1978) reported the presence of different amino acids in mycelia of *M. esculenta* whose concentration varies from 0.2 to 8.16 g/100g, however, protein content ranges from 0.22 to 9.12 g/100 g. Therefore, if supplemented with morels like *M. esculenta*, human diets would go a long way in overcoming the protein malnutrition problems in developing countries such as India and Bangladesh (Lakhanpal et al. 2010). Fruiting bodies of *Morchella* species are highly polymorphic in shape, taste and edibility (Masaphy et al. 2010), and the range of species within *Morchella* genus differs in colour, chemical composition and bioactivities (Tietel and Masaphy 2018).

Therefore, the present chapter deals with history of type *Morchella*, different Indian morels, traditional usages, phytochemistry, pharmacology and need for scientific intervention with special reference to *Morchella esculenta* growing in Indian Himalaya.

History on Discovery of Type Morchella

The genus *Morchella* Dill. ex Pers. was first established in 1719 by Johann Jacob Dillenius (1684-1747), and later on typified in 1974 by an early well known mycologist Christiaan Hendrik Persoon (1761-1836). The morel, *Morchella esculenta*, was designated as a type species for true morels. Among several earlier workers who took an interest in the genus were renounded mycologists of their era and several publications on new species and varieties accompanied by meticulously illustrated iconographic plates appeared in different journals and monographs (Fries 1822, Krombholz 1834, Cooke 1870, Saccardo 1889, Boudier 1897, Seaver 1928). Two morels, *Morchella elata* and *Morchella deliciosa*, whose true identity still remains unresolved (Loizides 2017) were described by Elias Fries (1794-1878) in 1822 from a fir forest of Sweden (Fries 1822). *Morchella vulgaris*, was recombined by Gray (1766-1828) as a distinct species following a *forma* of *Morchella esculenta* previously proposed by Persoon (Gray 1821). Going back to history, a large-spored species, *Morchella angusticeps*, was first described by American Mycologist Peck (1833-1917) in 1887 (Peck 1879). Another interesting purple morel, *Morchella purpurescens*, was first described by Jean Louis Émile Boudier (1828-1920) as a variety of *Morchella elata* in 1897 based on plate by Krombholz in 1834 and was recombined as a distinct species in 1985 by Emile Jacquetant (Miller 1981, Jacquetant and Bon 1985).

Distribution of Morels

Wild Morels reported to be harvested from China, India, Mexico, Turkey and United State of America (Pilz et al. 2007), however, different species of morels also reported from Australia, Canada, Cyprus, Israel, Japan and Spain. A map showing global distribution of *Morchella esculenta* is given in Fig. 1. and State-wise distribution of *Morchella esculenta* is given in Table 1.

Earlier history of morels discovery in India shows that Cooke (1870) described *Morchella gigaspora* from Kashmir valley, but currently this taxon is placed under the genus *Verpa* Sw. Majority of the Indian morels have been collected from Jammu & Kashmir, Himachal Pradesh, Uttarakhand, Assam and Punjab (Sydow and Butler 1911, Sohi et al. 1965, Hennings 1901, Bhattacharya and Baruah 1953, Ghurde and Wakode 1981, Kumar et al. 2014).

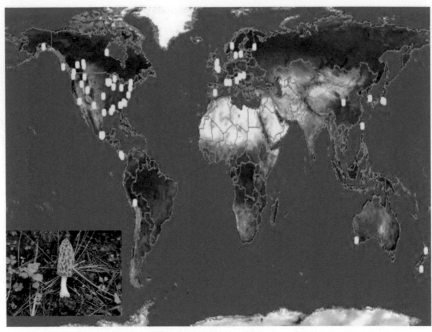

Fig. 1: Global distribution of *Morchella esculenta* worldwide
(*Courtesy*: www.google.com after slight modificaion)

Table 1: Various *Morchella* taxa in India and their distribution (modified after Lakhanpal et al. 2010)

S.No.	Botanical name	Habitat	Distribution
1.	*Morchella angusticeps* Peck. (Black morel)	Ground covered by dead and decaying organic matters of mixed coniferous forests	Himachal Pradesh, Jammu & Kashmir
2.	*Morchella conica* Pers. (Conic morel)	Occurs on ground covered by dead and decaying organic matters of forests rich in *Cedrus, Pinus* and *Quercus* species.	Maharashtra, Himachal Pradesh, Jammu & Kashmir, Uttarakhand
3.	*Morchella crassipes* (Vent.) Pers. ex Fr. (Thick stemmed morel)	Ground covered by dead and decaying organic matters of mixed coniferous forests rich in *Taxus, Cedrus, Juniperus, Pinus, Picea* and *Quercus* species.	Himachal Pradesh, Jammu & Kashmir
4.	*Morchella deliciosa* Fr. (Delicious morel)	Occurs on ground covered by dead and decaying organic matters rich in humus such as coniferous and mixed coniferous forests.	Maharashtra, Himachal Pradesh, Jammu & Kashmir, Uttarakhand
5.	*Morchella esculenta* Dill. ex Pers (Common morel)	Occurs on ground covered by dead and decaying organic matter rich in humus of forests such as	Maharashtra, Himachal Pradesh, Jammu & Kashmir, Uttarakhand

Contd.

		coniferous and mixed coniferous rich in *Quercus dilata* and *Pinus gerardiana.*	
6.	*Morchella semilibera* DC. ex Fr. (Half free morel)	Solitary and scattered on ground covered with dead decaying organic matter in humus in silver fir-spruce forests.	Western Ghats, Himachal Pradesh, Jammu & Kashmir
7.	*Morchella tibetica* Zang.	Occurs on ground covered by dead and decaying organic matter of coniferous and Oak forests	Kullu and Shimla (Himachal Pradesh); endemic
8.	*Morchella simlensis* Lakhanpal & Shad	Solitary on ground covered with dead decaying organic matter in Chir-Oak and mixed coniferous forests	Shimla, Kullu and Kinnaur of Himachal Pradesh; endemic

Habitat of *Morchella esculenta*

In general, morels are associated with live or dead trees, and disturbed on burnt soil, and they are tolerant to cold soils (Kuo 2005). Morels have adapted to a wide range of unusual habitats and environmental conditions, including river bottoms, dunes, garbage dumps, abandoned coal mines, cellars and basements, saw mills, wood piles, sand bars in rivers, excavations, deer trails, orchards, bomb craters and limed soils (Kaul 1975, Kuo 2002). Specifically, *Morchella esculenta* grows naturally on forest floor rich in humus. If the nutrient supply is sufficient, it collectively forms a compact mycelium on the surface soil. It has been found that *M. esculenta* stands solitary on the ground covered with dead organic matters rich in humus of conferous and mixed conferous forests having *Cedrus deodara, Pinus wallichiana* and *Quercus* species as the main species components. The ascocarps appear above the soil surface soon after the rains. However, the habitats are often distinguished by the dominance of tree species, *viz. Rhododendron arboreum, R. lepidotum, Taxus baccata, Pinus wallichiana, Cedrus deodara, Betula utilis, Cupressus juniperus*, and important medicinal and aromatic plants, such as *sinopodophyllum hexandrum, Dactylorhiza hatagirea, Picrorhiza kurroa, Rheum emodi, Pleurospermum angelicoides, Angelica glauca, Arnebia benthamii, Saussurea costus, Megacarpea polyandra, Selinum wallichianum, Nardostachys jatamansi, Aconitum* and *Polygonatum* species. It is noticed that this appears in a large scale during the month of March and its collection starts between March and April. Local people set the ground on fire every year during October and November, assuming that such a practice will improve morel yields.

Taxonomy and General Life Cycle of *Morchella esculenta*

Ascocarp: 6.0-8.0 cm long, pileate. Pileus: 3.0-4.5 x 1.8-2.5 cm at the base, ovate, apex obtuse, attenuated upwards but obtuse at the apex and adnate to the stipe at the base, pitted, pits irregular to somewhat rounded, yellowish-brown, on drying changes to brown-black. Stipe: 2.0-3.0 x 0.8-1.0 cm, cylindrical, soft, hollow, concorolous with pileus or slightly creamish yellow. Asci: 225.6-273.6 x 12.0 μm wide at the top and upto 6.0 μm at the base, eight-spored, cylindrical to sub-cylindrical, narrowed at the base. Ascospores: 18.4-25.6 x 9.6-10.4 μm, uniseriate, sub-hyaline, ellipsoidal, smooth. Paraphyses: upto 216 μm and 8 μmwide at the top and 4.8 μm wide at the base thin walled, septate, simple, yellowish in mass, sub-hyaline, singly. Pubescent hairs: upto 9.6 μm, septate, composed of 2 to 3 cells of variously shaped, thin walled, sub-hyaline (Fig. 2).

Fig. 2: Different habitat of *Morchella* species

Edibility: Edible as cullnary sapropyte

Specific distribution: Reported from various regions of India including Jammu and Kashmir, Punjab, Kumaon Himalayas (Cooke 1870, Bose and Bose 1940, Sohi et al. 1965, Kaul et al. 1978, Kaul 1981).

Fig. 3 explains the general life cycle of the genus *Morchella* which indicate that it has two kinds of life cycles (i) haploid fruiting and (ii) heterothallism. According to Volk and Leonard (1990), cytological studies revealed that the average number of nuclei per cellular compartment in vegetative hyphase of morchella is 10-15 and hyphal fusions are quit frequent. The resting structure called sclerotia form repeated branching. Fniting body is primordia and ascus development demonstrate autogamy rather than denova heterokaryan formation by hyphal fusion in the subhymenial layer of the fruiting body (Volk and Leonard 1990). Du et al. (2017) explains and illustrated life cycley of *Morchella* which is given in Fig. 3.

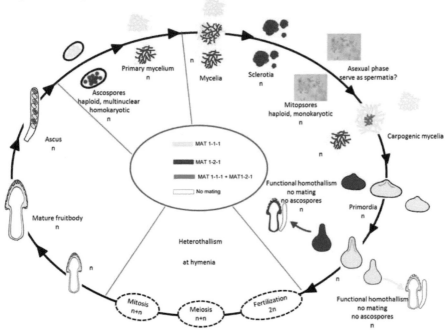

Fig. 3: Life cycle of the genus *Morchella* (Courtesy: Du et al. 2017. Illustration of two kinds of life cycles-haploid fruiting and heterothallism).

Traditional Usages in Western Himalaya

Many of species of morels (*Morchella angusticeps, M. conica, M. esculenta, M. elata, M. crassipes, M. rotunda, M. semilibera, M. tomentosa* etc.) are found in the forests of Himalayas (Cooke 1870, Kaul 1975, Waraitch 1976, Kaul et al. 1978, Lakhanpal et al. 2010, Kumar and Sharma 2011, Kumar et al. 2014).

Morel collection can be undertaken twice in a year between March-May and August-September. Local people set the ground on fire every year during October and November, assuming that such practice will improve their quality and yield. Due to lucrative returns, in many villages, almost all the families are actively involved in morel collection. On an average, each individual collect about 1-3 kg of fresh morels in a day. The various collectors sell their fresh or dried produce to local shopkeepers or middlemen. The middlemen buy morels directly from the villagers at the rate of rupees 2000-3000 per kg and further sell to the wholesellers at an exorbitant price of Rs. 7000-9000 per kg and these are finally retailed to the consumers @ rate of INR 10000-15000 per kg. Sometimes *Verpa conica* is used as an adulterant mushroom in the morel trade (Kumar et al. 2014).

Edible morels are commonly consumed by the residents of Uttarakhand, Himachal Pradesh and Jammu & Kashmir where they are grows in plenty. It is generally belief that morels comes out from ground after lightening strikes and prop up in greater abundance in burnt up areas. In high attitude mountaineous regions, the collection of morels are generally undertaken during March and April after the snow melts. The collection is done early in the morning, as there is intense competition for the mushroom gathering, especially for the morels because of their high commercial value and market demand. Mostly women and children from 'Gaddi and Shippi' tribes of J&K are frequently involved in these activities then men. Children frequently accompany the women, as they are good at locating mushrooms because of their sharp eyes and proximity to the ground and crevices where the occurrence of the mushroom is highest (Kumar and Sharma 2011). *Morchella* species are put up for sale in both fresh and dried forms. Several economically downtrodden natives also vend these mushrooms to well-to-do families in exchange for goods such as used woolen clothes, rice, flour etc. Ethnobotanical information collected from shepherds, herdsmen, rural women and rural school students revealed that *Morchella esculenta* known by different names in Western Himalaya. It is called as 'Guchhi' by several tribals of Himalaya, 'Chiaun' in Shimla district, 'Kiaun' in Kullu Valley, 'Chunchuru' in Kinnaur, 'Chamkaid' in Chamba districts of Himachal Pradesh. In Bhadarwah area of Jammu and Kashmir, *Morchella esculenta* is fondly called as 'Batt Kutch' (mushroom consumed with rice) and 'Thunthoo' in Kishtwari and Kashmiri jargon. Morels are generally cooked with rice and other vegetables including onion and tomatoes, and considered as highly nutritious then meat or fish. Several other prominent traditional recipes are 'Chaschni' (a local dessert), 'Thunthoo Pullow' (rice + morels), 'Thunthoo Kheer' (milk +morels), and 'Thunthoo Yakhni' (curd + morels) (Kumar and Sharma 2011). Negi (2006) discussed the nutritional value and medicinal uses of *Morchella* spp. from the Darma valleys of Pithoragarh district, Kumaun Himalaya, Uttarakhand.

In Himachal Pradesh, many traditional recipes such as pulao of *M. esculenta* with rice locally called as 'Chiaun ka bhaat' is prepared and it is believed that it is energy giving food and protect the body from extreme cold. As per Lakhanpal et al. (2010), traditional recipes of vegetables and soap such as 'Guchhi ka bhaat,' Guchhi Ka Biryani, Guchhi Pulao, Morchella fried rice, Sukhi Subji of Rangnu/Guchhi, Mixed vegetables of Morchella, Curry of Morchella, Morchella Shahi Paneer, Guchhi Masala, Taridar Subzi, Madra of Guchhi, Matar Guchhi,

Morchella Omelette, Kofta of Guchhi, Khadha of Guchhi, Mixed Non-Veg Soup, Stuffed Vegetarian Morels, Morchella Snacks with egg, Morels in local Wine and Morels Pickle are locally prepared by different tribes of Himalaya in India. They believe that it provide all essential nutrients required for growth of the body and protect body from severe cold and disease. In several villages, *Morchella conica* and *M. esculenta* are used as medicinal mushrooms that are usually given to mothers after childbirth to promote recovery (Kumar and Sharma 2011). Since the commercial cultivation of this mushroom has not yet been fully realized, its mycelium is extensively used as a flavouring agent.

PHYTOCHEMISTRY

Nutrient Analysis

Mushrooms are nutritionally appreciated for their high levels of protein, fibers and minerals. The nutritional value of several *Morchella* species including *M. crassipes, M. hortensis, M. conica* and *M. elata* is reported to contain 7.5–11.52 g protein, 2.2–3.9 g fat, 6.7–14.6 g ash and 74.55–80.5 g carbohydrates per 100 g dry weight (Litchfteld et al. 1963, Heleno et al., 2013, Vieira et al. 2016). Mushroom composition may vary as a result of growth areas, maturity stage, soil and environmental conditions (Liu et al. 2016). Nutrient composition and amino acid components in *Morchella esculenta* is presented in Table 2.

Table 2: Nutrient composition and amino acid components in *Morchella esculenta*

Nutrient composition			Amino acids	
Moisture weight (%)		84.15	Alanine	†
Crude Protein g/100 g dry weight		34.30	Aspartic	†
Elemental composition	Nitrogen	5.48	Cystine	†
(g/100 g dry weight)	Phosphorus	1.25	Glutamic	†
	Potassium	1.90	Glycine	†
	Calcium	0.60	Isoleucin	†
	Magnasium	0.17	Lysine	†
	Sulphur	0.37	Proline	†
	Zinc (PPM)	154	Seine	†
	Cupper (PPM)	44.0	Tyrosine	†
	Manganese (PPM)	52.0	Threonin	†
	Iron (PPM)	523.0	Tryptoph	†

Amino acid profile in *Morchella* species was recorded in several published works, and reports high levels of glutamic acid and alanine, while levels of other amino acids vary among varieties, with reports showing high levels of leucine, proline, aspartic acid, arginine, glycine and threonine (Rotzoll et al. 2006). Morels are also very rich in several kinds of minerals. According to Vandamme (2003), fatty acid profile of morels consists of C_{18}: $2n_6$ as the main fatty acid (linoleic acid, 63%–72%) followed by C_{18}:1n9 (oleic acid, 9.7%–

21%), C_{16}:0 (palmitic acid, 9.5%–11%) and C_{18}:0 (stearic acid, 1.5%–2.6%) (Vieira et al. 2016), and in one report C_{18}:3n_3 (linolenic acid) ranging between 0.2%–7.2% (Heleno et al. 2013). Morels also contain several organic acids, including oxalic acid (32.73–190 mg/100 g dry weight), quinic acid (0–880 mg/ 100 g dry weight), malic acid (0–199.1 mg/100 g dry weight), citric acid (0–233.4 mg/100 g dry weight) fumaric acid (17.38–561 mg/100 g dry weight) and ascorbic acid (13 mg/100 g dry weight), with total organic acids ranging between 279 and 1560 mg/100 g dry weight (Mau et al. 2004, Heleno et al. 2013).

Furthermore, *Morchella* sugar profile comprises of 0.21–0.71 g fructose, 0.99–11.54 g mannitol, 1.09–5.34 g trehalose, 43.07 g mannose, 0.086 g arabitol and 1.7–9.54 g glucose per 100 g DW (Rotzoll et al. 2006, Tsai et al. 2006, Beluhan and Ranogajec 2011, Heleno et al. 2013, Vieira et al. 2016).

Chemical Composition and Allied Chemistry

Several groups of phytochemicals have been reported from morels such as carotenoids polyphenols, steroids and tocopherols (Tietel and Masaphy 2018), although carotenoids are reported to occur in limited amount in mushrooms compared to plants (Kalac 2009). Tocopherols reported in morels (Shahidi and Ambigaipalan 2015), and this includes β-tocopherol (1.4–6.2 mg/100 γ dry weight), β-tocopherol (20 mg/100 g dry weight), α-tocopherol (12.4–20.3 mg/ 100 g dry weight) and δ-tocopherol (3.9–98.6 mg/100 g dry weight (Mau et al. 2004, Lakhanpal et al. 2010, Heleno et al. 2013, Vieira et al. 2016, Tietel and Masaphy 2018). Some major chemical constituents of *Morchella esculenta* is given in Fig. 4.

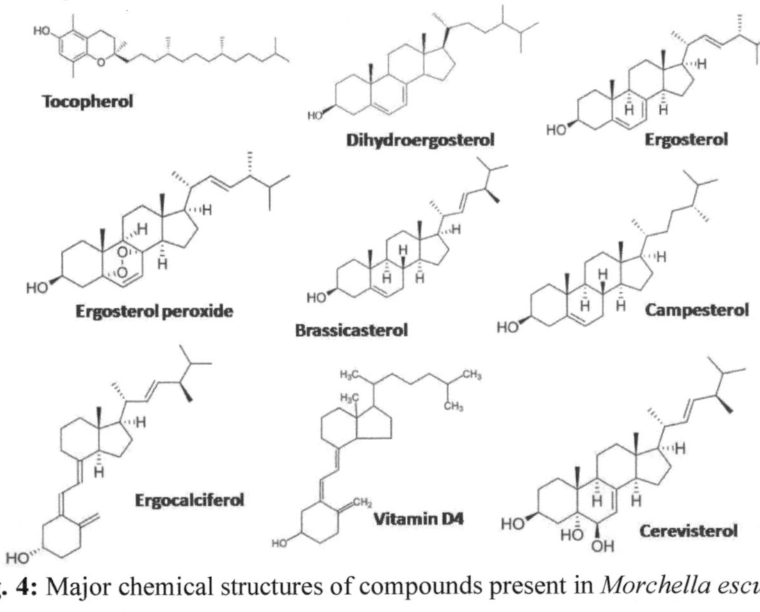

Fig. 4: Major chemical structures of compounds present in *Morchella esculenta*

Recently, new evidences on the roles of vitamin D on human health were published and specific attentions were paid to mushrooms as a significant dietary source of vitamin D_2, ergocalciferol (provitamin D) and of its provitamin, (Cashman et al. 2014). Mushrooms such as *Morchella esculenta* contain relatively high levels of ergosterol provitamin, ranging between 3000–7000 mg/kg dry weight and 0.3–59 mg/100 g fresh weight of vitamin D_2 (Teichmann et al. 2007, Phillips et al. 2011).

Morel sterol composition was not thoroughly investigated, although sporadic results are available: 13.4 mg/100 g dry weight of ergosterol peroxide in *M. esculenta*, 28.6 mg/100 g brassicasterol, 1.23–4.54 mg/100 g campesterol and 2.44–12.3 mg/100 g of an unknown sterol, 20.7– 32.6 mg/100 g fresh weight ergosterol, 4.98–7.13 mg/100 g fresh weight ergosta-5,7-dienol (22,23-dihydroergosterol), 4.39–6.26 mg/100 g fresh weight (ergocalciferol), and 2.15–2.36 mg/100 g vitamin D_4 (22,23-dihydroergocalciferol) (Krzyczkowski et al. 2009, Phillips et al. 2011). Sterol compounds, including 5-dihydroergosterol, ergosterol peroxide, ergosterol and cerevisterol were reported as anti-oxidative and anti-inflammatory activities in *M. esculenta*, with IC_{50} values for NF-kB inhibitory activity of 5.2, 4.6, 2.0, and 5.1 mM, respectively (Kim et al. 2011, Tietel and Masaphy 2018).

PHARMACOLOGY ASPECTS

Phenolic compounds present in morels are reported to possess anti-oxidative and anti-inflammatory activity. Literatures (Cheung et al. 2003, Taofiq et al. 2015) suggest that phenolic acids are the major compound responsible for main anti-oxidants in mushroom like morels (Kalac 2009). Total phenolics content (TPC) of water and methanolic extracts of *M. conica* was reported as 16.9 and 4.6 mg GAE/g dry weight, while that of *M. anguisticeps* showed values of 13.1 GAE/g dry weight and 2.6 GAE/g dry weight for water and methanolic extracts (Puttaraju et al. 2006). Ramirez et al. (2007) determined TPC values of 45.9 mg/g dry weight and 173.5 mg/g dry weight in methanol and water extracts of *M. esculenta*, respectively. Mau et al. (2004) reported TPC of 3.63 mg/g in methanolic extracts of *M. esculenta*.

MOREL TRADE IN WESTERN HIMALAYA

Morels are harvested from the wild on a commercial scale for sale in national and international markets. They have acquired global significance owing to their international trade which is extensive and profitable. Commercial cultivation or agro-technology of morels is very limited, though there are reports of morels cultivation from Israel, USA and China. Thus, wild-harvested morels are one of the most valuable forest products in the global market. In Western Himalaya of

India, commonly called as 'Guchchi' or morels are prime edibles with immense commercial importance and their strong demand among people as nutraceutical make them as an important Non-Timber Forest Produce (NTFP) as compared to all edible mushrooms collected from the wild. The sale of morels provides substantial monetary benefits to rural livelihoods and local traders. Morel collection, in northern state of India, is undertaken twice in a year between March-May and August-September. Due to lucrative returns, in many villages, almost all families are actively involved in morel collection. On an average, each individual can be able to collect about 1-3 kg of fresh morels in a day and sell them in fresh or dried form to the local shopkeepers or middlemen. Incidentally, middlemen usually buy morels directly from the villagers at very low prices (Rs. 2000-3000 per kg DW) and further sell them to the wholesellers and customers at an exorbitant price (Rs.10000-15000 per kg DW). This indicates that consumers pay 3-4 times more price than the initial price at the collectors' level. However, market sale price of dried morels may vary enormously from time to time (Kumar and Sharma 2010).

On the other hand, the official scenario of morel production is rather dismal. As per the data gathered from various forest divisions of the Western Himalaya, it is observed that morel output is not uniform in the region. In Kishtwar Forest Division of Jammu & Kashmir State, out-turn of morels production during 2003-2004, 2004-2005 and 2005-2006 were reported to be 46.50, 35.58 and 54.21 quintals, respectively while in Marwah Forest Division this figure stood at 10.27, 19.96 quintals during 2004-2005 and 2005-2006, respectively. Likewise, a similar division in Bhadarwah area of J&K state, 35.00, 35.00 and 45.00 quintals of morels were extracted from the wild during 2005-2006, 2006-2007 and 2007-2008 respectively (Kumar and Sharma 2010).

The actual morel collection is expected to be far greater than that depicted by the Forest Department since there is visibly no control over their collection and sale. Therefore, unorganized, rampant and unscrupulous trade of this valuable commodity cannot be ruled out in the region. Although the forests are occasionally leased out on contract basis by the State Forest Department for morel collection and the morel gathering by others is prohibited, yet the forests are freely accessible to the local inhabitants dwelling near the forest areas for morel collection thus inflicting a huge loss to the state exchequer every year. Dried morels are exported every year from India to the international markets and Jammu and Kashmir ranks second in the morel trade after Himachal Pradesh, Uttarakhand as the third largest producer of morels in India. Morels of the Jammu and Kashmir have far more cachet in Indian and foreign markets due to their extraordinary flavour and taste.

SCIENTIFIC INTERVENTION AND FUTURE PERSPECTIVES

Morchella esculenta and allied species of morels are highly valued worldwide, owing to their attractive organoleptic characteristics and high nutritional value. Additionally, they are consumed as a functional food, as they possess scientifically proven anti-oxidative, anti-inflammatory and immunostimulatory properties. Morels chemical composition were reported as nutritional value, phytochemical and taste soluble components, whereas, aroma was hardly investigated. Further research is required in order to establish understanding of their aroma profile and key-odorants for product development from morels.

No molecular identification of the reported species is mentioned along with the biological activities described in the reviewed reports. The recent multigene molecular phylogenetic assessment approach suggested a revision in morel species taxonomy, while increasing the number of morel species based on molecular identification (Kuo et al. 2012).

Another challenge in using morels as a functional food is that it is mostly obtained from wild growth, as morels are difficult to cultivate for mushroom production. Thus, an alternative way to exploit morel beneficial metabolites is by cultivation of *Morchella* species as fermented mycelia grown in liquid medium, and using their metabolites for consumption as functional food or for food flavoring. There is an increasing number of publications reporting investigations aiming to enhance the bioactive metabolites production or to reduce cultivation costs by using low cost substrates, using different *Morchella* species. Yet, the main challenges of this cultivation system are repeatability, isolation and characterization of the active metabolites produced in liquid cultures, to compare with the ones produced by fruiting bodies and to scientifically prove their activity *in vitro* and *in vivo*.

CONCLUSION

The morels are graded on the basis of their colour and size. Information gained from review of literatures is crucial for advancing morel conservation genetics and for formulating informed policies directed at insuring commercial harvests. *Morchella* species are well known across the globe as popular and prized edible fungi and unique culinary delights. Therefore, there is need of scientific intervention. More anti-oxidant, anti-inflammatory, anti-cancerous and immuno stimulatory properties of morels need to be studied for future development of value added nutraceuticals. Inspite of the high demand and the day-by-day increasing importance, cultivation of *M. esculenta* and other morels are limited. Phytochemicals including polysaccharide groups, phenolic compounds, tocopherols, ascorbic acid and vitamin D are the most active compunds. Literature revealed that no data on reproductive biology of *M. esculenta* are available till todate. Future investigation is required to study the life cycle and the ecological association of different morels in high altitude regions, and their association with conifer and broadleaved tree. There is urgent need to explore their reproductive modes as well as their basic biology needs to be elucidated for future conservation of this precious edible mushroom.

ACKNOWLEDGEMENTS

We are thankful to the entire team. The article bears publication number # IIIM/2251/2018

REFERENCES CITED

Beluhan S, Ranogajec A. 2011. Chemical composition and non-volatile components of Croatian wild edible mushrooms. *Food Chemistry* 124: 1076–1082.

Bhattacharya B, Barauh HK. 1953. Fungi of Assam. *Journal of University of Gauhati* 4: 287-312.

Bose SR, Bose AB. 1940. An account of edible mushrooms of India. *Science and Culture* 6: 141-149.

Boudier E. 1897. Révision analytique des morilles de France. *Bulletin trimestriel de la Société mycologique de France* 13: 130–150.

Cashman KD, Kinsella M, McNulty B A, Walton J, Gibney MJ, Flynn A, Kiely M. 2014. Dietary vitamin D2-A potentially underestimated contributor to vitamin D nutritional status of adults? *British Journal of Nutriention* 112: 193–202.

Cheung L, Cheung PC, Ooi VE. 2003. Antioxidant activity and total phenolics of edible mushroom extracts. *Food Chemistry* 81: 249–255.

Cooke MCMA. 1870. Kashmir Morels. *Transaction of the Botanical Society of Edinburgh* 10: 439-443.

Du XH, Zhao Q, Xia EH, Gao LZ, Richard F, Yang ZL. 2017. Mixed reproductive strategies, competitive mating-type, distribution and life cycle of fourteen black morel species. *Scientific Research* 7: 1493. DOI: 10.1038/s41598-017-01682-8.

Fries EM. 1822. Systema Mycologicum' Vol. 3/. Berlingiana, Lund, Sweden, pp. 238.

Ghurde VR, Wakode DD. 1981. A new report of Morchella from Central India. Indian Journal of Mycology. *Plant Pathology* 11: 314-315.

Gray SF. 1821. A Natural Arrangement of British Plants, according to their relations to each other. pp 662.

Heleno SA, Stojkovic D, Barros L, Glamoclija J, Sokovic M, Martins A, Queiroz MJRP, Ferreira ICFR. 2013. A comparative study of chemical composition, antioxidant and antimicrobial properties of *Morchella esculenta* (L.) Pers. from Portugal and Serbia. *Food Research International* 51(1): 236-243.

Hennings P. 1901. Fungi Indiae orientalis II, CW Gollama. 1900. Collecti: Hedw. 40: 323-342.

Jacquetant E, Bon M. 1985. Typifications et mises au point nomenclaturales dans l'ouvrage Les morilles (de E. Jacquetant), Nature-Piantanida 1984. Documents Mycologiques (in French). 14

Kalac P. 2009. Chemical composition and nutritional value of European species of wild growing mushrooms, a review. *Food Chemistry* 113: 9–16.

Kaul TN. 1975 Studies of the genus Morchella in Jammu and Kashmir I. Soil composition in relation to carpophore development. *Bulletin of the Botanical Society of Bengal* 29: 127–134

Kaul TN. 1981. Common edible mushrooms of Jammu and Kashmir. *Mushroom Science* 11: 79-82.

Kaul, TN, Kachroo, JL Raina, A. 1978. Common edible mushrooms of Jammu and Kashmir *The Journal of Bombay Natural History Society* 71: 26-31.

Kim J A, Lau E, Tay D, De Blanco EJC. 2011. Antioxidant and NF-kappa B inhibitory constituents isolated from Morchella esculenta. *Natural Product Research* 25: 1412–1417.

Krombholz JV von. 1834. Naturgetreue Abblidungen und Beschreibungen der essbaren, schädlichen und verdächtigen Schwämme, Heift 3. G. Calve, Praha J, 36 p., pl. XV-XXII.

Krzyczkowski W, Malinowska E, Suchocki P, Kleps J, Olejnik M, Herold F. 2009. Isolation and quantitative determination of ergosterol peroxide in various edible mushroom species. *Food Chemistry* 113: 351–355.

Kumar S, Kotwal M, Sharma YP. 2014. Morchellaceae from Jammu region of North-West Himalaya. *Mushroom Research* 23(1): 15-25.

Kumar S, Sharma YP. 2010. Morel trade in Jammu and Kashmir- Need for organized commercialization. *Everyman's Science* (ISCA) XLV (2):111-112.

Kumar S, Sharma YP. 2011. Diversity of wild edible mushrooms from Jammu and Kashmir (India). Proceedings of the 7[th]International Conference on Mushroom Biology and Mushroom Products (ICMBMP7) France, pp 568-577.

Kuo M, Dewsbury DR, O'Donnell K, Carter MC, Rehner SA, Moore JD, Moncalvo JM, Canfield SA, Stephenson SL, Methven AS. 2012. Taxonomic revision of true morels (Morchella) in Canada and the United States. *Mycologia* 104: 1159–1177.

Kuo M. 2002 When and where morels grow? http://www. mushroomexpert.com

Kuo M. 2005. Morels. MI: University of Michigan Press, Ann Arbor

Lakhanpal TN, Shad O, Rana M. 2010. Biology of Indian Morels. IK International Publishing House Pvt. Ltd., New Delhi, India, pp. 245.

Litchfield JH, Vely VG, Overbeck RC. 1963. Nutrient content of Morel Mushroom Mycelium: amino acid composition of the protein. *Journal of Food Science* 28(6): 741-743.

Liu C, Li P, Mao Q, Jing H. 2016. Antihyperlipidemic effect of endo-polysaccharide of *Morchella esculenta* and chemical structure analysis. *Oxidation Communication* 39: 968–976.

Loizides M. 2017. Morels: the story so far. *Field Mycology* 18(2): 42–53".

Masaphy S, Zabari L, Goldberg D, Jander-Shagug G. 2010. The complexity of Morchella systematics: A case of the yellow morel from Israel. *Fungi* 3:14–18.

Mau J L, Chang CN, Huang SJ, Chen CC. 2004. Antioxidant properties of methanolic extracts from *Grifola frondosa, Morchella esculenta* and Termitomyces albuminosus mycelia. *Food Chemistry* 87: 111–118.

Miller OK. 1981. Mushrooms of North America. Elsevier Dutton Company, New York, pp. 368.

Negi CS. 2006. Morels (*Morchella* spp.) in Kumaun Himalaya. *Natural Product Radiance* 5 (4): 306-310.

Peck CH. 1879. Report of the Botanist (1878). *Annual Report on the New York State Museum of Natural History* 32: 44.

Phillips KM, Ruggio DM, Horst RL, Minor B, Simon RR, Feeney MJ, Byrdwell WC, Haytowitz DB. 2011. Vitamin D and sterol composition of 10 types of mushrooms from retail suppliers in the United States. *Journal of Agriculture and Food Chemistry* 59: 7841-7853.

Pilz D, McLain R, Alexander S, Villarreal-Ruiz Berch S, Wurtz TL, Parks CG, McFarlane E, Baker B, Molina R, Smith JE. 2007. Ecology and Management of Morels Harvested from the Forests of Western North America. General Technical Report PNW-GTR-710. Portland: US Department of Agriculture, Forest Service, Pacific Northwest Research Station, 161pp.

Puttaraju NG, Venkateshaiah SU, Dharmesh SM, Urs SMN, Somasundaram R. 2006. Antioxidant activity of indigenous edible mushrooms. *Journal of Agriculture and Food Chemistry* 54: 9764-9772.

Ramýrez-Anguiano AC, Santoyo S, Reglero G, SolerRivas C. 2007. Radical scavenging activities, endogenous oxidative enzymes and total phenols in edible mushrooms commonly consumed in Europe. *Journal of the Science of Food and Agricuture* 87: 2272-2278.

Robinson RF, Davidson RS. 1959. The large scale growth of higher fungi. *Applied Microbiol*, 1: 216-278.

Rotzoll N, Dunkel A, Hofmann T. 2006. Quantitative studies, taste reconstitution, and omission experiments on the key taste compounds in morel mushrooms (*Morchella deliciosa* Fr.). *Journal of Agriculture Food Chemistry* 54: 2705–2711.

Saccardo PA. 1889. Sylloge Fungorum huscusque congitorum, Patavii. 8.

Samajpati N. 1978. Nutritive value of some Indian edible mushrooms. *Mushroom Science* 10: 695-703.

Seaver FJ. 1928. The North American Cup Fungi (Operculates). Seaver, New York, pp. 284.

Shahidi F, Ambigaipalan P. 2015. Phenolics and polyphenolics in foods, beverages and spices: Antioxidant activity and health effects-A review. *Journal of Functional Foods* 18: 820-897.

Sohi HS, Seth PK, Kumar S. 1965. Some interesting fleshy fungi from Himachal Pradesh-I. The *Journal of Indian Botanical Society* 44: 69-74.

Sydow H, Butler EJ. 1911. Fungi Indiae orientalis part III. *Annals of Mycology*, 9: 372-421.

Taofiq O, Calhelha R C, Heleno S, Barros L, Martins A, SantosBuelga C, Queiroz MJR, Ferreira IC. 2015. The contribution of phenolic acids to the anti-inflammatory activity of mushrooms: Screening in phenolic extracts, individual parent molecules and synthesized glucuronated and methylated derivatives. *Food Research International* 76: 821-827.

Teichmann A, Dutta PC, Staffas A, Jagerstad M. 2007. Sterol and Vitamin D2 concentrations in cultivated and wild grown mushrooms: Effects of UV irradiation. *LWT-Food Science and Technology* 40: 815-822.

Tietel Z, Masaphy S. 2018. True morels (*Morchella*)-nutritional and phytochemical composition, health benefits and flavor: A review. *Critical Reviews in Food Science and Nutrition* 58 (11): 1888-1901.

Tsai SY, Weng CC, Huang SJ, Chen CC, Mau JL. 2006. Nonvolatile taste components of Grifola frondosa, Morchella esculenta and Termitomyces albuminosus mycelia. LWT-*Food Science and Technology* 39:1066-1071.

Vandamme EJ. 2003. Bioflavours and fragrances via fungi and their enzymes. *Fungal Diversity* 13: 153-166.

Vieira V, Fernandes A, Barros L, Glamoclija J, Ciric, Stojkovi D, Martins A, Sokovi M, Ferreira C. 2016. Wild *Morchella conica* Pers. from different origins: A comparative study of nutritional and bioactive properties. *Journal of Science of Food and Agriculture* 96: 90–98.

Volk TJ, Lenard TH 1990. Cytology of the life cycle of Morchella. *Mycological Research* 94(3): 399-406.

Waraitch KS. 1976. The genus *Morchella* in India. *Kavaka* 4: 69-76.

2

Woodfordia fruticosa (L.) Kurz., a Potent Indian Medicinal Plant in Curing Peptic Ulcer

Anil Kumar Katare[1], Inshad Ali Khan[2], Durga Prasad Mindala[1] Prasoon Kumar Gupta[3], Surjeet Singh[4] and Bikarma Singh[5]

[1]cGMP Division, [2]Clinical Microbiology Division, [3]NPC Division, [4]Pharmacology Division, [5]Plant Sciences (Biodiversity and Applied Botany Division) CSIR-Indian Institute of Integrative Medicine, Jammu 180001, Jammu & Kashmir, INDIA Email: akkatare@iiim.ac.in, iakhan@iiim.ac.in, dpmindala@iiim.ac.in guptap@iiim.ac.in, ssingh@iiim.ac.in, drbikarma@iiim.ac.in

ABSTRACT

Traditional herbal drug preparations are known for centuries to protect several human diseases such as peptic ulcer, diarrhea, piles and dysentery. Current days knowledge about the underlying biochemical mechanism for most of the gastric ulcers and majority of the duodenal ulcers deserve appropriate consideration and due weightage while consolidating regarding the efficacy of a plant extract. The present investgation relates to a pharmaceutical composition comprising an effective amount of lyophilized extract or at least one bioactive fraction obtained from flowers of *Woodfordia fruticosa* (L.) Kurz, along with one or more pharmaceutically acceptable additives or carriers. The present results provides a composition for treating ulcers such as stress induced ulcer, peptic ulcer, cold restraint induced ulcer, drug induced ulcer and acid induced ulcer, and also the composition used as specific inhibitor of gastric H+, K+-ATPase. Herbal formulation in the form of capsules were conducted on marker-based stability studies as per ICH Q1 guidelines at accelerated and real time stability studies ($40\pm2°C$, $75\% \pm 5\%$ RH and $30\pm2°C$, $65\%\pm5\%$ RH, respectively), and also stability testing investigation was done as per stability specifications at intervals of 0, 3, 6 months for accelerated stability studies and 0, 3, 6, 9, 12, 18, 24, 36 months for real time stability. The present data reveals that this species is one of the potent Indian medicinal Plant which helps in curing ulcer and associated diseases.

Keywords: Medicinal Plant, *Woodfordia fruticosa*, Anti-ulcer, Herbal formulation, Lyophilised extract, Capsules, India.

INTRODUCTION

Peptic ulcer disease is one of the most common disease affecting millions of people (Kong et al. 2011), and a number of drugs are now available for treatment of this disease. The efficacy of anti-ulcer drugs are still debatable, and a number of anti-inflammatory drugs are fraught with adverse effects such as gastric ulceration and perforation. Hence, the research and the search for ideal drugs still continues and has been extended to herbal formulation. For commercialization of botanical products, there is responsibility of maintaining their quality falls to a large extent on the scientists, and to a certain extent on the manufacturers. In similar situation, the assurance of safety, quality and efficacy of medicinal herbs and botanical products becomes an important issues. The National Center for Complementary and Alternative Medicine, WHO or any regulatory bodies i.e. USFDA, EMA and Indian regulatory stress their importance of qualitative and quantitative methods for characterizing botanicals samples, quantification of the biomarkers and/or chemical markers and the fingerprint profiles (Baravalia et al. 2011). The herbal drug preparation itself as a whole is regarded as the active substance. Hence, the reproducibility of the total configuration of herbal drug constituents is important. Different approaches can be used for chemical standardization such as pre-treatment that involves drying and grinding, selection of a suitable method of extraction, analysis of compounds using suitable chromatographic or spectroscopic methods, analysis of data based on bioactive or marker compounds; quality control; elucidation of the properties of ADME and metabonomics evaluation of medicinal plants. There is also a need to approach scientific proof and clinical validation with chemical standardizations, biological assays, animal models and clinical trials for botanicals.

Woodfordia fruticosa (L.) Kurz (Syn. *Woodfordia floribunda* Salisb.) belongs to family Lythraceae. Frequently used english names for this species are 'Fire Flame Bush' and 'Shiranjitea', while Gujarat in India it is known as 'Dhavdi (Baravalia et al. 2011). All parts of this plant possess valuable medicinal properties *viz.* anti-inflammatory, anti-tumor, hepatoprotective and free radical scavenging activity (Chandan et al. 2008, Das et al. 2007), but flowers are in maximum demand (Baravalia et al. 2011). Development and commercialization of new bioactives and traditional preparations of *Woodfordia fruticosa*. (Lythraceae) is a single entity and has been collected, extracted by traditional method of using flower as a part, and targeted for the development of an effective anti-ulcer medicine as well as other pharmacodynamic activities and sufficient evidence about these activities exist in literatures (Plants 1968). Based on the available scientific data, the bioequivalent stable herbal formulation for clinical investigation studies and to develop it as a botanical drug for the treatment of anti-ulcer activity has been carried out at CSIR-IIIM, Jammu.

Plant Materials

The botanical raw material *Woodfordia fruticosa* (fresh flowers, Fig. 1) identified and voucher specimen RRLH18931 is deposited at the Janaki Ammal Herbarium of CSIR-Indian Institute of Integrative Medicine Jammu, India. This species is 'Least Concern' under the category of lower risk (Activite and Souris 2006) (IUCN red list of threatened species, 06 May 2015). The Plant materials were shade dried (below 35°C) up to moisture content not more than 10%w/w, and powdered (Sieve#40-65). The powdered raw materials packed in low density polythylene bags and kept in air tight containers, and later on analysed for its quality and taken up into further process in product development.

Fig. 1: Flowering portion of *Woodfordia fruticosa*

Chemicals and Reference Compounds

Standard marker compounds such as gallic acid, kaempferol and quercetin were purchased from Sigma-Aldrich (batch numbers # SLBL4700V, BCBQ1470V and SLBK4625V). The reference standards, reagents and chemicals used for toxicity studies were carboxy methyl cellulose (CMC) (lot number 0000149114 from Himedia Laboratories Pvt. Ltd., Mumbai and Rat pellet diet (M/S Altromin 1324, Germany) bearing batch no.: 040816/0823 supplied through M/S ATNT Laboratories, Mumbai. Acetonitrile, formic acid, hydrochloric acid (HCl) and sodium chloride (NaCl) were also used purchased from Lobal Chemie, Mumbai.

Apparatus

Agilent triple quad LC-MS/MS,HPLC (Shimadzu L), columns(YMCODS-A,3 mm, 150×4.6 mm; YMCODS-A, 5 mm, 250×4.6 mm;InertsilODS-3,3 mm, 150×4.6 mm; BDS Hypersil C18, 5 mm, 100×4.6 mm), Chromoliths high resolution RP-18 end capped, 50×4.6 mm, LC-MS (Waters), Oasis HLB1cc cartridge (Waters), sonicator and micropipettes were used for the study.

Extraction, Isolation and Qualitative Chemical Analysis of Markers

Markers gallic acid, kaempferol and quercetin were selected from flowers of *Woodfordia fruticosa*. These markers were first extracted from flowers using the optimised protocols and these isolated markers were characterised by comparison with their HPLC profile, 1H, 13C NMR and MS data of reference standards.

Composition

The raw material authentication was carried out following "Illustrated manual of herbal drugs in Ayurveda", published by Indian Council of Scientific & Industrial Research, New Delhi, India (Sarin et al. 1996). Macroscopic and microscopic characteristics of plant material and powders were studied in detail, and complete passport data were generated at CSIR-IIIM crude drug repository.

Method of Extraction

The plant materials of *W. fruticosa* were shade dried and then powdered by conventional method. The powdered materials were subjected to extraction with solvent ethanol (distilled from commercial grade ethyl alcohol): water (1:1), and left over-night (16 hrs) at room temperature (25-30°C). The contents were then filtered. The marc is reconstituted in the solvent mixture of ethanol: water (1:1) using lesser volume of the solvent mixture. The extraction process repeated three times more keeping same experimental conditions and worked up as described for first extraction. All the filtrates were then combined, passed through muslin cloth to remove insoluble/suspended material and then concentrated under reduced pressure 450 mm Hg (at 50°C) for recovery and recycle purpose of alcohol. Afterward remove maximum content of water through evaporation process under reduced pressure 450 mm Hg (< 100°C). The semi-concentrate is then subjected to freeze drying/spray drying and the dry extract powder passed through sieve # 40 mesh size and stored in air tight containers.

Standardization and Quality Control of Extract

Standardization and quality control of extracts were carried out in terms of Chemistry Manufacturing and Control (CMC) information. The extracts were evaluated for several parameters such as heavy metals (Pb, Cd, Hg and As), aflotoxins content (B1, B2, G1 and G2), acid insoluble ash, LOD at 105°C, total ash, water insoluble extractive and alcohol soluble extractive.

For determination of heavy metals approximately 0.25-1.0gm of test samples were digested with 5-8mL of nitric acid (concentrated) and the vessel was placed in the microwave oven. After completing digestion, vessels were removed from microwave and then cooled. The solution was transferred to 50mL centrifuge tube and made up to 50mL with elemental water. 100 μL of 10ppm internal standard was added to the sample in volume 50 mg (e.g. 20 ppb). The solution was injected directly into ICP-MS for determination of metals (AOAC 2002). Analysis of aflotoxins were performed as follows: To 50g of extract, 200mL of methanol and 50mL of 0.1N HCl was added. The resulting mixture was shaken at high speed for 5 min and filtered through Whatman Filter Paper 1. To 50mL of the filtrate, 50mL of 10% NaCl solution and hexane was added, and the mixture was shaken for 30 second. Hexane layer was discarded and aqueous layer was again partitioned using 25z dichloromethane. The lower layer of dichloromethane was collected and anhydrous sodium sulphate was added to remove water, if any. The partitioning was performed twice with dichloromethane as mentioned above. The collected elute was concentrated on steam bath and loaded on to silica gel column for separation of aflotoxins. Aflotoxins were eluted with 100mL of mixture of dichloromethane and acetone (9:1). Elute was evaporated on steam bath upto 6mL and was divided into 3 parts for further analysis. After transferring, 2mL of elute onto vial, it was evaporated to dryness using nitrogen. Derivatization of sample was carried out by adding 200 μL of hexane and 50 μL TFA. To this solution, 2mL of ACN-Water mixture (1:9) was added and vortexed for 30 second, 25 μL of lower aqueous layer was injected into LC system. For this, Shimadzu LC-10ATVP with auto sampler SIL-10 ADVP or equivalent; fluorescence detector Shimadzu RF-10 AXL EX360 nm, Em 450 nm fitted with PC windows 2000 was used. LC column-25 cm X 4.6 mm id, 5 μm RP-18 or equivalent with 20 cm X 1 cm id was used. The LC elution solvent was Water:Acetonitrile : Ethanol (700: 200: 200), individual aflatoxin concentration was calculated from standard calibration curve (AOAC, 1990). Microbial testing of extract was also performed to find out the total aerobic bacterial count; total yeast, mold counts; Enterobatecriace count, *Escherichia coli, Salmonella* spp. *Staphylococcus aureus* and *Pseudomonas aeruginosa.*

Passport Data of Botanical Raw Material (BRM)

- Plant source: *Woodfordia fruticosa* (L.) Kurz.

- Part used: Flowers

- Family: Lythraceae

- Habitat: Distributed throughout Northern parts India upto an elevation of 1500 m above mean sea level in sub-tropical forests.

- Season or time of collection: April

- Status: 'Least Concern' under the category Lower Risk. IUCN Red List of Threatened Species. Downloaded on 06 May 2015

- Harvest Location: CSIR-IIIM Experimental Farm, Chatha (*Latitude*: 32.664715; *Longitude:* 74.816203)

- Growth condition: Flowering

- Harvested time: April; 10: 00-14: 00hrs

- Collection: Fresh flowers were collected from the plants and dried in the shade below 35°C.

- Moisture contents of dried flowers: below 10% w/w.

Morphological Characterization of Flowers

Flowers of *Woodfordia fruticosa* is crimson red pedicels 1 cm long. Sepals 1-1.5 cm long, lobes 6, short, more or less triangular, alternate with small callous appendages. Petals 6, red, 3-4 mm long, lanceolate-acuminate. Stamens 12, inserted near the bottom of calyx tube. Ovary 4-6 mm long, oblong, 2-celled; ovules many; style 0.7-1.5 cm long. Capsules 0.6-1 cm long, 0.25-0.4 cm broad, ellipsoid, included in calyx. Seeds many.

Microscopic Characterization of Powder- form

Microscopic studies on powdered form of flowers of *Woodfordia fruticosa* showed unicellular trichomes, rosette and druse crystals, annular xylem vessels; pollen grains, ovules and cells filled with brown substances (Fig. 2).

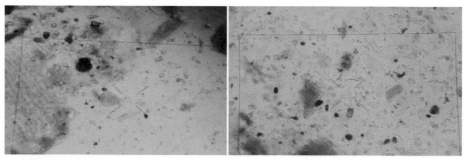

Fig. 2: Powder analysis of *Woodfordia fruticosa* showing unicellular trichomes, ovular parts, pollen grains and xylem vessels

Phytochemical studies of Woodfordia fruticosa have revealed the presence of different classes of chemicl compounds including steroids, fatty acids, flavonoids, tannins and flavonoids, Figure 1 represent the structure of the most important phytochemicals of the plant.

The major chemical constituents present in *W. fruticosa* are presented in Fig. 3-8.

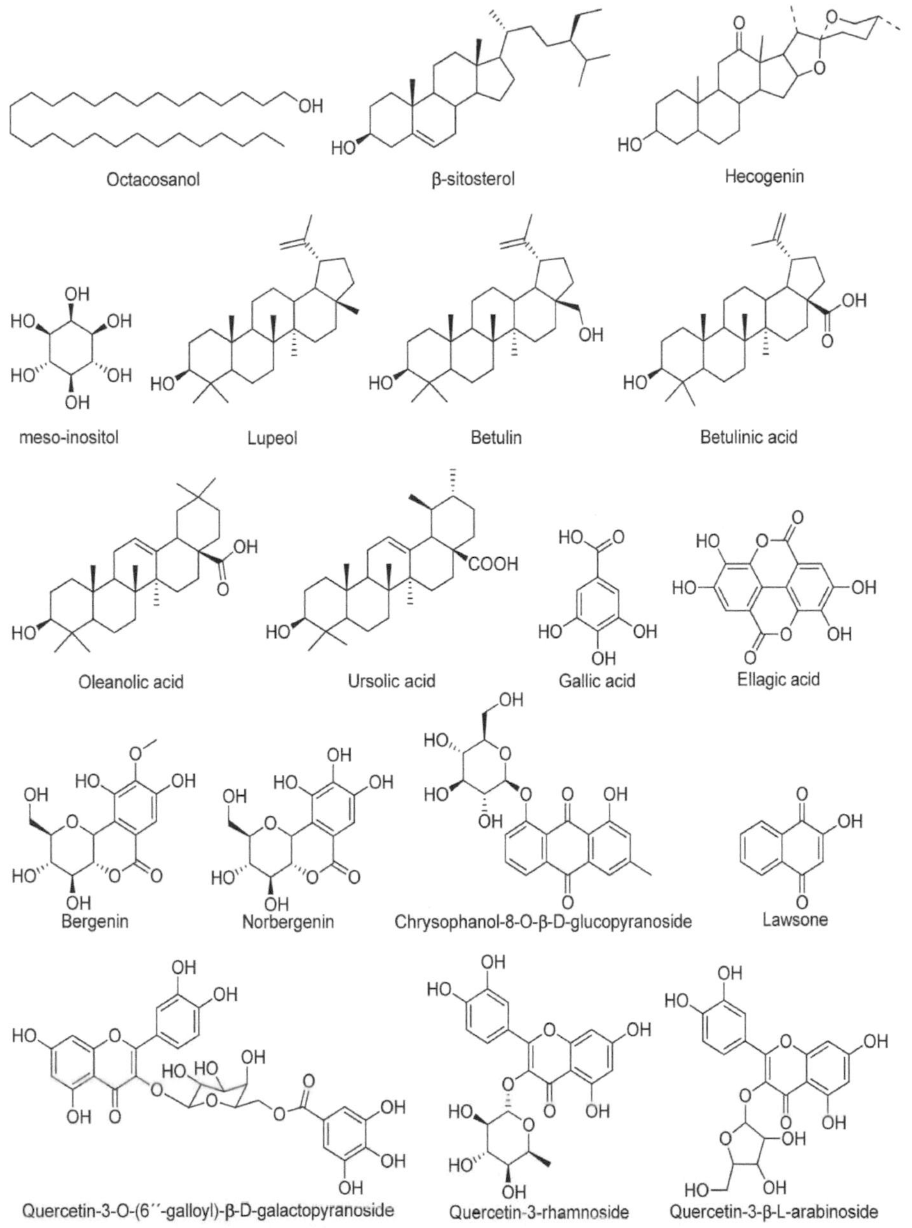

Fig. 3: Chemical constituents in *Woodfordia fructicosa*

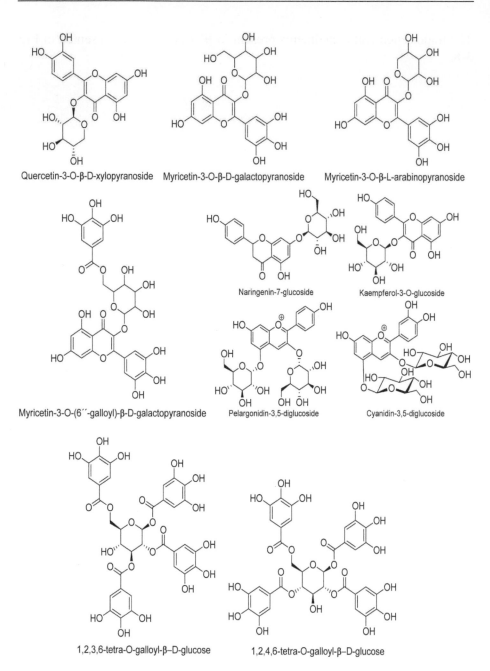

Quercetin-3-O-β-D-xylopyranoside   Myricetin-3-O-β-D-galactopyranoside   Myricetin-3-O-β-L-arabinopyranoside

Naringenin-7-glucoside

Kaempferol-3-O-glucoside

Myricetin-3-O-(6″-galloyl)-β-D-galactopyranoside

Pelargonidin-3,5-diglucoside

Cyanidin-3,5-diglucoside

1,2,3,6-tetra-O-galloyl-β–D-glucose

1,2,4,6-tetra-O-galloyl-β–D-glucose

Fig. 4: Chemical constituents in *Woodfordia fructicosa*

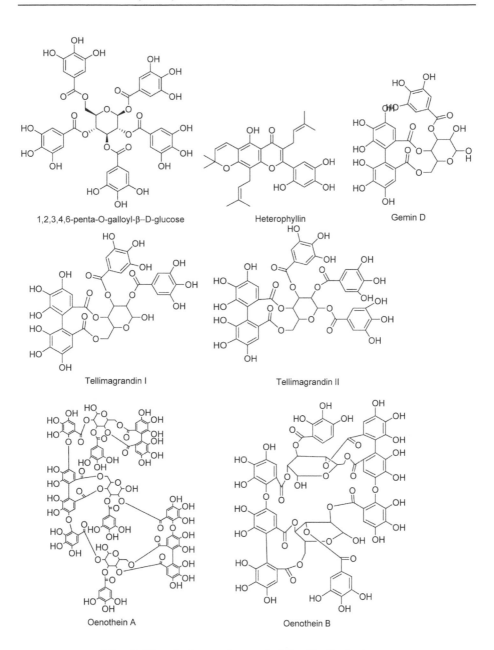

Fig. 5: Chemical constituents in *Woodfordia fruticosa*

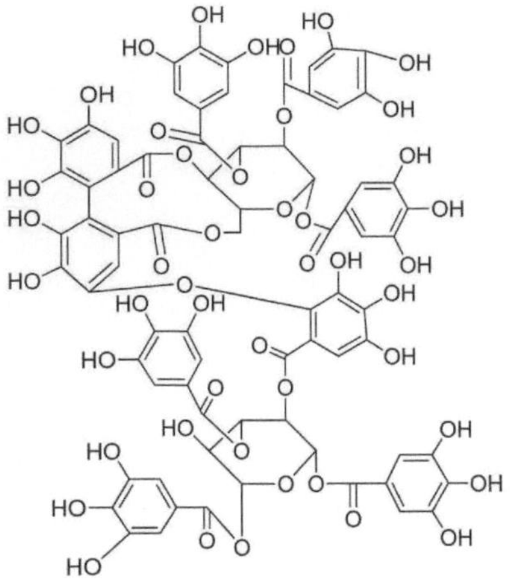

Woodfordin A

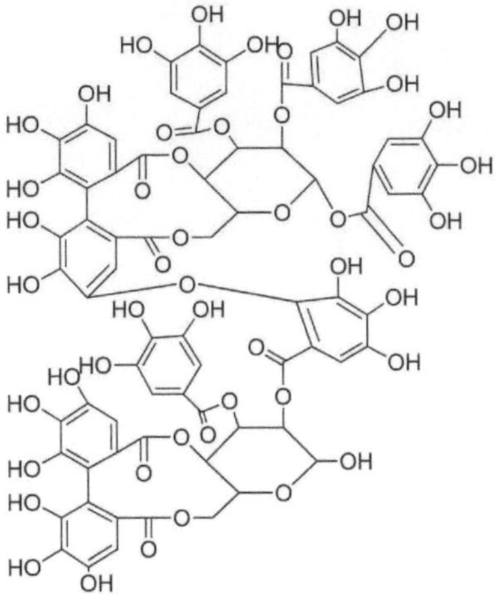

Woodfordin B

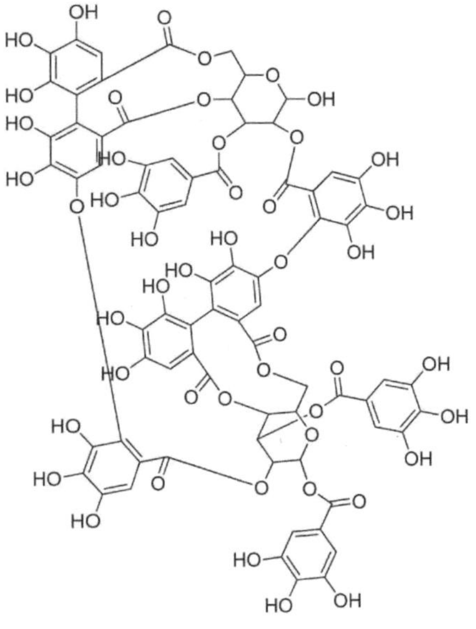

Woodfordin C

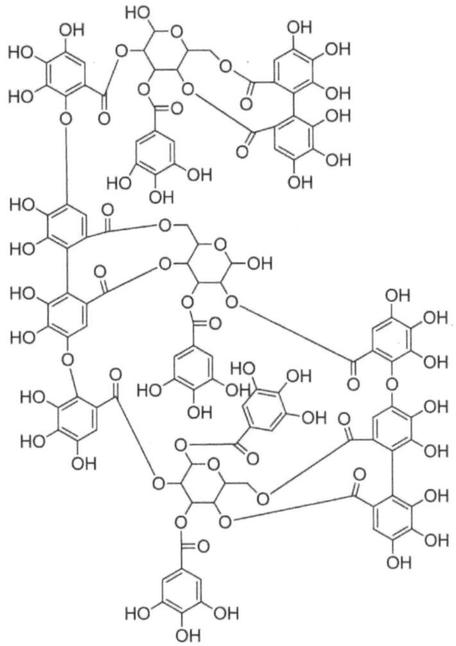

Woodfordin D

Fig. 6: Chemical constituents in *Woodfordia fruticosa*

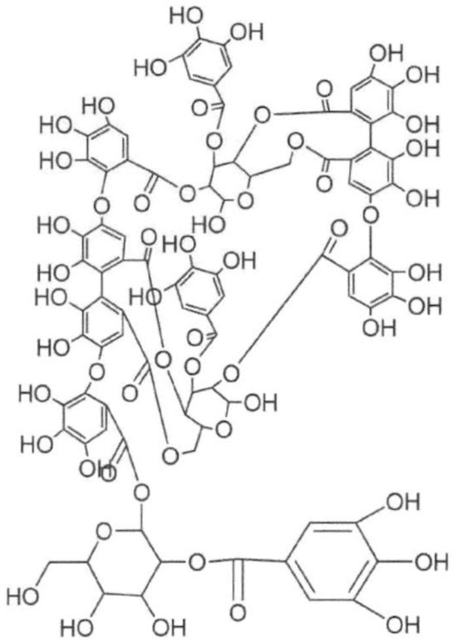

Woodfordin E

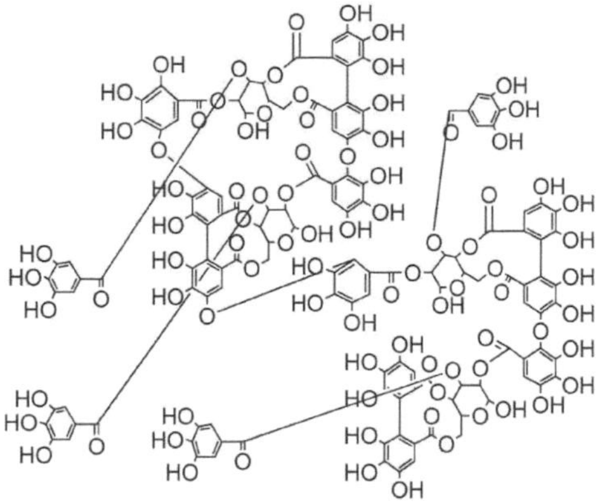

Woodfordin F

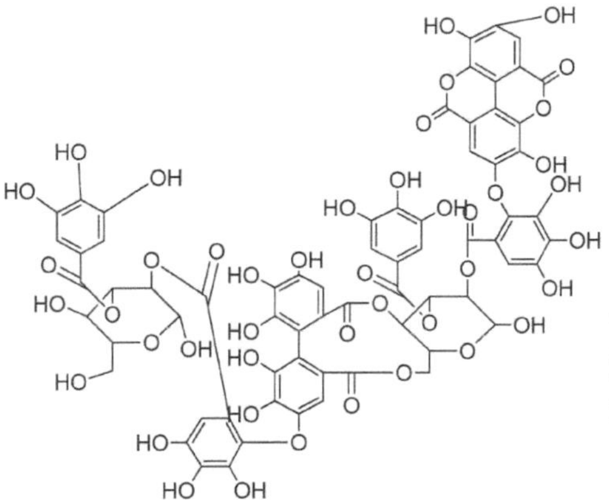

Woodfordin G

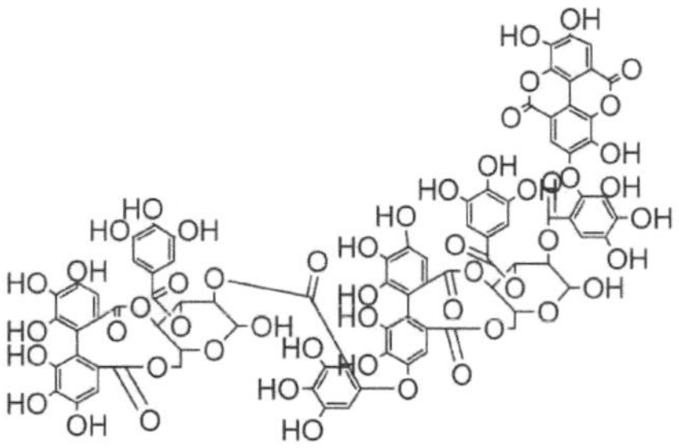

Woodfordin H

Fig. 7: Chemical constituents in *Woodfordia fruticosa*

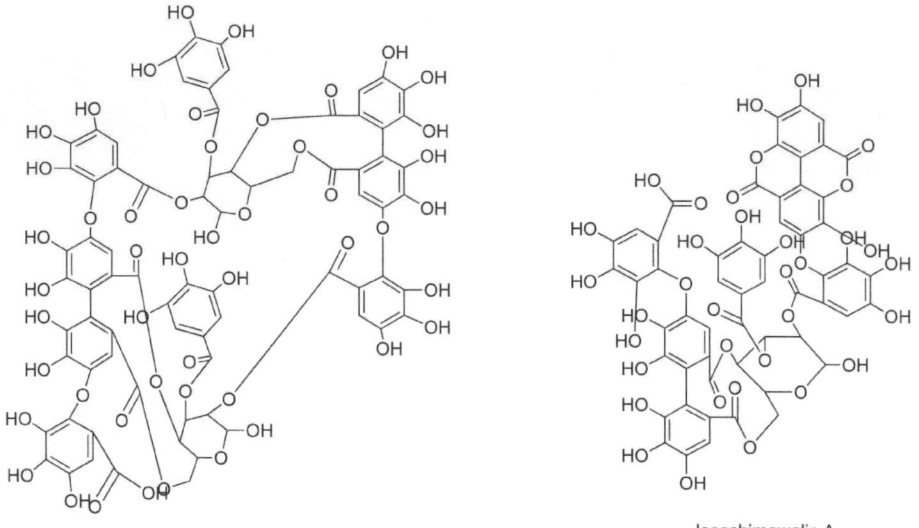

Woodfordin I

Isoschimawalin A

Fig. 8: Chemical constituents in *Woodfordia fruticosa*

Chemical Markers and Structure Elucidation

Following active markers were isolated and characterization has been done by the Mass, NMR, LC-Mass Spectroscopy:

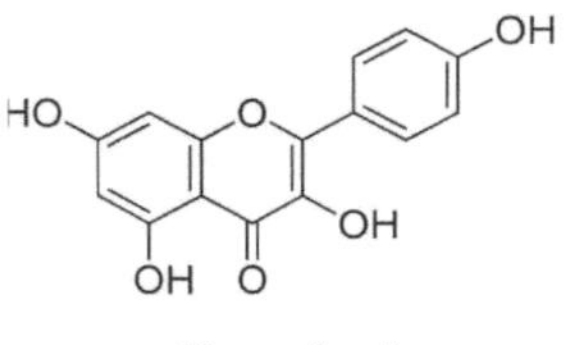

Kaempferol

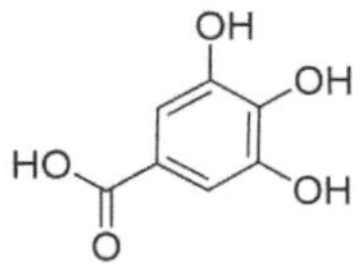

Gallic acid

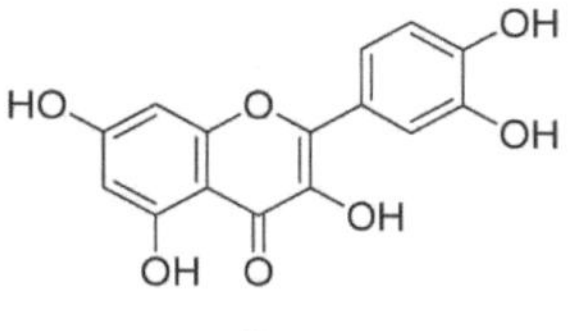

Quercetin

1. Kaempferol marker

Chemical name   : 2- (4- hydroxyphenyl)-3,5,7-trihydroxy-4H-chromen-4 one

Molecular formula   : $C_{15}H_{10}O_6$

Molecular weight   : 286.236 g/mol

HPLC purity   : 99.99%

^{1}H NMR and ^{13}C NMR spectrum (Fig. 9)

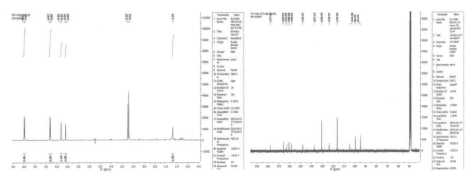

Fig. 9: Graphical representation of ¹H NMR and ¹³C NMR Spectrum of Kaempferol

2. Gallic Acid marker

Chemical Name   : Trihydroxybenzoic acid

Molecular Formula   : C7H6O5

Molecular Weight   : 170.12

HPLC Purity   : 89.18%

¹H NMR and ¹³C NMR Spectrum (Figure 10)

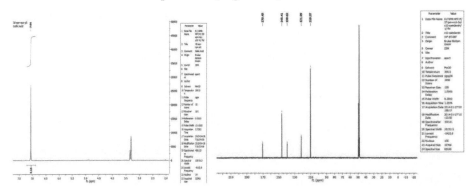

Fig. 10: Graphical representation of ¹H NMR and ¹³C NMR Spectrum of Gallic acid

3. Quercetin marker

Chemical Name   : 2-(3,4-dihydroxyphenyl)-3,5,7-trihydroxy-4H-
                       chromen-4-one

Molecular Formula   : C15H10O7

Molecular Weight   : 302.236 g/mol

HPLC Purity   : 99.99%

¹H NMR, ¹³C NMR Spectrum and HPLC prifiling (Fig. 11 and 12)

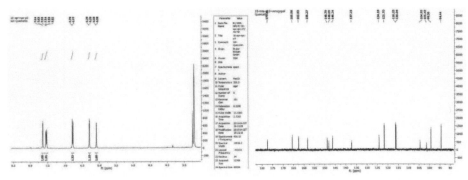

Fig. 11: Graphical representation of ¹H NMR and ¹³C NMR spectrum of Quercetin

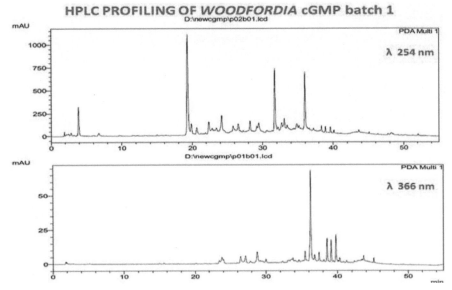

Fig. 12: Graphical representation of ¹H NMR and ¹³C NMR spectrum of Quercetin

PHARMACOLOGY ASPECTS

Evaluation of Anti-ulcer activity of ICB014 in Gastric Ulcer Models

Anti-ulcer activity of *Woodfordia fruticosa* was performed on wistar rats of either sex. The study protocols for pyloric ligation model (Acute model) and ethanol/ HCl induced chronic model were approved by institutional animal ethical committee CSIR-Indian Institute of Integrative Medicine, Jammu-180001: Animals were maintained under standard conditions in an animal house approved by Committee for the Purpose of Control and Supervision on Experiments on Animals (CPCSEA). The animals were housed in Poly propylene cages and maintained at 24°C ± 2°C under 12h light/ dark cycle and were feed ad libitum with standard pellet diet and had free access to water.

Acute Gastric Ulcer Model

Animals were divided into seven groups, each consisting of five rats. Control group received distilled water orally. Positive control group received Ranitidine at a dose of 100 mg/kg p.o. Remaining five groups rats received different doses of *Woodfordia fruticosa* the doses were 31.25, 62.5, 125, 250 and 500 mg/kg p.o. respectively. After 45 min of drug treatment, pyloric ligation was done by ligating the pyloric end of stomach of rats of respective groups under light ether anesthesia (Shay 1945).

Ligation was done without causing any damage to the blood supply of the stomach. Animals were allowed to recover and stabilize in individual cages and were deprived of water and food during postoperative period. After 4 h of surgery, rats were sacrificed, their stomach taken out, opened through the greater curvature, washed with normal saline and ulcer scoring was done as per the arbitrary scale and efficacy was compared with respect to control (Table 1).

Table 1: Anti-ulcer activity of ICB014 in pyloric ligation induced ulcer

Treatment	Ulcer Score	Inhibition (%)	ED50 (mg/kg)
CONTROL	3.6	0	—
Ranitidine 100 mg/kg	1.6 *	55.5*	—
ICB014, 31.25mg/kg	2.75	23.6	150
ICB014, 62.5 mg/kg	2.33*	35.18*	
ICB014, 125 mg/kg	1.66*	53.88*	
ICB014, 250 mg/kg	1.25*	65.27*	
ICB014, 500mg/kg	1.25*	65.27*	

N=5; *mean $p > 0.05$ using one way ANOVA when compared with control.

Ethanol / HCl-Induced Gastric Ulceration in Rats

The experiment was performed according to the method of (Mizui and Doteuchi, 1983) with certain modifications. Twenty-four hour fasted rats in the weight range of 180–200 g were taken. Drug and ulcerogenic agent were given according to the following schedule. 1.5 ml of ethanol/HCl mixture (70% ethanol and 5% HCl) were given on the first day, half its dose in the same volume on the second day and further half of the second dose in the same volume on the third day. Drug administration started from day 1 to 10. One group was kept as vehicle control which only received distilled water, group two received Ranitidine 100 mg/kg p.o and considered as positive control; whereas groups III, IV, V, VI, VII were treated with different doses i.e., 31.25, 62.5, 125, 250, 500 mg/kg p.o. respectively of hydro alcoholic extract of *Woodfordia fruticosa*. Feed was withdrawn 24 h before the last dose of drug. One hour after the last dose, 1.5 ml of ethanol/ HCl mixture (70% ethanol and 5% HCl) was given by intubation and the animals were sacrificed after 4 h.

The stomach was observed for percent protection of ulcer score according to the scale mentioned in Table 2.

Table 2: Anti-ulcer activity in 70% + 5% HCl induced ulceration

Treatment	Edema ± SEM	Inhibition (%)	ED50 (mg/kg
CONTROL	3.75±0.25	0	—
RANITIDINE 100 mg/kg p.o.	1.6±0.4*	57.33*	—
ICB014 500 mg/kg p.o.	0.8±0.34*	78.66*	162
ICB014 250mg/kg p.o.	0.9±0.29*	76.00*	
ICB014 125mg/kg p.o.	2.6±0.24*	30.66*	
ICB014 62.5mg/kg p.o.	2.8±0.37	25.33	
ICB014 31.5mg/kg p.o.	2.8±0.37	25.33	

*Values are Mean SEM, N=5; *mean $p>0.05$ using one way ANOVA when compared with control.

Stability Studies of Capsule Formulations

Three consecutive pilot batches were selected for conducting stability studies. These batches were manufactured by using method of manufacture and procedure that simulates the final process to be used for production batches. Herbal formulation (Capsule) was conducted the stability studies as per ICH Q1 guidelines at accelerated and Real time stability studies (40 ± 2°C, 75% ± 5% RH and 30 ± 2°C, 65%±5% RH respectively) (ICH 2003), and stability testing was done as per stability specifications at intervals of 0,3,6 months for Accelerated stability studies and 0,3,6,9,12,18,24,36 months for Real time stability. The stability analysis was done based on the estimation of bio marker i.e. gallic acid, quercetin and kaempferol (Table 3).

Table 3: Stability data of capsule prepared from *Woodfordia fructicosa*

Stability Data				
Accelerated Stability Condition: 40±2°C/75%±5%RH		Packing profile: Capsules are packed in food grade PVC white colour bottle.		
Capsule strength: 300 mg		Assay by HPLC (% w/w)		
		Gallic Acid	Quercetin	Kaempferol
	Specification Limit	0.06-3.0% w/w	0.01-1.0% w/w	0.01-1.0%w/w
Batch Numbers				
BEXTD 0001	0 Month	0.74	0.06	0.04
DEXTD 0001		0.84	0.06	0.04
DEXTD 0002		0.74	0.06	0.04
BEXTD 0001	3rd Month	0.69	0.01	0.01
DEXTD 0001		0.83	0.01	0.01
DEXTD 0002		0.82	0.01	0.01

Contd.

Real time stability conditions: 30±2°C/65%±5%RH

BEXTD 0001	0 Month	0.74	0.06	0.04
DEXTD 0001		0.84	0.06	0.04
DEXTD 0002		0.74	0.06	0.04
BEXTD 0001	3rd Month	0.85	0.01	0.02
DEXTD 0001		0.96	0.02	0.04
DEXTD 0002		0.87	0.01	0.02

Standardization, Characterization, Analytical Method Development and Quality Control Parameters

Standardization and quantitative estimation of markers in *W. fruticosa* and to established the HPLC method of this formulation with three markers. Three markers kaempferol, gallic acid and Quercetin were selected for quantification. The quantification of these markers in the flowers of said plant was determined using HPLC analysis. The markers (% w/w) quantified in *W. fruticosa* were 0.06-3.0%w/w of gallic acid, 0.01-1.0%w/w of quercetin and 0.01 – 1.0% w/w of kaempferol. Gallic acid, kaempferol and quercetin are the stable markers. Gallic acid is the bio active marker and it shows in the pharmacokinetic studies

CONCLUSION

Published works showed that *Woodfordia fruticosa* has been used pharmacologically in the treatment of various disease conditions. The main chemical constituents of this plant are woodfordins A, B, C, D, E, F, G, H and I and were identified from the flowers, and compound shows various pharmacological activities such as anti-inflammatory, anti-tumor, hepatoprotective and free radical scavenging. In summary, standardization, characterization and quality control of anti ulcer herbal *Woodfordia fruticosa* extract was carried out at QC/QA department. CSIR-IIIM, Jammu. Based on the available marker based (gallic acid, kaempferol and quercetin) stability data the proposed expiry date of herbal formulation is two years from the date of manufacturing. From the above enlisted three markers, gallic acid is the bioactive molecule for this plant. Hence, it can be considered to be safe botanical for the human use with an expiry date for two years.

Abbreviations

CMC   :   Chemistry Manufacturing control

USFDA :   United States Food and Drug Administration

EMA   :   European Medicines Agency

ICH   :   International Council for Harmonisation

WHO   :   World Health Organization

MS   : Material Safety

NMR   : Nuclear Magnetic Resonance

CSIR   : Council of Scientific and Industrial Research

RRLH   : Regional Research Laboratory Herbarium

IUCN   : International Union for Conservation of Nature

LOD   : Loss on Drying

ICP-MS : Inductively coupled plasma mass spectrometry

HPLC   : High Performance Liquid Chromatography

ACKNOWLEDGEMENTS

Authors would like to thank Dr. Ram A. Vishwakarma, Director IIIM Jammu, for encouragement and facilities. Authors are also thankful to QC/QA Department, CSIR-IIIM Jammu for the analysis of extract. It bears institutional publication No. IIIM/2251/2018.

REFERENCES CITED

Activite EDEL, Souris CLA. 2006. Woodfordia fruticosa 8235.

AOAC 1990. Official Methods of Analysis of AOAC International. AOAC Int. 1–25.

AOAC. 2005, 2002. AOAC Official Method 999.10 Lead, Cadmium, Zinc, Copper, and Iron in Foods Atomic Absorption Spectrophotometry after Microwave Digestion. *Gaithersburg, MD*, Ed. 18. 18–20.

Baravalia Y, Nagani K, Chanda S. 2011. Evaluation of pharmacognostic and physicochemical parameters of *Woodfordia fruticosa* Kurz. Flowers. *Pharmacognosy Journal* 2: 13–18.

Baravalia Y, Vaghasiya Y, Chanda S. 2011. Hepatoprotective effect of Woodfordia fruticosa Kurz flowers on diclofenac sodium induced liver toxicity in rats. *Asian Pacific Journal of Tropical Medicine* 342-346.

Chandan BK, Saxena AK, Shukla S, Sharma N, Gupta DK, Singh K. 2008. Hepatoprotective activity of *Woodfordia fruticosa* Kurz flowers against carbon tetrachloride induced hepatotoxicity. *Journal of Ethnopharmacology* 119(2): 218-224.

Chauhan JS, Srivastava SK, Srivastava SD. 1979. Chemical constituents of the flowers of Woodfodia fruticosa linn. *Journal of Indian Chemical Society* 56:1041.

Chauhan JS, Srivastava SK, Srivastava SD. 1979b. Chemical constituents of the flowers of Woodfordia fruticosa linn. *Journal of Indian Chemical Society* 56:1041.

Dan S, Dan SS. 1984. Chemical examination of the leaves of *Woodfordia fruticosa*. *Journal of Indain Chemical Society* 61:726-727.

Das PK, Goswami S, Chinniah A, Panda N, Banerjee S, Sahu NP. 2007. *Woodfordia fruticosa*: Traditional uses and recent findings. *Journal of Ethnopharmacology* 110(2): 189-199.

Desai HK, Gawad DH, Govindachari TR, Joshi BS, Kamat VN, Modi JD et al., 1971. Chemical investigation of some Indian plant. VI. *Indian Journal of Chemistry* 9:611-613.

ICH. 2003. Stability Testing of New Drug Substances and Products Q1A(R2). Int. Conf. Harmon. 24. https://doi.org/10.1136/bmj.333.7574.873-a

Kadota S. Takamori Y, Nyein KN, Kikuchi T, Tanaka K, Ekimoto H. 1972. Constituent of the leaves of Woodforida fruticosa Kurtz. Isolation, structure and proton and carbon 13 nuclear magnetic resonance signal assignments of Woodfruticosin (Woodforidin C), an inhibitor of deoxyribonucleic acid topoisomerase II. *Chemical and Pharmaceutical Bulletin* (Todyo). 38:2687-2697.

Kalidhar SB, Parthasarathy MR, Sharma P. 1981. Nobergenin, a new c-glycoside from Woodforida fruticosa Kurtz. *Indian Journal of Chem.* 20:720-721.

Kong S, Application F, Data P. 2011. United States Patent 2, 12–15. https://doi.org/10.1016/j.(73)

Mizui T, Doteuchi M. 1983. Effect of polyamines on acidified ethanol-induced gastric lesions in rats. *Japanese Journal of Pharmacology* 33: 939–945

Nair AGR, Kotiyal JP, Ramesh P, Subramanian SS. 1976. Polyphenols of the flowers and leaves of *Woodforida fruticosa*. *Indian Journal of Pharmacy* 38:110-111.

Plants I. 1968. Screening of Indian Plants 6.

Saoji AG, Saoji AN, Deshmukh VK. 1972. Presence of lawsone in Ammania baciferra Linn. And Woodfordia fruticosa salisb. *Current Science* 41-192.

Sarin YK. 1996. Illustrated Manual of Herbal Drugs Used in Ayurveda. Indian Council of Scientific & Industrial Research, New Delhi.

Shay H. 1945. A simple method for the uniform production of gastric ulceration in the rat. *Gastroenterology* 5: 43–61.

Yoshida T, Chou T, Haba K. Okama Y, Shingu T, Miyamoto K et al. 1989a. Macro cyclic ellagitanin dimmers and related dimmers and related dimmers and their anti tumor activity. *Chemical and Pharmaceutical Bulletin* 37:3174-3176.

Yoshida T, Chou T, Nitta A, Miyamoto K. Koshiura R, Okuda T. 1990. Woodfordin C. a macro cyclic hydolysable tannin dimer with antitumor activity and accompanying dimmers from Woodfordia fruticosa flowers. *Chemical and Pharmaceutical Bulletin* 38:1211-1217.

Yoshida T, Chou T, Nitta A, Okuda T. 1992. Tannins and related polyphenols of lythraceous plants III hydrolysable tannins oligomers with macro cyclic structures and accompanying tannins from Woodfordia fruticosa Kurz. *Chemical and Pharmaceutical Bulletin* 40: 2023-2030.

Yoshida T, Chou T. Matsuda M, Yasuhara T, Yazaki K, Hatano T et al. 1991. Trimeric hydrolysable tannins of macro ring structure with anti tumor activity. *Chemical and Pharmaceutical Bulletin* 39:1157-1162.

3

Tinospora cordifolia (Willd.) Miers ex Hook.f. & Thomson (Giloy), a Multifaceted Elixir Plant of India

Neha Sharma and Naresh Kumar Satti

Natural Product Chemistry Division, CSIR-Indian Institute of Integrative Medicine Jammu – 180001, Jammu and Kashmir, INDIA
Email: nehasharma.ns27@gmail.com, nksatti@iiim.ac.in

ABSTRACT

Tinospora cordifolia (Willd.) Miers ex Hook.f. & Thoms. is a versatile liana rich in various active compounds having diverse chemical structures. Limited research has been done on the biological properties and the probable medicinal applications of the individual isolated molecules actually responsible for the medicinal nature of the plant. Recent advances in the discovery of active constituents from *T. cordifolia* and their biological impact in disease cure has led to an increase interest on this plant across the globe. The therapeutic virtue of giloy used in traditional and Indian System of Medicine (ISM) has been well established through advance testing and evaluation, both pre-clinical and clinical trials in different diseased conditions. These studies make this indigenous drug plant a novel candidate for bio-prospection and drug development for the treatment of diseases such as cancer, ulcers, liver disorders, heart diseases, diabetes and post-menopausal syndrome where sufficient treatments are still not available. Over the past few decades, a noteworthy progress has been achieved in cancer prevention and treatment, and still there remains a need for effective treatment program. One of the best possible approaches with immense potential is chemoprevention, which involves the use of natural, synthetic or biological molecules to suppress, reverse or prevent cancer or the progression of cancerous cells. It is most likely that many agents present in specific herbs, herbal products and natural dietary factors will have an effect throughout the process of carcinogenesis. Thus, a detailed chemical investigation with the aim of isolating novel molecules, their structure elucidation, quantification, validation and pharmacological evaluation in terms of its anticancer potential was carried out on *T. cordifolia*. The present study was designed to isolate and identify new promising anti-cancer candidates from the aqueous alcoholic extract of *T. cordifolia* using bioassay-guided fractionation.

Keywords: Medicinal Plant, *Tinospora cordifolia*, DAPI, Apoptosis, MMP potential, ROS.

INTRODUCTION

Tinospora cordifolia (Willd.) Miers ex Hook.f. & Thomson (Family Menispermaceae) is the most versatile rejuvenating perennial liana used in Ayurvedic system of medicine, known for building up the immune system and the body's strife against various infecting organisms since ages (CSIR 2003, Aima 2003, Vaidya 1994). Apart from more than 100 common names, such as Guduchi, Amrit (Sanskrit) and Abb-e-Hyat (Urdu) meaning like Guduchi, Amrita, Amritavalli, Madhuparni, Guduchika, Chinnobhava, Gilo (Punjabi), Gulancha (Bengali), Chakralakshanika (Sanskrit), Gurcha (Hindi), Amrutavalli (Kannada), Galac (Gujarati), Gilo (Kashmiri), Gulvel (Marathi), Guluchi (Oriya), Seendal, Chittamrutu (Malayalam), Thippateega (Telugu), Amarlata (Assamese), Siddhilata, Heartleaf Moonseed, Seendil Kodi (Tamil), Tinospora (English), Amrit (Sanskrit) Guduchi, gurcha (Hindi), and Abb-e-Hyat in Urdu (meaning water of life), it is also known as Giloy. Giloy refers to a traditional term that alludes to the heavenly panacea which gives new life to the whole body, increases the human life span and keeps them perpetually young (Tirtha 2005, Singh et al. 2003).

Giloy has widely been employed as an ayurvedic remedy for the treatment of diabetes, gonorrhea, syphilis, rheumatism, cancer, jaundice, leprosy, general debility, fever, heart diseases, helmenthiasis, dysentery, chronic diarrhoea, dermatological diseases, bone fractures, liver, intestinal disorders and asthma (Sinha 2004, Khare 2007, Singh & Maheshwari 1983, Shah 1984, Bhatt & Sabnis 1987, Shah et al. 1983, Kirtikar &Basu 1975, Mhaiskar et al. 1980). In 1932, Indian Pharmacopoeia made this drug an official preparation. It also became a part of the Bengal Pharmacopoeia in 1844. According to the United States Dispensatory, it has an extensive history of its use as medication and in the preparation of palo or as gilaeka-sat, a highly nutritive and digestive starch used in the treatment of many diseases. Its therapeutic potential is attributed to its ability to revitalize, strengthen, detoxify and in purifying the whole system, specifically via liver (Zhao et al. 1991). Numerous monoherbal and polyherbal preparations prepared from various parts of the plant have been prescribed in Ayurveda as well as in other systems of medicine. It is also used as a combination remedy with other drugs, as an anti-dote for curing snake bite (Sudha et al.

2001, MHFW 2001). Although all parts of the plant have medicinal value, but Ayurvedic Pharmacopoeia of India has approved its stem for use in medicine due to high content of alkaloid than in the other parts of the plant (MHFW 2001).

The plant is used in numerous ayurvedic preparations such as Amritashtaka churana, Dashmoolarishta, Sanjivani vati, Kanta-kari avleha, Guduchyadi churna, Chyavanprasha, Guduchi saatva, Brihat, Guduchi taila, Stanyashodhana kashaya churana, Punehnimba churna, Guduchi ghrita, andAmrita guggulu (MHFW 2001). Myriad of potent phytoconstituents like alkaloids, tannins, phenolics, cardiac glycosides, flavanoids, sesquiterpenoids, saponins and steroids isolated from different parts of plant have been credited for ts numerous medicinal applications (Kiem et al. 2010, Fukuda et al. 1993, Rout 2006, Ahmed et al. 2006) (Table. 1). It's been a decade that the plant has been exploited for its extensive pharmacological, phytochemical, and clinical investigations, and has come out with interesting findings as to how it directly or indirectly affects the various metabolic cascades in case of various diseases. Medicinal value of this climbing shrub has become a matter of great significance particularly in case of some severe diseases like cancer as reported by various researchers (Adhvaryu et al. 2008), but still a detailed investigation on the chemical constituents, actually responsible for its anti-cancer potential is required. Hence, it can be chosen as a source of new lead anti-cancer molecules for the development of industrial products for treatment of various diseases.

DISTRIBUTION

T. cordifolia is distributed throughout the tropical Indian Sub-continent *viz*. China, as well as in Sri Lanka and Myanmar. There is evidence that the plant is also found in some tropical regions of Australia and Africa. In India, it grows right from the Kumaon Mountains to Kanyakumari. It is a very common plant of dry and deciduous forests.It can grow at any temperature and has a long life. It is also found in other tropical and subtropical environment of Sri Lanka and Myanmar ascending upto an altitude of 300 meters (Sinha 2004, Khare 2007).

BOTANICAL DESCRIPTION

T. cordifolia, a deciduous plant which is often found climbing up to the trunks of large neem trees, can grow in different types of soils, ranging from acidic to alkaline with moderate moisture. Stems of the *T. cordifolia* are succulent, having long filiform fleshy aerial roots, arising from the branches, varies in thicknesses, from 0.6 to 5 cm in diameter; young stems are green in colour with smooth surfaces and swelling at nodes, while circular lenticels impart warty protuberances in the light brown colored older stems. The plant is a succulent-stemmed, twinning, fast growing liana with grey-green branches becoming brown with age and tuberous roots, heart shaped, juicy, cordate and membranous leaves with circular petiole and mid-rib, 5 to 12 cm in diameter wood stem is soft, porous and white. The plant blooms in spring to tiny, unisexual, small and greenish flowers. Creamish white or greyish bark is thin with fleshy stem when peeled off. Male and female flowers are formed on different branches. Male flowers are small, green or yellow clusters, and female flowers are usually solitary. Fruits are fleshy, pea shaped, shiny, turn red when boiled and appear in winters (Singh et al. 2003, Sinha 2004, Khare 2007, Singh & Maheshwari 1983).

PHARMACOLOGY

Tinospora cordifolia is a medhya rasayana which enhances learning power and memory as per Ayurveda. Assorted classical texts of Ayurveda has mentioned Guduchi in Sushrut, Ashtang Hridaya, Charak and other treaties like Dhanvantari Nighantu and Bhava Prakash under different names such as Chinnarrhuha, Amritvalli, Amara, Vatsadani and Chinnodebha (Sinha 2004). In Sushurta Samhita, it has been used in the treatment of Kusth (leprosy), Svasa (asthma), Aruchi (anorexia) and a kind of fever known as Maha-jvara (MHFW 2001), while in Ashtang Hridaya and Charak Samhita. It has been employed in body conditions like Jvara (fever), Vat Rakta (gout) and Kamala (jaundice) (Upadhaya et al. 2010, Jeyachandran etal. 2003). It is also considered as a potent aphrodisiac, astringent, diuretic and bitter tonic, curative agent for skin infections, chronic diarrhoea, diabetes, jaundice and dysentery (Kumar et al. 2016). In Dhanvantari Nighantu, it has been claimed for curing bleeding piles, itching, erysipelas and promoting longevity (Tirtha2005). It is employed in curing many diseases causing debility due to its rejuvnating properties. The nutritious fecula is recommended in diarrhoea, cold fevers, in urinary infections, seminal weakness, skin diseases, jaundice, various forms of irritability of stomach and diabetes (Grover et al. 2000). It is traditionally believed that Guduchi Satva from *T. cordifolia* growing on *Azadirachta indica*, Neem tree possesses the medicinal properties of neem. It is more bitter but comparatively more efficacious than the normal tinospora herb (Jagetia & Rao2006). It was also included in the Bengal Pharmacopoeia of 1868 (Sinha 2004) when European physicians in India noticed the importance of this drug as an antiperiodic, tonic and for its diuretic properties (Upadhaya et al. 2010). It is widely employed in Unani system of medicine mostly as "Sat Giloe". Apart from that, "Arq Giloe" prepared from the fresh plant is considered a febrifuge, while "Arq Maul Laham Mako-kashiwala" is a general tonic (Singh et al. 2003). *T. cordifolia* has been acknowledged for its properties and uses in folk or tribal medicine in various regions of the country (Table 1).

Table 1: Uses of *Tinospora cordifolia* in folklore medicine

Tribals and areas	Diseases	Mode of application
Baiga, living in the interior areas of Naugarh and Chakia blocks of Varanasi district, Uttar Pradesh	Fever	The pills are prepared from the paste of stem of Guduchi (*T. cordifolia*) and the roots of Bhatkatiaya (*Solanum surattense*).
The tribals of Mumbai and its neighboring areas and the fishermen along the sea coast	Fever, jaundice, chronic diarrhea, periodic fever	The whole plant is used.
Tribals of Jammu (J & K) and Bigwada (Rajasthan)	Fever	Decoction of stem is administered orally.
The inhabitants of Bhuvneshwar (Orissa)	Fever	The warm juice of root of *T. cordifolia* orally.
Inhabitants of Banka (Bihar)	Balashosha (Emaciation in children), daha (burning)	Dyed shirt soaked in juice of Guduchi worm by children for Balashosha.Paste or juice of Amrita (*T. cordifolia*) leaves and Sarsapa beeja churna (seed powder of *Brassica campestris*) are used for relieving burning sensation.
Local people of Patiala (Punjab)	Fever	Juice or decoction of leaves is administered orally with honey.
The Muslim tribals of Rajouri, Jammu (Tawi) comprising Gujjars and Backwals	Bone fracture	Whole part is used.
Local women of Arjunpura (Rajasthan)	Raktapradara (leukorrhea)	Paste of Guduchi (*T. cordifolia*) and 5 seeds of Krishnamarich (*Piper nigrum*) is administered orally daily in morning.
The inhabitants of Badala (UP)	Swasa (Asthma)	Juice of stem orally with honey.
People of Dehrabara Kolaras, Sivpuri District of M.P.	Twak-roga (Skin disease)	Decoction of stem is administered orally.
Mundas of Chhota Nagpur	Fracture	Paste of whole plant used as plaster.
In certain parts of India	Bites of poisonous insects and venomous snake, eye disorders.	The paste of Guduchi is applied to the part bitten and administered internally through mouth at intervals of half an hour.Juice or decoction of the root is poured into the eyes.

Tinospora cordifolia is a miraculous liana which possesses anti-diabetic, hypolipidemic immunomodulatory, antimutagenic/ anticarcinogenic activity and several other medicinal properties (Table 2) (Adhvaryu 2008, Upadhaya 2010, Jeyachandran 2003, Kumar 2016, Grover 2000, Jagetia &Rao 2006, Premanath 2010). Polysaccharide rich fraction of *T. cordifolia* is very effective in reducing the metastatic ability of melanoma cells (Jagetia & Rao 2006). Water extract of the roots of *T. cordifolia*s shows a good hypolipidemic activity. Apart from being a powerful emetic, the roots are effective in treating visceral obstructions.The root extract prepared in water, exhibiting anti-diabetic effect is also used in leprosy. Strong antioxidant activities are possessed by the extracts of leaves, stem, roots and bark of this plant. This amazing Ayurvedic herb helps in increasing the efficiency of white blood cells, thereby increasing the ability to build up body's immune system. It's a boon for the tribal people and fishermen of India for the effective treatment of jaundice, fever, dysentery and chronic diarrhoea. Juice prepared from the leaves of Guduchi is effective in treating fever when it is orally administered along with honey (Singh 2003, Sinha 2004, Shah 1984). The powdered stem and root is taken along with milk for the treatment of cancer (Bhatt & Sabnis1987). Oral administration of decoction of stem is used for the treatment of various skin diseases, while the decoction with cold or hot water in morning on empty stomach is used as a tonic in general debility (Shah et al. 1983). The Muslim, Gujjar, and Backwal tribals of Jammu (India) use this plant in the treatment of bone fractures (Patel et al. 2010). Paste and juice of leaves is applied locally for relieve of burning sensation. Juice of stem with honey is used orally for treatment of asthma (Sinha 2004). Considerable response has been found in children for treating various mental deficit and behaviour disorders, along with improvement in IQ levels (Soni et al. 2011).

Table 2: Pharmacological activities of various parts of *Tinospora cordifolia*

Activity	Plant Part	Extract
Psychopharmacological	Whole plant (Agarwal et al. 2002)	Alcohol and aqueous
	Leaves (Murthy et al. 2010)	Petroleum ether and ethanol
	Stem (Patil et al. 1997,	Petroleum ether, aqueous
	Rawal et al. 2004)	and alcohol
Anti-diabetic	Whole plant (Wadood et al. 1992,	Aqueous, alcohol and
	Sengupta et al. 2009)	chloroform
	Leaf (Sai & Srividya, 2002)	Aqueous
	Roots (Prince et al. 1999,	Alcohol and aqueous
	Puranik et al. 2007)	
	Stem (Patel & Mishra 2011)	Aqueous, alcohol, hexane,
		ethyl acetate and methanol
Anti-inflammatory	Whole plant(Wesley et al. 2008)	Aqueous and alcohol
	Stem (Chopra et al. 2011)	Aqueous
	Leaves (Paval et al. 2011)	Alcohol
Anti-cancer	Whole plant (Muniyappan	Alcohol, aqueous,
	et al. 2009, Jagetia et al. 1998)	methylene chloride,CCl_4
		and methanol
Anti-allergic	Stem (Zalawadia et.al. 2009)	Aqueous
Anti-fertility	Stem(Gupta & Sharma 2003)	Hydromethanolic
Anti-oxidant	Root(Prince et al. 2004,	Aqueous
	Tyagi et al. 2009)	
	Leaf(Jadhao et al. 2009,	Acetone
	Sarla et al. 2011)	
Immunomodulatory	Stem(More & Pai 2011,	Methanol, aqueous, ethanol
	Chintalwar et al. 1999)	and hydromethanolic
	Whole plant(Kalikar et al. 2008)	Aqueous decoction
Radioprotective	Roots (Sharma et al. 2011)	Aqueous
	Whole plant (Tyagi et al. 2009)	Hydroalcoholic
Anti-microbial	Stems (Tiwari et al. 2011)	Aqueous and ethanol
	Stems and Leaves	CCl_4 fraction of methanol
	(Manjusha et al. 2011)	extract
	Roots (Rose et al. 2010)	Methanol extract
Anti-parasitic	Leaves (Jayaseelan et al. 2011)	Aqueous
	Whole plant (Singh 2005,	Aqueous
	Asthana et al. 2001)	
Osteoporotic	Stem (Kapur et al. 2008)	Ethanol
Anti-obesity	Stem (Dhingra et al. 2011)	Petroleum ether
Spasmolytic and anti-diarrheal	Stem (Tiwari et al. 2011)	Ethanol
Anti-ulcer	Whole plant (Bairy et al. 2002)	Alcohol
Cardioprotective	Whole plant (Rao et al. 2005)	Alcohol
Hepatoprotective	Whole plant (Panchabhai et al. 2008,	Alcohol
	Patel et al. 2009, Yang et al. 2010)	

PHYTOCHEMISTRY

A broad spectrum of chemical compounds with their well elucidated chemical structures including alkaloids, glycosides, sesquiterpenoids, aliphatic compounds, polysaccharides, phenolics, steroids such as tinosporaside, tinosporine, tinosporide, cordifolide, heptacosanol, cordifol, clerodane furano diterpene, tinosporidine, columbin and β-sitosterol have been reported to be isolated from this plant (Sriramaneni et al. 2010, Yang et al. 2010, Zhao et al. 2008, Kohno et al. 2002) (Fig. 1). The plant is rich in steroids, alkaloids, aliphatic compounds, glycosides, fatty acids, sesquiterpenoids, polysaccharides and essential oils. Alkaloids like Berberine, Tembetarine, Palmatine, Tinosporin, Magnoflorine Choline, etc., and glycosides like Tinocordifolioside, Tinocordiside, etc. have been isolated (Rout 2006, Upadhaya et al. 2010, Patel & Mishra 2011, Jagetia & Rao 2006, Patel et al. 2009, Gupta & Sharma 2011). X-ray emission technique based on particle-induction revealed high concentrations of Ca, Cl and K in the water extract of this medicinal plant and appreciable levels of Zn and Mn were identified from the leaves of the plant respectively. Leaves of the plant contain 11.2% of protein and a fair amount of phosphorus and calcium (Tirtha 2005). Studies on the chemical composition of the starch of *T. cordifolia* (extract) were conducted and the polysaccharide fraction was found to contain 1→4 linked glucan with occasional branching points (Soni et al. 2011, Padmapriya 2009). The water extract of Giloy contains arabinogalactan that has shown immunological activity. An arabinogalactan is obtained from extracts prepared from stems of *T. cordifolia*. The methanolic extract of giloy contains phenylpropanoids, diterpene furon glycosides, norditerpene furan glycosides and phytoecdysones (Lv et al. 2012, McKeown et al. 2012, Sundarraj et al. 2012).As per one of the post-harvest experiment conducted on this herb, the herb on mechanical drying at about 40°C provides the maximum percentage of alkaloid (tinosporin) content, which is 0.045%. This content decreases to 0.033% on drying either at a temperature of 60°C or on direct exposure to sunlight. Further, the small dried bits of stems packed in polyethylene bags are preferred as they remain consistent in the alkaloid content (0.042%) in comparison to the storage of the same under ambient condition (Padmapriya et al. 2009), which indicates the temperature sensitive or photosensitive nature of tinosporin.

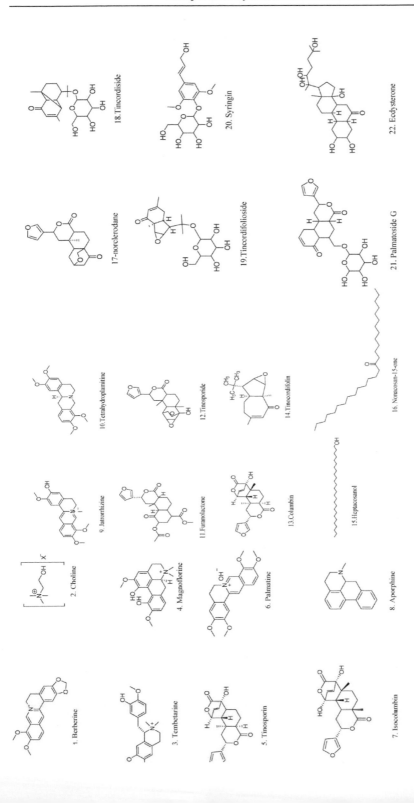

Fig. 1: Chemical structures of major molecules present in *Tinospora cordifolia*

EXPERIMENTAL PROCEDURES, CHEMICALS AND BIOCHEMICALS

Melting points were recorded on digital melting point apparatus B-542 (Buchi). UV spectras were measured with Shimadzu UV-2600 UV-Vis Spectrophotometer and the IR spectras were obtained with Perkin Elmer FT-IR, Spectrum Two. ^1H and ^{13}C NMR spectroscopic data were recorded on Bruker-Advance DPX FT-NMR 500 and 400 MHz instruments (125 MHz for $^{13}CNMR$). LCESIMS data was acquired on Agilent UHD-6540 LCMS/MS (HRMS) system. All chromatographic purifications were performed on silica gel (#60–120 or #100–200 from E. Merck, Germany), Supelco Diaion HP-20SS, USA and Sephadex LH-20 were purchased from Sigma Aldrich respectively.

PLANT MATERIAL

Fresh stems of *Tinospora cordifolia* were collected from Indian Institute of Integrative Medicine (IIIM), Jammu, India. A voucher specimen (RJM/0010) was identified by the Scientist of botany division and voucher deposited in the herbarium of the Institute.

EXTRACTION AND ISOLATION

Fresh stems (1.5kg) of *Tinospora cordifolia* were crushed, soaked in 4.5 L of ethyl acetate:water in the ratio of 1:1 and mechanically stirred for 2 h. The mixture was double filtered through musclin cloth and distilled on rota vapour at 55°C. 59.2 g of the aqueous portion (coded as TCE) obtained was freeze dried, which was divided into two parts, 20 g of which was used for column chromatography on Diaion HP-20 (400g, column size - 90 cm x 6 cm). The column was eluted successively with 100% water (9×500 ml), 25% MeOH/H_2O (6×500ml), 50% MeOH/H_2O (4×500ml), 100% MeOH (15×500ml) and 50% EtOAc/MeOH (3×500ml). A total of 47 fractions, each of 500ml volume were collected. Fractions 11-14 yielded 125 mg of residue with one major spot on the TLC. On repeated column chromatography of the pooled 11-14 fractions on Sephadex LH-20, 14mg of white amorphous compound was obtained which was identified as cordifolioside A (coded as TC-1) on the basis of spectral data obtained with that reported in the literature. 560mg of fraction 20-22(coded as TCFR) was taken up for further purification on flash chromatography using RP-18 silica gel (33.5g). Column was eluted with water (3×100ml), 10% ACN in water (2×100ml), 20% ACN in water (2×100ml), 30% ACN in water (2×100ml), 40% ACN in water (2×100ml), 50% ACN in water (2×100ml), 100% ACN (2×100ml), 100% MeOH (3×100ml) and finally with isopropyl alcohol. On the basis of similar TLC pattern obtained, fractions from 8-12 (TCFR-3) were pooled and dried. Total 121mg of the residue so obtained was

Table 3: Major and sub groups of natural products present in different parts of *Tinospora cordifolia* and their biological activities

Active component	Compound	Plant part	Biological Activity	References
Alkaloids	Berberine (1), Choline (2), Tembetarine (3), Magnoflorine (4), Tinosporin (5), Palmatine (6), Isocolumbin (7), Aporphine (8), Jatrorrhizine (9), Tetrahydropalmatine (10)	Stem, Root	Anti-viral infections, Anti-cancer, anti-diabetes, inflammation, Neurological, immunomodulatory, psychiatric conditions	(Upadhaya et al. 2010, Rout 2006, Patel & Mishra 2011, Patel et al. 2009, Gupta & Sharma 2011.)
Diterpenoid Lactones	Furanolactone (11), Clerodane derivatives[(5R,10R) -4R-8R- dihydroxy-2S-3 R:15,16- diepoxy-cleroda-13 (16), 14-dieno-17,12S:18,1S- dilactone], Tinosporon, Tinosporide (12), Jateorine, Columbin (13)	Whole plant	Vasorelaxant: relaxes norepinephrine induced contractions, inhibits Ca++ influx, anti-inflammatory, anti-microbial, anti- hypertensive, anti-viral. Induce apoptosis in leukemia by activating caspase-3 and bax, inhibits bcl-2.	(Sriramaneni et al. 2010, Yang et al. 2010, Zhao et al. 2008, Kohno et al. 2002, Dhanasekaran 2009)
Sesquiterpenoids	Tinocordifolin (14)	Stem	Antiseptic	(Maurya & Handa 1998)
Aliphatic compounds	Octacosanol, Heptacosanol (15), Nonacosan-15-one (16),	Whole Plant	Anti-nociceptive and anti- inflammatory. Protection against 6- hydroxydopamine induced parkinsonisms in rats. Down regulate VEGF and inhibits TNF-α from binding to the DNA.	(De-Oliveria et al.2012, Wang et al. 2010, Thippeswamy et al. 2008)

Contd.

Active component	Compound	Plant part	Biological Activity	References
Glycosides	18-norclerodane (17), glucoside, Furanoid diterpene glucoside, Tinocordiside (18), Tinocordifolioside (19), Cordioside, Cordifolioside Syringin (20), Syringin- apiosylglycoside, Pregnane glycoside, Palmatoside G (21), Cordifolioside A, B, C, D and E	Stem	Treats neurological disorders like ALS, Parkinsons, Dementia, motor and cognitive deficits and neuron loss in spine and hypothalamus, Immunomodulation, Inhibits NF-k Band act as nitric oxide scavenger to show anticancer activities.	Ly et al. 2007, Karpova et al. 1991, Kapil & Sharma 1997, Chen et al. 2001, Badwin 2001, Yang et al. 2010, Kim et al. 2008)
Steroids	β–sitosterol, δ–sitosterol, 20 β–hydroxyecdysone, Ecdysterone (22), Makisterone A, Giloinsterol	Shoot	IgA neuropathy, glucocorticoid induced osteoporosis in early inflammatory arthritis, induce cell cycle arrest in G2/M phase and apoptosis through c-Myc suppression. Inhibits TNF- α, IL-1 β, IL-6 and COX-2.	(Lv et al. 2011, McKeown et al. 2012, Sundarraj et al. 2012)
Others	3,(a,4-di hydroxy-3- methoxy-benzyl)-4-(4- compounds hydroxy-3- methoxy-benzyl) - tetrahydrofuran, Jatrorrhizine, Tinosporidine, Cordifol, Cordifelone, Giloinin, Giloin, N-trans- feruloyltyramine as diacetate, Tinosporic acid.	Root, Whole plant	Protease inhibitors for HIV and drug resistant HIV.	(Ghosh et al. 2008, Mukherjee et al. 2010)

re-chromatographed on RP-18 silica gel. Elution with 100% ACN and repeated crystallization in methanol led to the isolation of 76mg of creamish colored pure compound, coded as TC-2.

Remaining 39.2g of the freeze dried extract was dissolved in water and extracted with butanol. Butanol extract was chromatographed over silica gel column (100-200 mesh, column size - 90 cm × 15 cm) and eluted with a combination of 1-100% MeOH/CHCl$_3$. A total of 112 fractions were collected. Initial fractions eluted in chloroform yielded octacosanol, TC-3 (10mg) followed by the isolation of β-sitosterol, TC-4 (25 mg). Fractions 33-39 (eluted with 10% MeOH/CHCl$_3$) and 48-56 (eluted with 15% MeOH/CHCl$_3$), crystallized in methanol and acetone afforded clerodadiene-diolide, TC-5 (15 mg) and Ecdysterone, TC-6 (10mg). 82-91 of the total 112 fractions on repeated column chromatography and crystallization in acetone yielded Tinosporaside, TC-7 (32 mg) respectively. The structures of the known isolated compounds were determined on the basis of spectroscopic data interpretation and in comparison with the data reported in the literature.

ACETYLATION OF TC-2

TC-2 was reacted with acetic anhydride in catalytic amount of pyridine to obtain acetate crystals of the isolated molecule for X-ray structure determination and confirmation. 10mg of TC-2 was dissolved in 1ml of pyridine, and 2ml of acetic anhydride was added to the solution. The reaction mixture was heated on a steam bath for 2h under dry conditions. Usual work up followed by crystallization yielded hexaacetate, TC-2a. In ^{1}H NMR (500Hz, CDCl$_3$), signal shifts from δ 3.41, 3.53, 3.84 and 3.778 to higher delta valuesand increase in number of 9 protons at δ 2.1 confirmed the formation of three acetate groups.The triacetate so formed was confirmed by X-ray crystallography.

X-RAY CRYSTAL STUDIES OF TC-2

Crystal structure determination and refinement

X-ray intensity data of 7288 reflections (of which 4004 unique) were collected on X'calibur CCD area-detector diffractometer equipped with graphite monochromated MoKa radiation (λ = 0.71073 Å). The crystal used for data collection was of dimensions 0.30 x 0.20 x 0.20 mm. The intensities were measured by w scan mode for q ranges 3.55 to 26.00. 1149 reflections were treated as observed (I >2σ(I)). Data were corrected for Lorentz, polarisation and absorption factors. The structure was solved by direct methods using SHELXS97 (Sheldrick 2008) All non-hydrogen atoms of the molecule were located in the best E-map. Full-matrix least-squares refinement was carried out using SHELXL97 (Sheldrick 2008).The final refinement cycles converged to an R= 0.0614 and wR(F2) = 0.1204 for the observed data. Residual electron densities ranged from -0.175<Δα< 0.409 eÅ-3. Atomic scattering factors were taken from International Tables for X-ray Crystallography (1992, Vol. C, Tables 4.2.6.8 and 6.1.1.4).The crystallographic data has been summarized in table-4 (Fig. 2) represents the OTEP view of the TC-2 molecule.

Table 4: Crystallographic data of TC-2 acetate.

Chemical formula	$C_{36}H_{46}O_{17}$
M_r	750.73
Temperature (K)	293
a, b, c (Å)	12.4621 (15), 7.4751 (6), 21.069 (3)
α, β, γ (°)	90, 105.277 (13), 90
V (Å³)	1893.3 (4)
Z	2
Radiation type	Mo $K\acute{a}$
μ (mm⁻¹)	0.11
No. of measured, independent and observed [$I> 2\sigma(I)$] reflections	7288, 4004, 2173
R_{int}	0.049
$(\sin \theta/\lambda)_{max}$ (Å⁻¹)	0.617
Refinement	
$R[F^2> 2\sigma(F^2)]$, $wR(F^2)$, S	0.061, 0.147, 0.99
No. of reflections	4004
No. of parameters	486
No. of restraints	1
H-atom treatment	H atoms treated by a mixture of independent and constrained refinement
$\Delta\theta_{max}$, $\Delta\tilde{n}_{min}$ (e Å⁻³)	0.41, "0.18
Absolute structure	Flack H D (1983), Acta Crystal. A39, 876-881
Absolute structure parameter	10 (10)

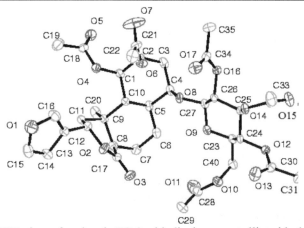

Fig. 2: ORTEP view of molecule TC-2 with displacement ellipsoids drawn at 40%. H atoms are shown as small spheres of arbitrary radii.

CHROMATOGRAPHIC CONDITIONS AND APPARATUS

The HPLC system consisted of an Agilent series 1100 instrument equipped with a binary pump, an autosampler, an automatic electronic degasser, an automatic thermostatic column oven, a diode array detector, and a computer with Chemstation software (version 06.03 [509]) for data analysis. The LC separations were optimized using RP-18e, Merck column (4 X 250mm, 5 ìm) where mobile phase consisted of acetonitrile and water. The mobile phase was delivered at a flow rate of 0.8 ml/min. The column temperature was maintained at 30°C to provide sharpness to the eluting peaks. The UV chromatograms were recorded at 215nm. TC-1, TC-2, TC-5, TC-6 and TC-7 exhibited peaks at retention times of 16.2, 22.28, 23.08, 18.20 and 20.0 min. respectively.

PREPARATION OF STANDARD SOLUTIONS AND SAMPLE SOLUTIONS

Standard Solutions

1 mg of each of the five pure U.V active molecules (Figs. 1a) i.e. TC-1, TC-2, TC-5, TC-6 and TC-7 were properly weighed and dissolved in HPLC grade methanol. The standard solutions were sonicated and filtered through Millipore micro filters (0.45 μm) before injecting into the system. Working solutions in the concentration range of 1.0-100 μg/mL were prepared by diluting with methanol. These working solutions of all the marker compounds were mixed together and were injected in different concentrations (1.25μg/ml, 2.5μg/ml, 5μg/ml, 10μg/ml, 25μg/ml, 50μg/ml and 100μg/ml) respectively.

Sample Solutions

Accurately weighed quantities (20mg) of dried extract samples from aqueous extract of *T. cordifolia* were dissolved 1 ml of HPLC grade methanol. The solutions were filtered through Millipore (0.45 μm) filters before injection into the HPLC system.

BIOLOGICAL EVALUATION

Cell Culture, Growth and Treatment Conditions

Human lung carcinoma cell lines (A549), Prostate (PC-3), SF-269 (CNS), MDA-MB-435 (Melanoma), HCT-116 (Colon) and Breast (MCF-7) were procured from National Cancer Institute (NCI), USA. The human cancer cell lines were grown in tissue culture flasks in complete growth medium (RPMI-1640 medium supplemented with 10% fetal calf serum, 100 μg/ml streptomycin and 100 units/ml penicillin) in carbon dioxide incubator (New Brunswick, Galaxy 170R, Eppendorf) at 37°C, 5% CO_2 and 98% RH.

In vitro Cytotoxicity Assay

Sulforhodamine B (SRB) assay was performed, in which cell suspension of optimum cell density (7500-15000 cells/100 μL) was seeded and exposed to 1 μM, 10 μM, 30 μM and 50 μM concentrations of test materials in complete growth medium (100μL) were added after 24 hours of incubation along with known cytotoxic agents paclitaxel, mitomycin, 5-fu and doxorubicin as positive controls. After further 48hr incubation cells were fixed with ice-cold TCA for 1hr at 4 °C. After 1 h, the plates were washed five times with distilled water and allowed to air dry followed by the addition of 100 μL of 0.4% SRB dye for 0.5 h at room temperature. Plates were then washed with 1% v/v acetic acid to remove the unbound SRB. The bound dye was solubilized by adding 100 μL of 10 mM Tris buffer (pH 10.4) to each well. The plates were put on the shaker for 5 min to solublize the dye completely, and finally the reading was taken at 540 nm on microplate reader (Thermo Scientific). IC_{50} was determined by plotting OD against concentration from Graph PAD Prism version 5.

Nuclear Morphology Studies by DAPI Staining

The presence of apoptotic cells were determined by staining HCT-116 cells with DAPI (4' -6-Diamidino-2-phenylindole). Human colon cancer HCT-116 cells ($2x10^5$/ml/well) were seeded in 60 mm culture dishes. After 24 h, cells were incubated with different concentrations of TC-2 and paclitaxel (positive control) for 24 h. Media was collected and cells were rinsed with PBS, detached by trypsinization, and added back to the conditioned media to ensure that floating and poorly attached cells were included in the analysis. Air dried smears of HCT-116 cells were fixed in methanol at -20°C for 20 min, air dried and stained with DAPI at 1μg/mL in PBS at room temperature for 20 min in the dark and the slides were mounted in glycerol–PBS (1:1) and examined in an inverted fluorescence microscope (Olympus, 1X81) (Rello 2005).

Detection of Apoptosis by Annexin V-FITC and PI

Annexin V-FITC and propidium iodide (PI) dual staining is usually used to detect the early and late apoptotic cells. For measuring apoptosis, human colon cancer (HCT-116) cells were seeded in six-well plates (2 × 105 cells) and treated with TC-2 for 24 h. Paclitaxel (1μM) was used as a positive control. After 24 h treatment cells were collected, washed twice with PBS and resuspended in binding buffer. Thereafter, the cells were stained with Annexin V/FITC and PI for 15 min in dark and analyzed by laser scanning confocal microscope (Olympus Fluoview FV 1000) (Zhang 2013, Bai 2015, Dai 2008, Munafo & Colombo 2001, Acharya 2009, Kumar 2016).

Detection of Intracellular Reactive Oxygen Species (ROS) Accumulation

Intracellular ROS levels were monitored by fluorescence microscopy after staining with DCFH-DA (dichlorodihydro-fluorescein diacetate). Human colon cancer (HCT-116) cells (2x105/ml/well) were seeded in 60 mm culture dishes. After 24 h, cells were incubated with different concentrations of TC-2 for 24 h and observed under fluorescence microscope using 40X lens (Olympus, 1X81). H_2O_2 (0.05%) was used as positive control (Bai 2015, Kumar 2016).

Loss of Mitochondrial Membrane Potential (MMP)

Loss in mitochondrial membrane potential ($\Delta\psi m$) as a result of mitochondrial perturbation was studied using confocal microscopy after staining with Rhodamine 123 (Rh123). Human colon cancer (HCT-116) cells (2x105/mL/well) were seeded in six well plate and treated with different concentrations of TC-2 and paclitaxel (positive control) for 24 h. Cells untreated and treated with test material (s) were trypsinized and washed twice with PBS. The cell pellets were then suspended in 2 ml fresh medium containing Rh123 (1.0 µM) and incubated at 37°C for 20 min with gentle shaking. Cells were collected by centrifugation and washed twice with PBS, then analyzed by laser scanning confocal microscope (Olympus Fluoview FV1000) (Dai 2008, Kumar 2016).

Analysis of Autophagy by Monodansylcadaverine (MDC) Staining

A fluorescent compound, monodansyl cadaverine (MDC), has been proposed as a special tracer for autophagic vacuoles. Human colon cancer (HCT-116) cells (2x10^5/ml/well) were seeded in 60 mm culture dishes. After 24 h, cells were incubated with different concentrations of 2 for 24 h. The autophagic vacuoles were labelled with MDC by incubating the cells with 0.05 mM MDC in PBS at 37°C for 1 hr. After incubation, cells were washed three times with PBS and immediately analyzed by fluorescence microscope using 40X lens (Munafo & Colombo 2001, Kumar 2016).

Immunofluorescence Microscopic Studies for Detection of Cytochrome C and LC3B

Cultured HCT-116 cells were seeded on cover slips in 35 mm culture dishes, incubated in the presence of different concentrations of compound for 24 h. Cells were washed twice in PBS and fixed in 4% paraformaldehyde for 15 min at room temperature. Cells were permeabilized in PBS with 0.1% TritonX-100 at room temperature for 10 min. Nonspecific binding sites were blocked by incubating the cells in 10% BSA. Cells were then incubated with Cytochrome c and LC3B antibody diluted 1:100 in 0.1% Triton X-100 in PBS for 1hr at room temperature and respective Alexa Fluor 555and 488 conjugated secondary antibodies diluted 1:500 in PBS for 1 h at room temperature. Cells were then washed three times in PBS and stained with 4', 6-diamidino-2-phenylindole (DAPI) 1µg/ml in PBS. The cover slips were mounted over glass slides and cells were imaged by a laser scanning confocal microscope (Olympus Fluoview FV1000) by using 60X oil immersion objective lens (Kumar 2016, Campos et al. 2006).

PHYTOCHEMICAL STUDIES

Isolation and Characterization of Isolated Molecules

Through bioassay guided fractionation and isolation of the aqueous alcoholic stem extract of *T. cordifolia*, we were successful in obtaining a new clerodane furano diterpene glycoside (TC-2) along with six more compounds which are already known, i.e. cordifolioside A (β-D-Glucopyranoside, 4-(3-hydroxy-1-propenyl)- 2,6-dimethoxyphenyl 3-O-D-apio-β-D-furanosyl) (TC-1), octacosanol, (TC-3), β-Sitosterol (TC-4), 2β,3β:15,16-Diepoxy- 4β, 6β-dihydroxy-13(16), 14-clerodadiene-17,12:18,1-diolide (TC-5), ecdysterone (TC-6) and tinosporoside (TC-7).

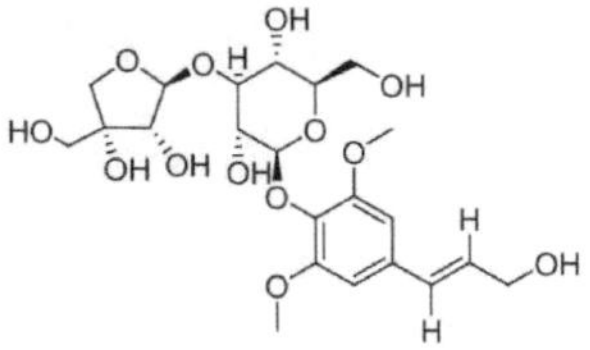

Cordifolioside A, TC-1

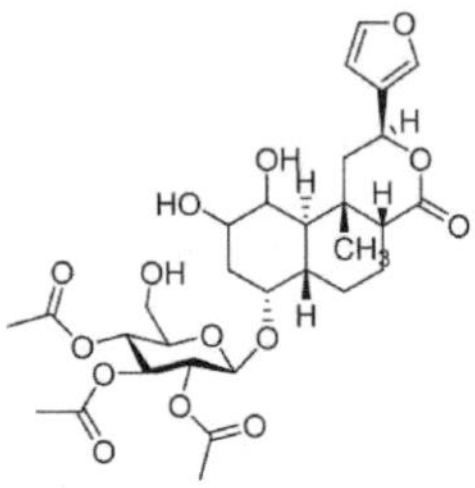

New clerodane diterpenoid, TC-2

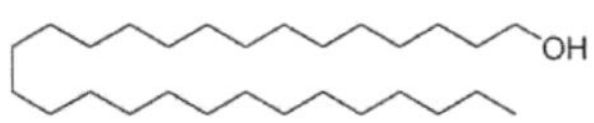

Octacosanol, TC-3

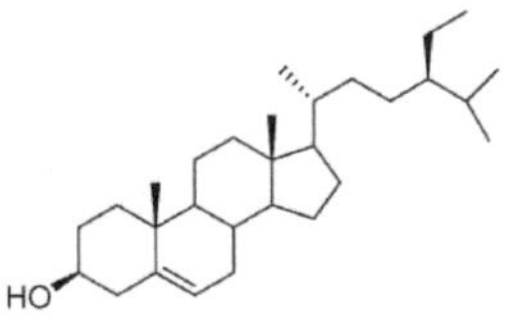

β-sitosterol, TC-4

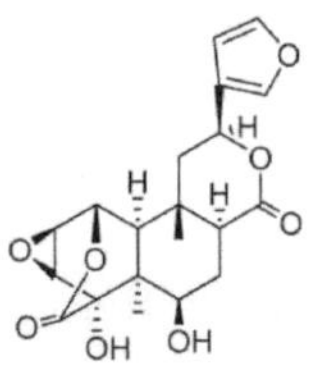

Diepoxy diolide, TC-5

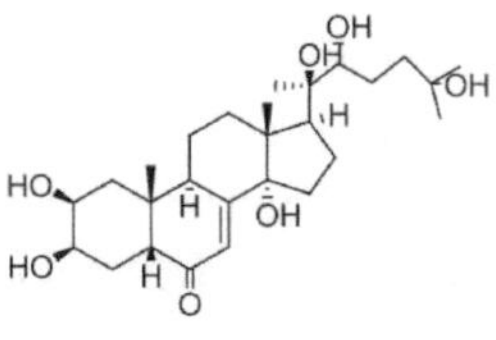

Ecdysterone, TC-6

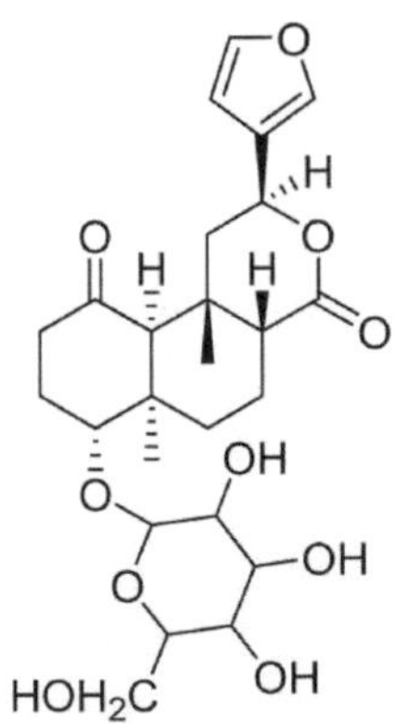

Fig. 3: Chemical structures of the compounds isolated from *Tinospora cordifolia*

STRUCTURE ELUCIDATION OF TC-2

Spectral analysis of new clerodane diterpenoid, TC-2

TC-2 was obtained as an amorphous powder, creamish white in colour. Analysis of the ^{13}C NMR and DEPT-135 spectra revealed resonances along with a pseudo molecular ion peak [M-H3O]$^+$ at m/z 605. Its IR spectrum showed absorptions corresponding to hydroxyl and furan moieties at 3384 cm^{-1} and 771.62 cm^{-1}. Signals at 1729.05 cm^{-1} indicated the presence of lactone which was confirmed by the presence of resonance at 172.062 ppm in the ^{13}C NMR spectrum. The UV absorption at 209 nm supported the presence of an α, β-unsaturated ketone group. ^{1}H-NMR of TC-2 displayed signals at δ 7.42 (t, ^{1}H, J=3.3 Hz), δ 7.43 (d,1H, J=0.6 Hz) and δ 6.40 (dd, ^{1}H, J= 0.9 Hz) suggested the presence of protons of the β-substituted furan moiety commonly reported in clerodanes isolated from different Tinospora species (32-36). One angular methyl group at C-9 was observed as three proton singlet at 1.09 respectively. The signal at δ 3.40 (m) was allocated to the C-12 proton carrying the β-substituted furan moiety. The signals at the aliphatic region δ 1.94 (m) and δ 1.74 (m) were attributed to the C-3 methylene protons (Fig. 4). The HMBC correlations observed from H-14 (δ 6.4) to C-13 (δ 125.38), C-15 (δ 139.22) & C-16(δ 143.90); H-15 (δ 7.44) to C-13 (δ 125.38), C-14 (δ 108.16) & C-16(δ 143.90); H-16 (δ 7.43) to C-13 (δ 125.38), C-14 (δ 108.16) & C-15 (δ 139.22) and CH$_3$-9 (δ 1.09) methyl protons to C-2 (δ 70.47), C-8 (δ 51.10) and C-11 (δ 43.53) augmented the above argument. Of the three hydroxyl resonances, the one at δ 3.10 was allocated to the carbon at position 1. The other deshielded hydroxyl peak at δ 3.53 was assigned to the C-2 based on HMBC correlations (Fig. 4). The relative configuration of TC-2 was deduced based on NOESY experiment. A cross peak was found from H-8 (δ 2.31) to CH3-9 (δ 1.09); H-

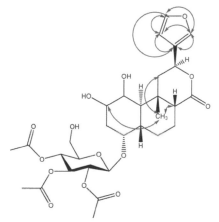

Fig. 4: Major HMBC correlations (blue arrows) for compound TC-2

10 (δ 1.67) to CH3-9 (δ 1.09), H-5' or 6' (δ 3.83), H-2' (δ 5.46), H-4'(δ 4.96); CH3-9 (δ 1.09) to H-8 (δ 2.31) and H-10 (δ 1.67) indicated similar configurations between different groups which was confirmed by the X-ray crystallography. The other compounds,TC-1, TC-3 to TC-7isolatedfromthe bioactive fraction were identified on the basis of comparing their spectral data with the ones already reported in the literature.

BIOLOGICAL EVALUATION

Anticancer activity of different fractions of the aqueous extract

Using SRB assay, we evaluated the cytotoxicity of the different fractions of the aqueous extract of *T.cordifolia* against three different cancer cell lines; Colon (HCT-116), Lung (A549) and Prostate (PC-3) cell lines respectively (Table 5). Out of all the evaluated fractions from the aqueous extract of *T. cordifolia*, TCFR fractioncame out to be the most potent one. The bioactive fraction was subjected to repeated column chromatography for the expected isolation of more potent molecules.

Anticancer activity of the isolated molecules

All the isolated molecules were screened for their anti-cancer potential where TC-2 exhibited more promising results than the other molecules. TC-2 was examined on a panel of human cancer cell lines of various origins for a period of 48h incubation. The following cell lines were used: Lung (A549), Prostate (PC-3), SF-269 (CNS), MDA-MB-435 (Melanoma), HCT-116 (Colon) and Breast (MCF-7).To calculate the IC_{50} values, cells were given treatments with different concentrations of the new compound. The results exhibited that the incubation of different cancer cells with 1 μM, 10 μM, 30 μM and 50 μM concentrations of TC-2 for 48 h imparted varied effects on cellular viability. IC_{50} values were of the order of 33 μM (A549) in lung cancer; 23 μM (SF-295) in CNS carcinoma, 10.4 μM (PC-3) Prostate, 14.8 μM (MDA-MB-435) for melanoma, 8 μM (HCT-116) in colon carcinoma and 40 μM (MCF-7) in breast carcinoma. TC-2 revealed highest activity against colon cancer (HCT-116) cells and least activity against breast cancer (MCF-7) cells with IC_{50} values of 8 and 40 μM respectively (Table 6)

Table 5: NMR Data for TC-2 in CDCl$_3$ (δ in ppm, J in Hz in parentheses), bs- broad signal

Position	δ_C	δ_H	HMBC Correlation observed in TC-2	COSY	NOESY
1	76.31	3.41 m	-	3.4(C-12)	
2	70.47	3.53 m	-	-	
3	30.49	1.94b, 1.74a m	-	I) 2.4(C-11), 2.10(3',4'), II)2.2(C-7)	
4	75.51	3.31 m	-	3.57(C-2)	
5	34.58	Q -	-		
6	25.17	1.74a,1.447b m	-		
7	21.39	2.36a,1.73b m	-		
8	51.10	2.31 m	-		CH3-9
9	35.14	Q -			
10	39.06	1.67 m	-		CH3-9, 5' or 6', 4', 2'
11	43.53	2.42a,1.73b d at 2.42(3.25) & m at 1.73	-		
12	73.28	3.40 m	-	3.54(C-1), 4.27(C-1')	
13	125.38	Q -			
14	108.16	6.40 d,1H, (0.9)	C-13,C-15,C-16	6.4(C-14)	
15	139.22	7.44 d,1H, (0.6)	C-13,C-14,C-16		
16	143.901	7.43 in HSQC t,1H,(3.3)	C-13,C-15,C-14		
17	172.062	Q -	-		
1'	99.91	4.26 d,1H, (7.7Hz)	-	3.38(C-4)	
2'	68.61	5.46 dd,1H,(3.15) and (2.95)	-		

Position	δ_C	δ_H		HMBC Correlation observed in TC-2	COSY	NOESY
3'	71.02	5.61	dd, 1H, (3.10) and (3.20)	-	1.76 (C-3 or C-6)	
4'	73.37	4.96	dd, 1H, (3.25) and (3.25)	-		
5'	73.79	3.84	m, bs	-		
6'	62.30	3.83b,3.78a	m, bs			
9-CH3	21.53	1.09	s,3H	C-11, C-8,C-2	1.65 (C-10)	H8, H10
2'-OCOCH3	21.06	1.94	s, 3H	-	2.42 (C-11)	
3'-OCOCH3	20.94	2.13	s, 3H	-		
4'-OCOCH3	21.147	2.36	s, 3H	-		
2'-OCOCH3	172.06	Q	-	-		
3'-OCOCH3	170.60	Q	-	-		
4'-OCOCH3	169.57	Q	-	-		

Table 6: *In vitro* cytotoxicity of the extract, bioactive fraction and the isolated compounds against human cancer cell lines

Cell line type Tissue			HCT-116 Colon	A549 Lung	PC-3 Prostate
Sr No.	Code	Conc. (μg/ml)	% Growth inhibition		
1.	T.C.E	50	23	0	27
		100	52	5	48
2.	TCFR-3	50	62	2	72
		100	94	9	97
3.	TC-1	50	0	12	5
		100	20	19	14
4.	TC-2	50	94	77	71
		100	97	78	78
5.	TC-3				
6.	TC-4	50	7	13	11
		100	24	16	20
7.	TC-5	50	12	0	0
		100	27	2	10
8.	TC-6	50	0	0	3
		100	0	0	17
	5-Fluorouracil	20μM	52	-	-
	Paclitaxel	1 μM	-	76	-
	Mitomycin	1 μM	-	-	66

Effect of TC-2 on human colon HCT-116 cancer cell line

To understand the interesting potency revealed by new clerodane diterpenoid TC-2, detailed mechanistic studies were carried out. The differential cytotoxicity exhibited by the compound may be due to the varying molecular characteristics of these cells. These findings substantiate the findings in *Polyalthia longifolia* (Verma 2008) and *Ocimum basilicum* (Manosroi & Manosroi 2006). To examine, whether TC-2 treatment killed cancer cells by inducing apoptosis, we analysed the human colon cancer (HCT-116) cells for nuclear morphological changes, by staining nuclei with DAPI. TC-2 induced chromatin condensation and fragmentation of nuclei of few cells in concentration dependent manner, typical of apoptosis (Fig. 5). Annexin V/PI dual staining suggested the significant externalization of PS (phosphatidylserine) in the events of early cell death after 24 h in the present studies. The concentration dependent increase in percentage of early and late stage apoptosis with treatment of TC-2 was observed (Fig. 6 & 7). The property of cancer cells is their resistance to apoptosis induction (Hanahan & Weinberg 2000). Therefore, inducing apoptosis is the aim of many anticancer therapeutic approaches as it makes it possible to kill cancer cells without causing inflammation (Reed 2002).

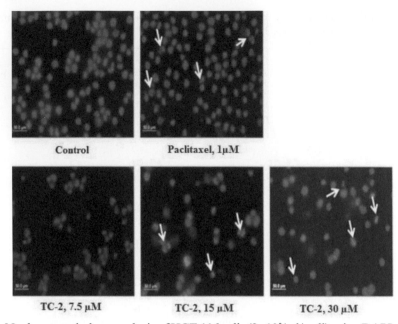

Fig. 5: Nuclear morphology analysis of HCT-116 cells (2×10^5/ml/well) using DAPI. After treatment with indicated concentrations of TC-2 for 24 h and examined using fluorescence microscopy (40X). Paclitaxel (1μM) was used as positive control. With increase in concentration of TC-2 there is significant increase in nuclear condensation and formation of apoptotic bodies.

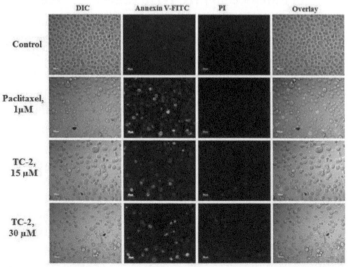

Fig. 6: The representative images of TC-2 treatment on the exposure of phosphatidylserine (PS) in HCT-116 cells after 24 h treatment. Phosphatidylserine exposure was assessed by the Annexin V/propidium iodide assay and analysed by confocal microscopy using 40X oil immersion lens.

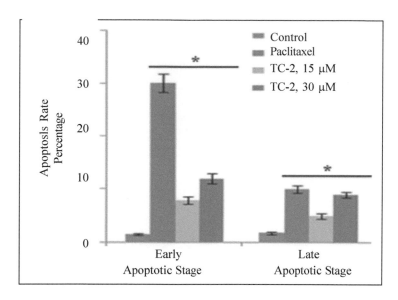

Fig. 7: Histogram showing the percentage of cells in early and late stages of apoptosis obtained by analysis of the cell images. Data are mean ± S.D. of three similar experiments; statistical analysis was done with *p < 0.05.

Mitochondria are intermediate to the intrinsic pathway of apoptosis and therefore are targets of choice for cancer therapy (Fulda 2010). The mitochondrial pathway of apoptosis is portrayed by mitochondrial membrane permeabilization (MMP) and release of pro-apoptotic proteins (e.g. cytochrome c) from the intermembrane space of the mitochondria to the cytosol. These events lead to the activation of the initiator caspase 9, triggering the caspase cascade causing DNA condensation/fragmentation and ultimately cell death (Kroemer 2009). MMP involves pore formation such as Bax/Bak oligomers and the permeability transition pore complex (PTPC) (Kroemer 2007). Mitochondria play a considerable role in energy production and are also significant sensors for apoptosis. Experiments on the measurement of the changes in mitochondrial membrane potential (MMP) with laser scanning confocal microscope revealed considerable loss of mitochondrial membrane potential (MMP) in human colon cancer (HCT-116) cells treated with different concentrations of TC-2 (Fig. 8).

Further, Cytochrome c localization was determined by immunofluorescence with a cytochrome c specific antibody. Cytochrome c was colocalized with mitochondria in control cells. In contrast, TC-2 and paclitaxel treatments were associated with a decrease in functional mitochondria and release of cytochrome c to the cytosol respectively (Fig 9).

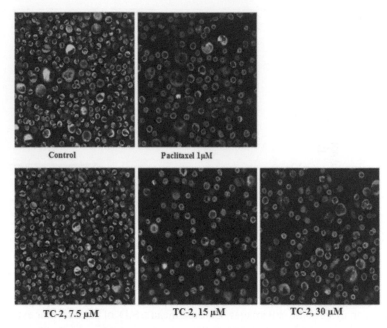

Fig. 8: Loss of mitochondrial membrane potential ("øm) was measured in human colon cancer (HCT-116) cells (2×10^5/mL/well) treated with indicated concentration of TC-2 in 6 well plates for 24 h and incubated with Rodamine-123 (1.0 µM) in serum free media for 20 min at 37ÚC and washed with PBS. The loss of mitochondrial membrane potential (MMP) in HCT-116 cells was observed under laser scanning confocal microscope using 40X lens (Olympus Fluoview FV1000)

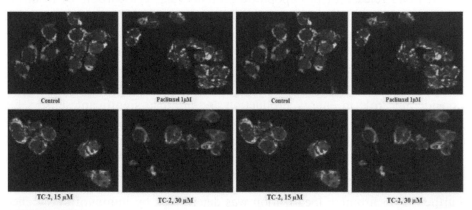

Fig. 9: Colocalization of cytochrome c and mitochondria was determined by confocal microscopy using 60 × oil immersion lenses. Human colon cancer (HCT-116) cells were immunostained for cytochrome c release (green) and the mitochondria of the cells were stained with MitoTracker (red). HCT-116 cells were treated with different concentrations of TC-2 and paclitaxel (1 µM) for 24 h and stained with anticytochrome c antibody and Alexa Fluor 488-labeled secondary antibody

Reactive oxygen species (ROS) is generated during cellular metabolism through leakage of electrons by mitochondrial electron transport and is an intermediary in apoptosis. To gain insight into the mechanism by which TC-2 results in cell death we next examined ROS production, since excessive generation of ROS results in cells injury and death. TC-2 increased ROS/oxidative stress in HCT-116 cells in concentration dependent manner (Fig. 10). Mitochondria play an important role in apoptosis. Mitochondria-mediated reactive oxygen species (ROS) generation is a major source of oxidative stress in the cells. The apoptosis induction in the present studies causes the activation of the mitochondrial pathway and is subsequently associated with ROS production, Cyt c release, and nuclear fragmentation.

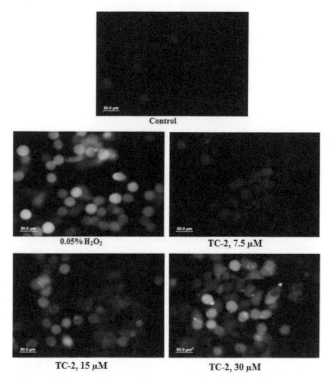

Fig. 10: Intracellular ROS level was detected by fluorescence microscopy using 22, 72-dichlorofluorescein diacetate (DCFH-DA) after 24 h treatment. HCT-116 cells were treated with indicated concentrations of TC-2 and 0.05% H_2O_2 and incubated with 5 µM 22 ,72 -dichlorofluorescein diacetate and examined by fluorescence microscope (40X). We also detected changes in autophagic activity by monitoring the MDC fluorescence which is a known marker for autophagic vacuoles. The number of autophagic vacuoles stained by MDC in the TC-2 treated HCT-116 cells was much higher than in the control (Fig. 11).

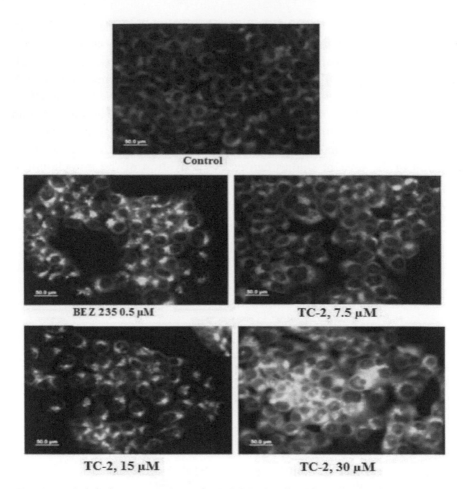

Fig. 11: TC-2 induces autophagy in HCT-116 cells. The autophagic vacuoles were observed under fluorescence microscope (40×) with MDC staining. The treatment of compound and BEZ235 (positive control group) induced concentration-dependent formation of autophagic vacuoles in HCT-116 cells after 24 h.

Next, to further confirm the induction of autophagy by TC-2, a set of autophagy-related factors including LC3-I and LC3-II in the HCT-116 cells after treatment with different concentrations of TC-2 for 24 h were investigated by immunofluorescence microscopy. The microtubule associated protein light chain 3 (LC3) is another signature marker of autophagosomes. Cleavage of the 18 kDa full length LC3, known as LC3-I, to a 16 kDa form, known as LC3-II, results in recruitment of LC3-II to double layered membrane of autophagosomes and this is a key step in autophagy. The immunofluorescence confocal microscopic studies of TC-2 treated HCT-116 cells also showed concentration dependent induction of autophagy (Fig. 12).

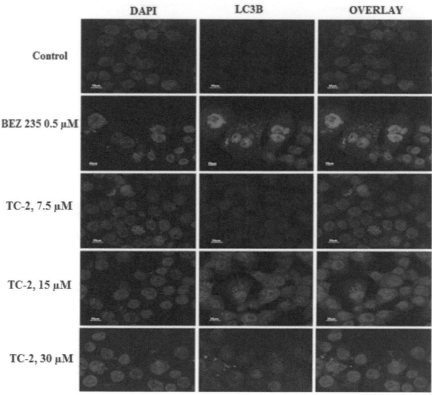

Fig. 12: Detection of autophagy with LC3b antibody by confocal microscopy using 60X oil immersion lens. Immunocytochemical staining was conducted using anti-LC3b antibody and Alexa Flour-555-labelled secondary antibody. Nuclei were stained with DAPI.

STANDARDIZATION OF THE ACTIVE EXTRACTS

Quantification of Isolated Molecules in Aqueous Extract

A rapid, simple and reliable High performance liquid chromatography method was developed for the quantification of five pure molecules (Figure 13) namely, cordifolioside A (b-D-Glucopyranoside,4-(3-hydroxy-1-propenyl)-2,6-dimethoxyphenyl 3-O-D-apio-b-D-furanosyl) (TC-1), a new clerodane furano diterpene glycoside (TC-2),2b,3b:15,16-Diepoxy- 4a, 6b-dihydroxy-13(16),14-clerodadiene-17,12:18,1-diolide (TC-5), ecdysterone (TC-6) and tinosporoside (TC-7) in aqueous extract of stem of *T. cordifolia*.TC-3 and TC-4 being UV inactive were not included in the quantification part.4.02 % of Cordifolioside A, 1.04 % of Diepoxy- clerodadiene-diolide, 6.10 % of Ecdysterone and 1.34 % of Tinosporaside was found to be present in the active aqueous extracts of *T. cordifolia*. The percentage of potent new clearodane diterpenoid, TC-2 in the extract was found to be 0.74 %.

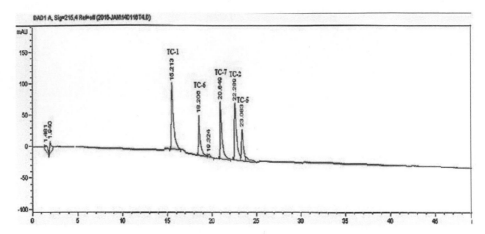

Fig. 13: HPLC chromatogram depicting the separation of five markers; viz. TC-1, TC-2, TC-5, TC-6 and TC-7.

METHOD VALIDATION

Specificity

The method was found to be specific. Mobile phase when compared with mobile phase spiked with chemical markers did not show any interference at the respective retention times of each chemical markers.

Linearity and range

The calibration curves of TC-1, TC-2, TC-5, TC-6 and TC-7were linear over the concentration range of 1.25μg – 100μg (R^2> 0.999). Results are shown in Table 7 and Figure 14.

Table 7: LOD and LOQ Calibration curve range: 1.25 – 100 μg /ml, LOD: Limit of Detection, LOQ: Limit of quantification

Analyte	Retention Time (min)	LOD(μg)	LOQ(μg)	Regression equation (y = mx + b)	Correlation coefficient
TC-1	16.2	0.02	0.75	y = 15226x + 3651.7	0.999
TC-2	22.2	0.06	0.84	y = 8004.6x + 7408.5	0.999
TC-5	23.0	0.42	1.20	y = 11982x + 23446	0.999
TC-6	18.2	0.13	1.18	y= 32531.33x + 15365.18	1.000
TC-7	20.0	0.05	0.88	y= 9216.6x + 1631.7	0.999

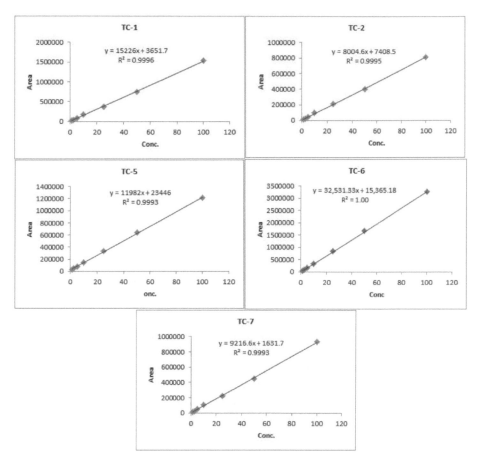

Fig. 14: Calibration curves for TC-1, TC-2, TC-5, TC-6 and TC-7.

Limit of detection (LOD) and Limit of quantitation (LOQ)

The LOD and LOQ for TC-1, TC-2, TC-5, TC-6 and TC-7are explained. The marker compounds in the aqueous extract of *Tinospora cordifolia* were quantified using above calibration curves. Presence of marker compounds in aqueous extract of *Tinospora cordifolia* was confirmed by comparison of their retention times and overlying of UV spectra with those of standard compounds.

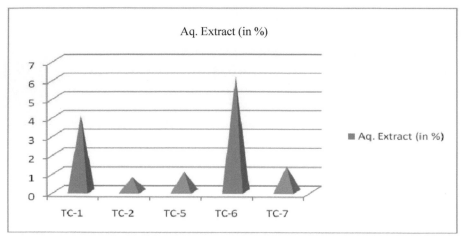

Fig. 15: Bar graph representing the percentage of five marker compounds in different geographical locations

Recovery, Accuracy and precision

The combined recovery for TC-1, TC-2, TC-5, TC-6 and TC-7 was carried out in LQC, MQC and HQC samples. The % recovery of TC-1 was 95.50 (from LQC), 91.05 (from MQC), and 94.97 (from HQC). The recovery % of TC-2 was 105.25 (from LQC), 107.80 (from MQC), and 103.25 (from HQC). For TC-5, it was 98.00 (from LQC),97.30 (from MQC), 103.13 (from HQC),for TC-6, 81.00 (from LQC), 107.25 (from MQC), 95.28 (from HQC)and for TC-7, the recovery % was found to be 105.25 (from LQC), 98.90 (from MQC), 101.32 (from HQC) (Table 8). The intra-assay accuracy in terms of % RE was in the range of -2.0 to 2.5 for TC-1, 2.2 to 3.0 for TC-2, 0.4 to 2.7 for TC-5, 1.8 to 3.0 for TC-6 and 2.1 to 3.0 for TC-7. Inter-assay % RE was in the range of -1.0 to 2.4 for TC-1, - 0.4 to 1.0 for TC-2, 1.5 to 2.0 for TC-5, 0.5 to 1.9 for TC-6 and 0.5 to 2.5 for TC-7. Inter- assay precision (% RSD) was in the range of 0.5 to 1.7 for TC-1, 1.9 to 2.5 for TC-2, 1.4 to 2.0 for TC-5, 1.0 to 1.9 for TC-6 and 1.0 to 2.5 for TC-7. Intra- assay precision (% RSD) was in the range of 0.5 to 2.9 for TC-1, 1.0 to 2.7 for TC-2, 0.7 to 2.1 for TC-5, 1.1 to 1.6 for TC-6 and 1.4 to 2.4 for TC-7 (Table 9).

Table 8: Recovery percentage

Analyte	Actual Concentration (μg/ml)	Detected concentration (Mean, n=3)	Recovery (%)
TC-1	4.0	3.82	95.50
	20.0	18.21	91.05
	80.0	75.98	94.97
TC-2	4.0	4.21	105.25
	20.0	21.56	107.80
	80.0	82.6	103.25
TC-5	4.0	3.92	98.00
	20.0	19.46	97.30
	80.0	82.51	103.13
TC-6	4.0	3.24	81.00
	20.0	21.45	107.25
	80.0	76.23	95.28
TC-7	4.0	4.21	105.25
	20.0	19.78	98.90
	80.0	81.06	101.32

Table 9: Interday and intraday precision (RSD %) and accuracy (RE %)

Analyte	Actual Concentration (μg/ml)	Detected concentration (Mean ± SD, n=6)		RSD (%)		RE (%)	
		Interday	Intraday	Interday	Intraday	Interday	Intraday
TC-1	2.0	1.98 ± 0.03	1.96 ± 0.05	1.79	2.91	-1.0	-2.0
	12.0	11.96± 0.07	12.16 ± 0.15	0.58	1.27	-0.3	1.3
	30.0	30.73± 0.52	30.75 ± 0.68	1.71	0.55	2.4	2.5
TC-2	2.0	2 ±0.05	2.06 ±0.04	2.24	1.58	2.56	2.70
	12.0	11.95 ± 0.08	12.27± 0.31	1.95	1.05	1.0	3.0
	30.0	30.19 ± 0.58	30.86± 0.32	-0.4	2.2	0.6	2.8
TC-5	2.0	2.04 ±0.02	2.01 ± 0.04	1.40	2.11	2.05	0.75
	12.0	12.18± 0.25	11.95± 0.08	1.64	1.45	2.0	0.5
	30.0	30.54± 0.50	30.83± 0.44	1.5	0.4	1.8	2.7
TC-6	2.0	2.01 ± 0.18	2.06 ± 0.02	1.07	1.27	1.51	1.1
	12.0	12.21± 0.69	12.28± 0.50	1.99	1.61	0.5	3.0
	30.0	30.58± 0.61	30.54± 0.49	1.7	2.3	1.9	1.8
TC-7	2.0	2.05 ± 0.05	2.05 ± 0.05	2.58	1.72	1.08	2.40
	12.0	12.06± 0.13	12.26± 0.29	2.06	1.41	2.5	3.0
	30.0	29.71± 0.61	30.84± 0.43	0.5	2.1	0.9	2.8

CONCLUSION

Due to the innumerable medicinal properties and extensive medical application of Giloy, chemical investigation on this plant were carried out in order to isolate the active compounds responsible for its anti-cancer activity. Bioassay guided isolation resulted in the isolation of six known and one new clerodane diterpenoid, potent against HCT-116 colon cancer cells. We demonstrated that the new clerodane diterpenoid, TC-2 induces apoptosis of colon cancer (HCT-116) cells mainly by triggering ROS production. This new natural compound thus shows potential for the treatment of colon cancer. Autophagy was also observed after the treatment. Taken together, this study identified a new clerodane furano diterpenoid that exhibited anticancer activity via induction of mitochondria mediated apoptosis and autophagy in HCT116 cells. The results from the present studies will be very advantageous in further development of new chemotherapeutic agents. Apart from this, the active aqueous extract of *T. cordifolia* was also standardized as per the ICH guidelines. The standardized extract can be safely consumed as herbal remedy for the treatment of colon cancer and the method so developed can be employed for the consistent chemical profile and quality control of the herbal formulations prepared from this medicinal herb.

Abbreviations

μg/mL : microgram per mililiter

μM   : micromolar

ALS   : Amyotrophic lateral sclerosis

CAN   : Acetonitrile

CCD   : Charge coupled device

CCl_4 : Carbon tetrachloride

$CDCl_3$: Deuterated chloroform

$CHCl_3$: Chloroform

cm   : centimeter

COSY : Correlation spectroscopy

COX-2 : cyclooxygenase-2

DAPI   : 4', 6-diamidino-2-phenylindole

DCFH-DA: Dichlorodihydro-fluorescein diacetate

EtOAc : Ethyl acetate

FT-IR   : Fourier transform- Infra red

g   : gram

H_2O : Water

HIV   : human immunodeficiency virus

HMBC: Heteronuclear Multiple Bond Correlation

ACKNOWLEDGEMENTS

Dr. Neha is thankful to Prashant Kunawat for motivating to keep reaching for excellence and to my parents for making what she is today.

REFERENCES CITED

ABalazs, Hunyadi A, Csabi J,Jedlinszki N, Martins A, Simon A, Toth G. 2013.^{1}H and ^{13}C NMR investigation of 20-hydroxyecdysone dioxolane derivatives, a novel group of MDR modulator agents. *Magn. Reson. Chem*. 830-6.

Acharya BR. 2009.Vitamin K3 disrupts the microtubule networks by binding to tubulin: a novel mechanism of its antiproliferative activity. *Biochem.* 6963-6974.

Adhvaryu MR, Reddy N, Parabia MH. 2008. Anti-tumor activity of four ayurvedic herbs in Dalton Lymphoma Ascites bearing mice and their short term in vitro cytotoxicity on DLA-cell line.*Afr. J. Trad. CAM* 409 - 418.

AgarwalA, Malini S, Bairy KL, Rao MS. 2002.Effect of *Tinospora cordifolia* on learning and memory in normal and memory deficit rats. *J.Pharmacol.* 339-349.

Ahmed SM, Manhas LR, Verma V, Khajuria RK. 2006. Quantitative determination of four constituents of *Tinospora* sps. by a reversed phase HPLC-UV-DAD method. Broad based studies revealing variation in content of four secondary metabolites in the plant from different eco geographical regions in India. *J. Chromatogr Sci.* 504-509.

Aima RK. 2003. Pictorial Guide to Plants. Natraj Publishers,1, 454-455.

Asthana JG, Jain S, Mishra A, Vijaykant MS. 2001. Evaluation of antileprotic herbal drug combinations and their combination with dapsone. *Indian Drugs* 82-86.

Badwin AS. *J.* 2001. Control of oncogenesis and cancer therapy resistance by the transcription factor NF-kappa B. *Clin. Invest.* 241-246.

Bai J. 2015.Down-Regulation of deacetylase HDAC6 inhibits the melanoma cell line A375.S2 growth through ROS-dependent mitochondrial pathway. *PLoS One* 1371-1382.

Bairy KL, Malini S, Rao CM. 2002. Protective effect of *Tinospora cordifolia* on experimentally induced gastric ulcers in rats. *J. Nat. Remed.* 49-53.

Bhatt RP, SabnisSD, 1987, Contribution of the ethnobotany of khedbrahma region of north Gujarat. *J. Econ. Taxon. Bot.* 139-144.

Campos CBL, Paim BA, Cosso RG, Castilho RF, Rottenberg H, Vercesi AE. 2006. *Cytometry Part A*: 515-523.

Chakraborty B, Sharma GD, Sengupta M. 2009. Immunomodulatory properties of Tinospora cordifolia in carbon tetrachloride intoxicated swiss albino mice. *Int. J. Biol. Sci.* 35-39.

Chaudhary R, Jahan S, Goyal PK. 2008.Chemoprotective potential of an Indian medicinal plant (*Tinospora cordifolia*) on skin carcinogenesis in mice. *J Environ Pathol. Toxicol. Oncol.* 233-243.

Chen S, Wu K, Knox R. 2001.Structure-function studies of DT-diaphorase (NQO1) and NRH: Quinone oxidoreductase (NQO2). *Free Radic. Biol. Med.* 276-284.

Chintalwar G, Jain A, Sipahimalani A, Banerji A, Sumariwalla R, Ramakrishnan R. 1999. An immunologically active arabinogalactan from *Tinospora cordifolia.Phytochem.*: 1089-1093.

Chopra A, SalujaM, Tillu G,Venugopalan A, Sarmukaddam S, Raut AK. 2011. A randomized controlled exploratory evaluation of standardized ayurvedic formulations in symptomatic osteoarthritis knees: A government of India NMITLI project. Evidence based complementary and alternative medicine. *Evi.Complement. Alt. Med.* 724291.

Chougale AD, Ghadyale AD, Panaskar SN, Arvindekar AU. 2009. Alpha glucosidase inhibition by stem extract of *Tinospora cordifolia*. *J. Enzyme Inhib. Med. Chem.* 998-1001.

Dai J, 2008.Scutellaria barbate extract induces apoptosis of hepatoma H22 cells via the mitochondrial pathway involving caspase-3. *World J. Gastroenterol.*7321-7328.

De-Oliveria AM, Conserva LM, Ferro JND, Brito FD, LyraLemos RP, Barreto E. 2012.Antinociceptive and anti-inflammatory effects of octacosanol from the leaves of sabiceagrisea var. Grisea in mice. *Int. J. Mol. Sci.* 1598-1611.

Dhanasekaran M, Baskar AA, Ignacimuthu S, Agastian P, Duraipandiyan V. 2009. Chemopreventive potential of Epoxy clerodane diterpene from *Tinospora cordifolia* against diethyl nitrosamine-induced hepyocellular carcinoma. *Invest. New Drugs* 347-355.

Dhingra D, Goyal PK. 2008. Evidences for the involvement of monoaminergic and GABAergic systems in antidepressant like activity of *Tinospora cordifolia* in mice.*Intl. J. Pharmaceut. Sci.* 761-767.

Dhingra D,Jindal V, Sharma S, Kumar HR. 2011.Evaluation of antiobesity activity of *T. cordifolia* stems in rats. *Int. J. Res. Ayurv. Pharmac.* 306-311.

Farrugia LJ. 1997. ORTEP-3 for Windows - a version of ORTEP-III with a Graphical User Interface (GUI).*J. Appl. Cryst.* 565.

Farrugia LJ. 2012. WinGX and ORTEP for Windows: an update. *J. Appl. Cryst.* 849-854.

Fukuda N, Yonemitsu M, Kimura T. Studies on the Constituents of the Stems of *Tinospora tuberculata*, IV. Isolation and Structure Elucidation of the Five New Furanoid Diterpene Glycosides Borapetoside C-G.1993. *Ann. Chem.* 491-495.

Fulda S. 2010. Targeting mitochondria for cancer therapy. *Nat. Rev. Drug Discov.* 447-464.

Gacche RN, Dhole NA, 2011. Profile of aldose reductase inhibition, anti cataract and free radical scavenging activity of selected medicinal plants: An attempt to standardize the botanicals for amelioration of diabetes complications. *Food Chem Toxicol.* 1806-1813.

Ghosh AK, Chapsal BD, Webar IT, Mitsuya H. 2008.Design of HIV protease inhibitors targeting protein backbone: An effective strategy for combating drug resistance.*Acc. Chem. Res.* 78-86.

Grover JK, Vats V, Rathi SS. 2000. Anti-hyperglycemic effect of Eugenia jambolana and *Tinospora cordifolia* in experimental diabetes and their effects on key metabolite enzymes involved in carbohydrate metabolism. *J. Ethnopharmacol.* 461-470.

Gupta R, Sharma V. 2011. Ameliorative effects of *Tinospora cordifolia* root extract on histopathological and biochemical changes induced by aflatoxin-b (1) in mice kidney. *Toxicol. Int.* 94-98.

Gupta RS, Sharma A. 2003.Antifertility effect of *Tinospora cordifolia* (Wild.) stem extract in male rats. *Intl. J. Experim. Biol.* 885-889.

Hamsa TP, Kuttan G. 2010. *Tinospora cordifolia* ameliorates urotoxic effect of cyclophosphamide by modulating GSH and cytokine levels. *Exp. Toxicol. Pathol.* 1-8.

Hanahan D, Weinberg RA. 2000. The hallmarks of cancer. *Cell.* 57-70.

Jadhao KD, Badwe DV, Wadekar MP, Iqbal M. 2009. Evaluation of antioxidants from *Tinospora cordifolia. Biosci.Biotechnol. Res. Asia.* 803-806.

Jagetia GC, Nayak V, Vidyasagar MS. 1998. Evaluation of the antineoplastic activity of guduchi (*Tinospora cordifolia*) in cultured HeLa cells. *Cancer Lett.* 71-82.

Jagetia GC, Rao SK. 2006. Evaluation of cytotoxic effects of dichloromethane extract of guduchi (*Tinospora cordifolia* Miers ex Hook F and THOMS) on cultured HeLa cells. *eCAM*: 267-272.

Jagetia GC, Rao SK. 2006. Evaluation of the Antineoplastic Activity of Guduchi (*Tinospora cordifolia*) in Ehrlich Ascites Carcinoma Bearing Mice. *Biol. Pharm. Bull.* 460-466.

Jana U, Chattopadhayay RN, Shu BP. 1999. Preliminary studies on anti-inflammatory activity of Zingiber officinale Rosc, Vitex negundo Linn and *Tinospora cordifolia* (Wild) in albino rats. *Intl. J. Pharm.*: 232-233.

Jayaseelan C, Rahuman AA, Rajkumar G, Kirthi A, Santhoshkumar VT, Marimuthu S. 2011. Synthesis of pediculocidal and larvicidal silver nanoparticles by leaf extract from heartleaf moonseed plant, *T. cordifolia* Miers. *Parasitol. Res.*185-194.

Jee V, Dar GH, Bhat GM. 1984.Taxoethnobotanical studies of the rural areas in district Rajouri (Jammu). *J. Econ. Taxo. Bot.* 831-834.

Jeyachandran R, Xavier TF, Anand SP. 2003. Antibacterial activity of stem extracts of *Tinospora cordifolia* (Willd) Hook. f& Thomson. *Ancient Science of Life* 40-43.

Joladarashi D, Chilkunda ND, Salimath PV, 2011. Glucose-uptake stimulatory activity of *T. cordifolia* stem extracts in Ehrlich ascites tumour model system. *J. Food Sci Tech.,* 480-3.

Kalikar M, Thawani V, Varadpande U, Sontakke S, Singh R, Khiyani R. 2008. Immunomodulatory effect of *Tinospora cordifolia* extract in human immune-deficiency virus positive patients. *Intl. J. Pharmacol.*107-110.

Kapil A, Sharma S.1997. Immunopotentiating compounds from *Tinospora cordifolia. J. Ethopharmacol.* 89-95.

Kapur P, Jarry H, Wuttke WB, Pereira MJ, Wuttke DS. 2008. Evaluation of antiosteoporotic potential of *Tinospora cordifolia* in female rats. *Maturitas* 329-338.

Kapur PB, Pereira MJ, Wuttke W, Jarry H. 2009. Androgenic action of *Tinospora cordifolia* ethanolic extract in prostate cancer cell line LNCaP. *Phytomed.* 679-682.

Karpova EA, Voznyi YV, Dudukina TV, Tsvetkva IV. 1991. 4-Trifluoromethylumbelliferyl glycosides as new substrates form revealing diseases connected with hereditary deficiency of lysosome glycosidases. *Biochem. Int.* 1135-1144.

Khare CP, 2007. Indian Medicinal Plants - An Illustrated Dictionary. 1st Ed. *Springer Science and Business Media.*

Kiem PV, Minh CV, Dat NT, Kinh LV, Hang DT, Nam NH, Cuong NX, Huong HT, Lau TV.2010. Aporphine alkaloids, clerodane diterpenes, and other constituents from *Tinospora cordifolia. Fitoterapia.* 485-489.

Kim SK, Kim HJ, Choi SE, Park KH, Choi HK, Lee MW. 2008. Antioxidative and inhibitory activities on nitric oxide (NO) and prostaglandin E2 (COX-2) production of flavonoids from seeds of prunustomentosa Thunberg. *Arch Pharm. Res.* 424-428.

Kirtikar KR, Basu BD, 1975. Indian Medicinal Plants, *Ind. Med. Plants*: II.

Kohno H, Maeda M, Tanino M, Tsukio Y, Ueda N, Wada K. 2002. A bitter diterpenoid furano lactone columbine from calumbae Radix inhibits azoxy methane-induced rat colon carcinogenesis. *Cancer let.* 131-139.

Kroemer G. 2007. Mitochondrial membrane permeabilization in cell death. *Physiol. Rev.* 99-163.

Kroemer G. 2009. Classification of cell death: recommendations of the Nomenclature Committee on Cell Death.*Cell Death Differ.* 3-11.

Kumar A. 2016. A novel colchicine-based microtubule inhibitor exhibits potent antitumor activity by inducing mitochondrial mediated apoptosis in MIA PaCa-2 pancreatic cancer cells.*Tumor Biol.* 13121-13136.

Kumar A. 2016. A novel microtubule depolymerizing colchicine analogue triggers apoptosis and autophagy in HCT-116 colon cancer cells. *Cell Biochem.Funct.* 69-81.

Kumar V, Mahdi F, Singh R, Mahdi AA, Singh RK. 2016. A clinical trial to assess the antidiabetic, antidyslipidemic and antioxidant activities of *Tinospora cordifolia* in management of Type-2 diabetes. *IJPSR.* 757-764.

Kunkuma VL, Kaki SS, Rao BVSK, Prasad RBN, Devi BLAP. 2013. A simple and facile method for the synthesis of 1 octacosanol. *Eur. J. Lipid Sci. Technol.* 921-927.

Leyon PV, Kuttan G. 2004. Inhibitory effect of a polysaccharide from *Tinospora cordifolia* on experimental metastasis. *J. Ethnopharmacol.* 233-237.

Lv J, Xu D, Perkovic VMX, Johnson DW, Woodward M. 2012. Corticosteroid therapy in IgA nephropathy. *J. Am. Soc. Nephrol.* 1108-1116.

Ly PT, Singh S, Shaw CA. 2007. Novel environmental toxins: Steryl glycosides as a potential etiological factor for age- related neurodegenerative diseases. *J. Nrurosci. Res.* 231-237.

Manjrekar PN, Jolly CI, Narayanan S. 2000. Comparative studies of the immunomodulatory activity of *Tinospora cordifolia* and *Tinospora sinensis. Fitoterapia.* 254-257.

Manjusha GV, Rajathi K, Alphonse JKM, Meera KS. 2011. Antioxidant potential and antimicrobial activity of Andrographis paniculata and *T. cordifolia* against pathogenic organisms. *J. Pharma. Res.* 452-4555.

Manosroi JP, Manosroi DA. 2006. Anti-proliferative activity of essential oil extracted from Thai medicinal plants on KB and P388 cell lines.*Cancer Lett.* 1120-1140.

Mathew S, Kuttan G. 1999. Immunomodulatory and antitumor activities of *Tinospora cordifolia.Fitoterapia.* 35-43.

Maurya R, SS, 1998. Tinocordifolin, a sesquiterpene from *Tinospora cordifolia. Phytochem.* 1343-1346.

Maurya R, Wazir V, Kapil A, Kapil RS. 2006. Cardifoliosides A and B, two new phenylpropene disaccharides from *Tinospora cordifolia* possessing immunostimulant activity. *Nat. Prod. Lett.* 7-10.

Maurya R, Wazir V, Tyagi A, Kapil RS. 1995. Clerodane diterpenoids from Tinospora cordifolia. *Phytochem.* 659-661.

McKeown E, Bykerk VP, Deleon F, Bnner A, Thorne C, CA. 2012.Quality assurance study of the use of preventative therapies in glucocorticoid-induced osteoporosis in early inflammatory arthritis: Result from the CATCH cohort. *Rheumatol.* 1662-1669.

Mhaiskar VB, Padya DC, Karmakar KB, 1980. Clinical evaluation of *Tinospora cordifolia* in amavata and sandhigata vata. *Rheumatism.* 35-39.

More P, PaiK. 2011. Effect of *T. cordifolia* (Guduchi) and LPS on release of H_2O_2, O_2 and TNF-á from murine macrophages in vitro. *J. Pharma. Biomed.Sci.* 1-6.

Mukherjee R, De UK, Ram GC. 2009. Evaluation of mammary gland immunity and therapeutic potential of *T. cordifolia* against bovine subclinical mastitis. *Trop. Anim. Health Prod.* 645-651.

Mukherjee R, De UK, Ram GC. 2010. Evaluation of mammary gland immunity and therapeutic potential of *Tinospora cordifolia* against bovine subclinical mastitis. *Trop. Anim. Health Prod.* 645-651.

Munafo DB, ColomboMI. 2001. A novel assay to study autophagy: regulation of autophagosome vacuole size by amino acid deprivation. *J .Cell Sci.* 3619-3629.

Muniyappan D, Baskar AA, Jagnacimuthu S, Agastian P, Duraipandyan V. 2009. Chemoprotective potential of epoxy clerodane diterpene from *Tinospora cordifolia* against diethylnitrosamine induced hepatocellular carcinoma. *Invest. New Drugs.* 347-355.

Murthy JR, Meera, Venkataraman RS, Satpute B, Chidambaranathan N,Devi P. 2010. Phytochemical investigation and anticonvulsant activity of leaves of *Tinospora cordifolia* Miers. *J Pharm. Sci.* 522-527.

Nagarkatti DS, Rege NN, Desai NK, SA. 1994. Modulation of kuppfer cell activity by *T. cordifolia* in liver damage.*J. Postgrad. Med.* 65-67.

Nardelli M. 1995. PARST95 - an update to PARST: a system of Fortran routines for calculating molecular structure parameters from the results of crystal structure analyses. *J. Appl. Cryst.* 659.

More P, Pai K. 2011.Immunomodulatory effects of *T. cordifolia* (Guduchi) on macrophage activation. *Biol. Med.* 134-140.

Padma P, Khosa KL. 2002. Anti-stress agents from natural origin. *J. Nat. Remed.*: 21-27.

Padmapriya S, Kumanan K, Rajamani K. 2009. Optimization of post-harvest techniques for *Tinospora cordifolia. Acad. J. Plant Sci.* 128-131.

Panchabhai TS, Ambarkhane SV, Joshi AS, Samant BD, RegeNN. 2008. Protective effect of *Tinospora cordifolia*, Phyllanthus emblica and their combination against antitubercular drugs induced hepatic damage: an experimental study. *Phytother. Res.* 646-650.

Parveen TD, Nyamathulla S. 2011.Antihyperlipidemic activity of the methanolic extract from the stems of *T. cordifolia* on Sprague dawley rats.*J. Pharm. Sci.* 104-109.

Patel JP, Gami B, Patel K. 2010. Evaluation of in vivo schizonticidal properties of acetone extract of some Indian medicinal plants. *Adv. Biol. Res.* 253-258.

Patel MB, Mishra S. 2011. Hypoglycemic activity of alkaloidal fraction of *T. cordifolia*. Phytomedicine. *Phytomed.* 1045-1052.

Patel SS, Shah RS, Goyal RK. 2009. Anti-hyperglycemic, anti-hyperlipidemic and antioxidant effects of Dihar, a poly herbal ayurvedic formulation in streptozotocin induced diabetic rats. *Indian J. Exp. Biol.* 564-570.

Patil M, Patki P, Kamath HV, Patwardhan B. 1997. Antistress activity of *Tinospora cordifolia*. *Indian Drugs*. 211-215.

Paval J, Kaitheri SK, Kumar A, Govindan, S. 2011.Anti-arthritic activity of the plant T. cordifolia Willd. *J. Herb. Med. Toxicol*. 11-16.

Pierre LL, Moses MN, 2015.Isolation and Characterization of Stigmasterol and betasitosterol from Odontonema Strictum (Acanthaceae). *JIPBS*. 88-95.

Premanath R, Lakshmidevi N. 2010. Studies on Anti-oxidant activity of *Tinospora cordifolia* (Miers.) Leaves using *in vitro* models. *J. Am. Sci.* 736-743.

Prince PS, Venugopal MP, Menon GG.1999.Hypolipidaemic action of *Tinospora cordifolia* roots in alloxan diabetic rats. *J. Ethnopharmacol.* 53-57.

Prince PSM, Menon VP. 1999. Antioxidant activity of *Tinospora cordifolia* roots in experimental diabetes. *J. Ethnopharmacol.* 277-281.

Prince PSM, Menon VP. 2001. Antioxidant action of *Tinospora cordifolia* root extract in alloxan diabetic rats. *Phytother. Res.* 213-218.

Prince PSM, Menon VP. 2003. Hypoglycaemic and hypolipidaemic action of alcohol extract of *Tinospora cordifolia* roots in chemical induced diabetes in rats. *Phytother. Res.* 410-413.

Prince PSM, Padmanabhan M, Menon VP. 2004. Restoration of antioxidant defence by ethanolic *Tinospora cordifolia* root extract in alloxan-induced diabetic liver and kidney. *Phytother. Res.*785-787.

Puranik N, Kammar KF, Devi S. 2010. Anti-diabetic activity of *Tinospora cordifolia* (Willd.) in streptozotocin diabetic rats; does it act like sulphonylureas? *Turk. J. Med. Sci.* 265-270.

Puranik NK, Kammar KF, Devi S. 2007. Modulation of morphology and some gluconeogenic enzymes activity by *Tinospora cordifolia* in diabetic rat kidney.*Biomed Res.* 179-183.

Pushp P, Sharma N, Joseph GS, RP Singh, 2011. Antioxidant activity and detection of (-) epicatechin in the methanolic extract of stem of *T. cordifolia*. *J. Food Sci Technol*. 567–572.

Rajalakshmi M, Eliza J, Priya CL, Nirmal A, Daisy P. 2009. Antidiabetic properties of *Tinospora cordifolia* stem extracts on streptozotocin-induced diabetic rats. *Afr. J. Pharma. Pharmcol.* 171-180.

Rao PR, Kumar VK, Viswanat RK, Subbaraju GV. 2005. Cardioprotective activity of alcoholic extract of *Tinospora cordifolia* in ischemia-reperfusion induced myocardial infarction in rats. *Biol. Pharm. Bull.* 2319-2322.

Rawal A, Muddeshwar M, Biswas S. 2004. Effect of *Rubia cordifolia*, Fagonia cretica linn and *Tinospora cordifolia* on free radical generation and lipid peroxidation during oxygen-glucose deprivation in rat hippocampal slices. *Biochem.Biophys. Res. Commun.*: 588-596.

Reed JC. 2002. Apoptosis-based therapies. *Nat. Rev. Drug Discov*. 111-121.

Rello S. 2005. Morphological criteria to distinguish cell death induced by apoptosis and necrotic treatments. *Apoptosis*. 201-208.

Rose MF, Noorulla KM, AsmaM, Kalaichelvi R, Vadivel K, Thangabalan B. 2010. *In vitro* antibacterial activity of methanolic root extract of *T. cordifolia* (Willd.). *J. Pharma Res. Develop*. 1-5.

Rout GR. *Z.* 2006. Identification of *Tinospora cordifolia* (Willd.) Miers ex Hook F & Thomas using RAPD markers. *NaturforschC.* 118-122.

S.S. Tirtha, 2005. The Ayurveda Encyclopedia-Natural Secrets to Healing, Prevention and Longevity. 2nd Ed. New York: Ayurveda Holistic Centre.

Sai KS, Srividya N. 2002. Blood glucose lowering effect of the leaves of *Tinospora cordifolia* and Sauropus androgynous in diabetic subjects. *J. Nat. Remed.* 28-32.

Sangeetha MK, Raghavendran BRH, Veeraraghavan G,Vasanthi HR. 2011. *T. cordifolia* attenuates oxidative stress and distorted carbohydrate metabolism in experimentally induced type 2 diabetes in rats. *J. Nat. Med.* 544-550.

Sarla M, Velu V, Anandharamakrishnan C, Singh RP. 2011. Spray drying of *T. cordifolia* leaf and stem extract and evaluation of antioxidant activity, Journal of Food Science and technology. *J. Food Sci. Technol.* 119-122.

Sengupta S, Mukherjee A, Goswami R, Basu S. 2009. Hypoglycemic activity of the antioxidant saponarin characterized as á-glucosidase inhibitor present in *Tinospora cordifolia. J. Enzyme Inhib. Med. Chem.* 684-690.

Shah GL, 1984. Some economically important plants of salsette island near Bombay. *J. Econ. Taxon. Bot.* 753-756.

Shah GL, Yadav SS, Badari N. 1983. Medicinal plants from Dahanu forest division in Maharashtra state. *J. Econ. Taxon. Bot.*141-144.

Sharma P, Parmar J, Sharma P, Verma P, Goyal PK. 2011. Radiation induced testicular injury and its amelioration by *T. cordifolia* extract. *Evi. Complement. Alt. Med.* 1-9.

Sharma, Pandey D. 2010. Protective role of *Tinospora cordifolia* against lead-induced hepatotoxicity. *Toxicol. Int.* 12-17.

Sheldrick GM. 1997. SHELX97, Program for the Refinement of Crystal Structures from Diffraction Data, University of Göttingen, Göttingen.

Sheldrick GM. 2008. A short history of SHELX Acta Cryst A.*Acta Crystall.* 112.

Singh KK, Maheshwari JK, 1983. Traditional phytotherapy amongst the tribals of Varanasi district U.P. *J. Econ. Taxon. Bot.* 829-32.

Singh NS, Singh M, Shrivastava P. 2005. Effect of *Tinospora cordifolia* on the antitumour activity of tumor-associated macrophages derived dendritic cells. *Immunopharmacol. Immunotoxicol.* 1-14.

Singh RK. 2005. *T. cordifolia* as an adjuvant drug in the treatment of hyper-reactive malarious splenomegaly-a case report. *J. Vector Borne Dis.*36-38.

Singh SM, Singh N, Shrivastava P. 2006. Effect of alcoholic extract of ayurvedic herb *Tinospora cordifolia* on the proliferation and myeloid differentiation of bone marrow precursor cells in a tumor-bearing host. *Fitoterapia.*1-11.

Singh SS, Pandey SC, Srivastava S, Gupta VS, Patro, B, Ghosh AC. 2003. Chemistry and medicinal properrties of *Tinospora cordifolia* (Guduchi). *Indian J. Pharmacol.* 83-91.

Sinha K, 2004. *Tinospora cordifolia* (Guduchi), a reservoir plant for therapeutic applications: A review. *Indian J. Tradit.Know.* 257-270.

Sivakumar V, Rajan MSD. 2011. Hypoglycemic and antioxidant activity of *T. cordifolia* in experimental diabetes.*Int. J. Pharmac. Sci. Res.* 608-613.

Soni HP, Nayak G, Patel SS, Mishra K, Singh RP.2011. Pharmacognostic studies of the leaves of *Tinospora cordifolia. IJPI's J. Pharmacog. Herb. Form.*1-6.

Spek AL. 2009. Structure validation in chemical crystallography Acta Cryst D. *Acta Crystallogr. a*: 148-155.

Sriramaneni RN, Omar AZ, Ibrahim SM, Amirin S, Mohd ZA. 2010. Vasorelaxant effect of diterpenoid lactones from and rographis paniculata chloroform exract on rat aortic rings. *Pharmacog. Res.* 242-246.

Subramanian M, Chintalwar G, Chattopghay S. 2007. Antioxidant properties of a *Tinospora cordifolia* polysaccharide against iron-mediated lipid damage and γ-ray induced protein damage. *Redox Rep.* 137-143.

Sudha P, Zinjarde SS, Bhargava SY, Kumar AR. 2011. Potent α-amylase inhibitory activity of Indian Ayurvedic medicinal plants.*BMC Complement.Altern. Med.* 5.

Sundarraj S, Thangam R, Sreevani V, Kaveri K, Gunasekaran P, Achiraman S. 2012.Ò-Sitosterol from acacia nilotica L. induces G2/M cell cycle arrest and apopyosis through c-Myc suppression in MCF-7 and A549 cells. *J. Ethnopharmcol.* 803-809.

Thatte UM, Kulkarni MR, Dahanukar SA.1992. Immunotherapeutic modification of E. coli peritonitis and bacteremia by *T. cordifolia. J. Postgrad. Med.* 13-15.

The Ayurvedic Pharmacopoeia of India. Part I. New Delhi: Department of Ayush, Ministry of Health and FW, 2001, I 53-55.

Thippeswamy G, Sheela ML, Salimath BP. *Eur.* 2008. Octacosanol isolated from *Tinospora cordifolia* downregulates VEGF gene expression by inhibiting nucular translocation of NF<kappa>B and its DNA binding activity. *J. Pharmcol.* 141-150.

Tiwari P, Kumar B, Kaur M, Kaur G, Kaur H, Dayal R. 2011. Spasmolytic, antidiarrhoel and intestinal modulatory activities of ethanolic extract of stem of *T. cordifolia* on isolated rat ileum.*Int. Pharmac. Sci.* 123-132.

Tiwari P, Kumar B, Kumar M, Kaur M, Debnath J, Sharma P. 2011. Comparative anthelmintic activity of aqueous and ethanolic stem extract of *T. cordifolia* (Willd). *Intl. J. Drug Develop. Res.*70-83.

Tomar A, Singh A, Thakur G, Agarwal AK, SinghVK. 2010. *In vitro* and *in vivo* study of *Tinospora cordifolia* as an antidiabetic agent in rat. *Biochem. Cell Biol.* 175-177.

Tyagi S, Singh L, Devi MM, Goel HC, Rizvi MA. 2009. Augmentation of antioxidant defence system by *T. cordifolia*: Implications inradiation protection. *J. Complem. Integr. Med.* 36.

Uddin MH, Hossain MA, Kawsar MH. 2011. Antimicrobial and cytotoxic acivities of *T. cordifolia. Int. J. Pharma. Sci. Res.* 656-658.

Upadhaya AK, Kumar K, Kumar A, Mishra, HS. 2010. *Tinospora cordifolia* (Willd.) Hook. F. and Thoms. (Guduchi)-validation of theAyurvedic pharmacology through experimental and clinical studies. *Int. J. Ayurveda Res.* 112-121.

Vaidya DB, 1994. Materia Medica of Tibetan Medicine. Delhi: Sri Satguru Publications, 163.

Verma M. 2008. *In vitro* cytotoxic potential of Polyalthia longifolia on human cancer cell lines and induction of apoptosis through mitochondrial-dependent pathway in HL-60 cells.*Chemico-Biol. Interact.* 45-56.

Vinutha B, Prashanth D, Salma, K, Sreeja SL, Pratiti D, Padmaja R. 2007. Screening of selected Indian medicinal plants for acetylcholinesterase inhibitory activity. *J. Ethnopharmacol.* 359-363.

Wadood N, Wadood A, Shah S.A.W. 1992. Effect of *Tinospora cordifolia* on blood glucose and total lipid levels of normal and alloxan diabetic rabbits. *Planta Med.*131-136.

Wang T, Liu YY, Wang X, Yang N, Zhu HB, Zuo PP. 2010. Protective effects of octacosanol on 6-hydroxydopamine-induced Parkinsonism in rats via regulation of ProNGP and NGF signalling. *Acta Pharmacol. Sin.* 765-774.

Wealth of India 2003. A dictionary of Indian Raw Materials and Industrial Products. 2003, New Delhi: CSIR. 251-252.

Wesley J, Christina AJM, Chidambaranathan N, Livingston RNR, Ravikumar K. 2008. Effect of alcoholic extract of *Tinospora cordifolia* on acute and subacute inflammation. *Pharmcology Online.* 683-687.

Yang JH, Kondratyuk TP, Marler LE, Qiu X, Choi. 2010. Isolation and evaluation of kaempferol glycosides from the fern neocheiropterispalmatopedata.Y. *Phytochem.* 641-647.

Yang S, Evens A, Prachands M, Singh AT, Bhalla S, Devid K. 2010. Diterpenoid lactone and rographolide, the active component of and rographis paniculata.*Clin. Cancer Res.* 4755-4768.

Zafar Z, Talkad MS, Bandopadhyay C, Sinha M, Sarkhel J. 2011. Antioxidant activity of five selective medicinal plants. *Afr. J. Sci Res.* 127-147.

Zalawadia R, Gandhi C, Patel V, Balaraman R. 2009. The protective effect of *Tinospora cordifolia* on various mast cell mediated allergic reactions. *Pharma. Biol.* 1096-1106.

Zhang H. 2013. The ClC-3 chloride channel associated with microtubules is a target of paclitaxel in its induced-apoptosis. *Sci. Rep.* 2615.

Zhao F, He EQ, Wang L, Liu K. 2008. Anti-tumor activities of and rographolide, a diterpene from And rographis paniculata, by inducing apoptosis and inhibiting VEGF level. *J. Asian Nat. Prod. Res.* 467-473.

Zhao TF, Wang X, Rimando AM, Che C. 1991. Folkloric medicinal plants: *Tinospora sagittata* var. cravaniana and Mahonia bealei. *Planta Med.* 505.

4

Modelling the Environmental Niche and Potential Distribution of *Magnolia campbellii* Hook.f. & Thomson for its Conservation in the Indian Eastern Himalaya

[1]Dibyendu Adhikari, [1]Prem Prakash Singh, [1]Raghuvar Tiwary and [1&2]Saroj Kanta Barik[*]

[1]Department of Botany, North-Eastern Hill University, Shillong - 793022 Meghalaya, INDIA
*[2]*CSIR-National Botanical Research Institute, Lucknow - 226001, Uttar Pradesh INDIA*
Corresponding author email: sarojkbarik@gmail.com

ABSTRACT

Magnolia campbelii Hook.f. & Thomson is a deciduous tree species belonging to Magnoliaceae and is highly valued for its ornamental uses. The species is threatened by habitat degradation as well as its utility as timber. It is distributed in the eastern Himalaya, southern Myanmar, and Yunnan province of China. However, its exact distribution in the eastern Himalaya is unknown owing to rough terrain and remote location of the Himalayas. We modelled the environmental niche and delineated the potential habitats and distribution areas of the species in this study using ENVIREM dataset and Maxent software. Thermicity Index and Topographic Wetness were the critical environmental variables determining the species distribution in the eastern Himalaya as revealed by Jackknife analysis. Geographical projection of the modelled environmental niche indicates that most suitable areas for its distribution are the western part of Arunachal Pradesh, few pockets in Sikkim, and the central part of Bhutan. The results of the study have implications for in situ conservation of the species as well as identifying suitable areas for reintroduction.

Keywords: *Magnolia campbelii*, Ecological Niche Modelling, Conservation, Northeast India

INTRODUCTION

Native plant diversity in the tropics is under threat owing to habitat destruction, invasion by alien plant species, and changing climatic conditions (Chitale et al. 2014, Adhikari et al. 2015, Roy et al. 2015, Barik et al. 2018a). Putatively, these agents of environmental change have brought at least one-fifth of the total plant species to the brink of extinction (Brummitt and Bachman 2010). Considering the gravity of the issue, biotechnological interventions seems to be the only way, to curb the extinction process and aid threatened plant conservation (Barik et al. 2018b).

Numerous methods, tools, techniques, and protocols have been developed/ proposed in this respect, most of which are available online. Recently, Barik et al. (2018b) proposed a 10-step protocol that should help in threatened plant conservation in India *viz.*, (i) population inventory, characterization and mapping using ecological niche modelling, (ii) metapopulation modelling of selected species populations to determine the conservation status, minimum viable population size and to assess extinction risk, (iii) identification of factors responsible for depleting species populations and developing a species-specific recovery strategy, (iv) molecular characterization of the selected species populations to identify those with greater diversity for genetic enrichment based on source–sink concept, (v) characterization of active principles in selected species in different habitats/populations, (vi) standardizing the macro- and micropropagation techniques for mass multiplication, (vii) reproductive biology of the selected species to address the regeneration failure, (viii) production of planting materials for reintroduction of the species in the areas identified through ENM, (ix) herbarium and establishment of field gene banks at appropriate ecological zones, and (x) memorandum of understanding with the Forest Department and communities, and reintroduction with post-introduction monitoring protocol. It is important to mention, that studying all the aspects requires a lot of time and resources. Nonetheless, we opine that modeling the ecological niche and predicting geographical distribution of species, which helps in prioritizing conservation areas, may be an effective way to protect the species and prevent extinction (Adhikari et al. 2018).

Prioritizing areas for biodiversity conservation have been in practice for a long time. Here, the emphasis is primarily on protection of undisturbed natural habitats for a wide range of species through the creation of protected area networks. However, when the concern is to identify conservation areas for a single species, such method may be inefficient. Nonetheless, recent technological developments in the field of computing, satellite remote sensing and geographical information system (GIS) have made it possible to model the ecological niches of individual species as well as anticipate their geographical distribution with high accuracy and precision. Subsequently, the modelled niche is projected in geography to anticipate the potential distribution of the species.

Ecological niche modelling (ENM) is a robust computational tool, which relates point distributional data of a species with rasterized environmental data to reconstruct its ecological niche (Thuiller et al. 2005, Hijmans and Graham 2006). Subsequently, the reconstructed niche may be translated in the form of a predictive distributional map, which is essentially a distributional hypothesis of the species. The resultant distributional hypothesis is evaluated using various statistical tests to construct a formal distributional model of the species (Peterson et al. 1999). ENM tool has been successfully used in planning conservation measures for several threatened species (Adhikari et al. 2018).

We modeled the environmental niche and potential distribution area of *Magnolia campbellii* Hook.f. & Thompson (Magnoliaceae), a medium to large sized threatened deciduous tree species in the Eastern Himalayan states of Arunachal Pradesh, Sikkim and Bhutan. Reportedly, the species is distributed in eastern Nepal; Sikkim and Assam in India; southern Xizang, Yunnan, and southern Sichuan in China and south to northern Myanmar (Khela 2014). However, it is not clear how the environmentally suitable areas of the species are distributed in these regions. Moreover, the natural population of the species is currently under threat because of habitat degradation and harvesting for fuelwood; although the species has been classified as 'least concern' in the IUCN Red list (ver. 3.1). Therefore, there is a need to conserve the species through technological interventions.

MATERIALS AND METHODS

Occurrence Data

A total of 38 occurrences records were compiled of which 22 were first-hand data and 16 were from online sources. Primary data on the species occurrences were collected through field surveys in Tawang and West Kameng districts of Arunachal Pradesh. Secondary data on species occurrences in Sikkim and Bhutan were obtained from online sources (www.indiabiodiversity.org and www.biodiversity.bt/species).

Environmental Predictors

We used ENVIREM variables (Environmental Rasters for Ecological Modeling, http://envirem.github.io/) to model the environmental niche and predict the geographical distribution of *M. campbelii*. The dataset consists of a set of 16 climatic and 2 topographic variables that are relevant to species ecological and physiological processes (Title and Bemmels 2017). Generic grids of the dataset with a spatial resolution of 30 arc seconds (~1 km) were downloaded from the archives of the University of Michigan's Deep Blue Data (https:// deepblue.lib.umich.edu/data/concern/generic_works/gt54kn05f). The data was then resampled to a spatial resolution of 500 meters to match with the terrain complexity in the study area. Thereafter, we performed correlation analysis on the dataset to eliminate the redundant variables (ENM tools, Warren et al. 2010).

The variables retained for ENM were climatic moisture index, thermicity index, PET coldest quarter, PET seasonality, PET wettest quarter, Emberger Q, Thornthwaite aridity index, topographic wetness and topographic roughness (Table 1).

ECOLOGICAL NICHE MODELING

We used Maximum entropy modeling software (Maxent 3.4.1, Phillips et al. 2017) to reconstruct the environmental niche and predict the geographical distribution of the species. Model parameterization was done using 10,000 background points, 500 iterations, and a convergence threshold of 0.00001. Hinge, linear and quadratic feature types were used to optimize model fitting. Ten bootstrap runs were employed to derive an optimized model and opted for complementary log-log (cloglog) output.

Model Evaluation

Model quality was evaluated based on AUC value and was graded following the conservative estimate of Thuiller et al. (2005) as poor (AUC<0.8), fair (0.8<AUC<0.9), good (0.9<AUC<0.95) and very good (0.95<AUC<1.0). Field surveys were undertaken in the predicted areas for independent validation of the model.

Determinants of Environmental Niche

The important determinants of the environmental niche of *M. campbellii* were identified through ranking of the predictor variables and jackknife analysis (Phillips et al. 2006). Ranking of the predictor variables was done based on two metrics *viz.*, percent contribution and permutation importance. To determine percent contribution, in each iteration of the training algorithm, the increase in regularized gain is added to the contribution of the corresponding variable or subtracted from it if the change to the absolute value of lambda is negative. For determining permutation importance, for each environmental variable in turn, the values of that variable on training presence and background data are randomly permuted. The model is reevaluated on the permuted data, and the resulting drop in training AUC is shown in the table, normalized to percentages.

Table 1: Estimates of relative contributions of the environmental variables to the Maxent model. Values shown are averages over replicate runs.

Variable	Variable description	Percent contribution	Permutation importance
Climatic moisture index	A metric of relative wetness and aridity	34.1	2.9
Thermicity index	Compensated thermicity index: sum of mean annual temperature, minimum temperature of coldest month, maximum temperature of the coldest month, x10, with compensations for better comparability across the globe	21.9	47.2
PET coldest quarter	Mean monthly PET of coldest quarter	12.2	21.5
PET seasonality	Monthly variability in potential evapotranspiration	10.2	15.6
Emberger Q	Emberger's pluviothermic quotient: a metric that was designed to differentiate among Mediterranean type climates	8.9	8.4
Topographic wetness	Topographic wetness index	5.7	1.1
Thornthwaite aridity index	Thornthwaite aridity index: Index of the degree of water deficit below water need	5.1	1.1
Topographic roughness	Terrain roughness index	1.7	1.7
PET wettest quarter	Mean monthly PET of wettest quarter	0.1	0.5

Distribution of Environmentally Suitable Areas

Suitable areas of the species were mapped by importing the averaged probability model output in ArcGIS software. The predicted probability distribution was grouped into five suitability classes viz., very low (0-0.2), low (0.2-0.4), medium (0.4-0.6), high (0.6-0.8), and very high (0.8-1.0).

RESULT AND DISCUSSION

Model Evaluation

Tests of model performance yielded 'very good' results for ROC (mean AUC: 0.962) indicating that the Maxent could discriminate the true presences from false presences with high rates of accuracy (Fig. 1). It shows very good model consistency.

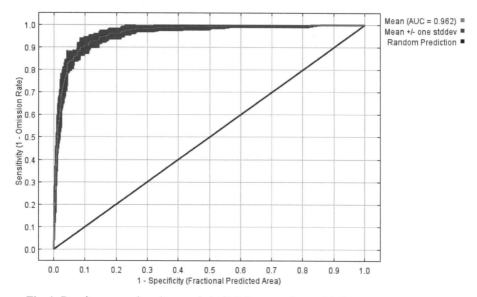

Fig. 1: Receiver operating characteristic (ROC) curve along with the area under the curve (AUC) showing model performance.

Analysis of Variable Contribution

Climatic moisture index, thermicity index, potential evapotranspiration of the coldest quarter and potential evapotranspiration seasonality had ~78 percent cumulative contribution, indicating their dominant role in defining the environmental niche of *M. campbellii* (Table 1). The relatively high values of permutation importance for thermicity index and potential evapotranspiration of coldest quarter indicate their role in influencing the model fitness. Permuting these variables caused a substantial drop in the training AUC, which signifies the reduced ability of the Maxent model to discriminate known presences from the background points.

To determine percent contribution, in each iteration of the training algorithm, the increase in regularized gain is added to the contribution of the corresponding variable or subtracted from it if the change to the absolute value of lambda is negative. For determining permutation importance, for each environmental variable in turn, the values of that variable on training presence and background data are randomly permuted. The model is reevaluated on the permuted data, and the resulting drop in training AUC is shown in the table, normalized to percentages.

Jackknife test shows that thermicity index and topographic wetness index were the most effective variables in predicting the distribution of *M. campbellii* in the Eastern Himalaya (Fig. 2). These variables had the useful information content by themselves that are not present in the other variables, and therefore may be most relevant in the distribution of the species. Overall, the environmental determinants of the *M. campbellii* niche indicate its cold climatic preference.

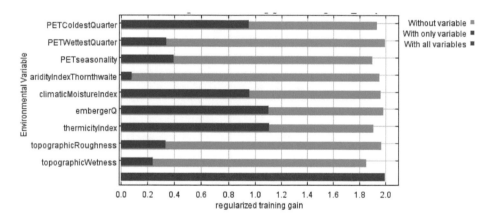

Fig. 2: Jackknife of regularized training gain for *M. campbellii* elucidating the variable importance. The environmental variable with the highest gain, when used in isolation, is thermicity index, which therefore appears to have the most useful information by itself. The environmental variable that decreases the gain the most when it is omitted is topographic wetness, which therefore appears to have the most information that is not present in the other variables. Values shown are averages over replicate runs.

Potential Distributional Areas

About eight percent of the total area in the Eastern Himalaya i.e., ~10,725 km² has high to very high environmental suitability for *M. campbellii* (Fig. 3 and 4). Substantial portions of these suitable areas are anticipated in the Western parts of Arunachal Pradesh i.e., the West Kameng, Tawang and Subansiri districts, and the central and Eastern parts of Bhutan. Nonetheless, areas with medium to high suitability are also expected in the state of Sikkim.

In general, the model predictions show that river valleys are environmentally more suitable compared to other areas. Field surveys done in West Kameng and Tawang districts of Western Arunachal Pradesh corroborate this finding, where the species were found growing in the valley slopes. A review of the literature also revealed that the species mostly grows in sheltered valleys.

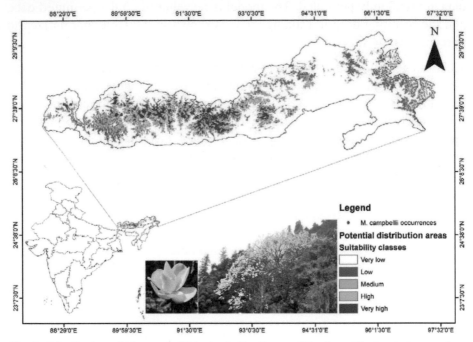

Fig. 3: Distribution of *M. campbellii* in the Indian Eastern Himalaya. The inset pictures show individual flower and the habitat of the species.

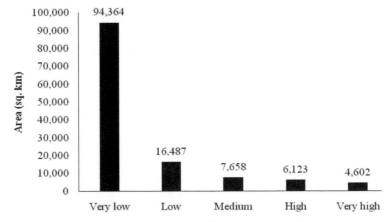

Fig. 4: Area under different environmental suitability classes in the Eastern Himalayan states of Arunachal Pradesh, Sikkim, and Bhutan. Numbers on top of the bars are the areas under different suitability classes.

Status of habitat and population

Most habitats of the species were disturbed and had signs of cultivation, harvesting of fuelwood, and developmental activities. During field visits, we enumerated about 378 mature individuals in Western Arunachal Pradesh. However, no saplings and seedlings of the species were observed. This indicates its seemingly poor regeneration in nature.

IMPLICATIONS FOR CONSERVATION

The study reveals that *Magnolia campbellii* have a highly restricted potential distribution area in the Eastern Himalayas. In addition, most of such areas fall within areas impacted by a range of anthropogenic activities. Considering these, it is anticipated that the species may become Threatened in near future from the current threat classification of 'Least Concern'. Thus, it is imperative to undertake appropriate conservation steps such as mass multiplication through vegetative means and/or biotechnological interventions followed by reintroduction of mass propagated species to areas identified as suitable by ENM. In addition, the local communities should be sensitized and involved in the conservation process. It is important to mention that species because of its showy white flowers have great ornamental value. Therefore, they can be used for landscaping and beautification in the hill stations.

ACKNOWLEDGEMENT

We acknowledge the financial support received from Water and Power Consultancy Services (WAPCOS), Ministry of Water Resources, Government of India, and Government of Arunachal Pradesh.

REFERENCES CITED

Adhikari D, Reshi Z, Datta BK, Samant SS, Chettri A, Upadhaya K, Shah MA, Singh P, Tiwary R K, Majumdar K, Pradhan A, Thakur ML, Solan N, Zahoor Z, Mir SH, Kaloo ZA, Barik SK, 2018. Pradhan A. 2018. Inventory and characterization of new populations through ecological niche modelling improve threat assessment. *Current Science* 114(3): 519-531.

Adhikari D, Tiwary R, Barik, SK. 2015. Modelling hotspots for invasive alien plants in India. *PLoS One* 10(7): e0134665.

Barik SK, Tiwari ON, Adhikari D, Singh PP, Tiwary R, Barua S. 2018a. Geographic distribution pattern of threatened plants of India and steps taken for their conservation. *Current Science* 114(3): 470-503.

Barik SK, Chrungoo NK, Adhikari D. 2018b. Conservation of threatened plants of India. *Current Science* 114(3): 468-469.

Brummitt NA, Bachman SP. 2010. Plants under pressure a global assessment: the first report of the IUCN sampled red list index for plants. *Kew, UK: Royal Botanic Gardens.*

Chitale VS, Behera MD, Roy PS. 2014. Future of endemic flora of biodiversity hotspots in India. *PLoS One* 9(12): e115264.

Hijmans RJ, Graham CH. 2006. The ability of climate envelope models to predict the effect of climate change on species distributions. *Global Change Biology* 12: 2272–2281. doi: 10.1111/j.1365-2486.2006.01256.x

Khela S. 2014. *Magnolia campbellii*. The IUCN Red List of Threatened Species 2014: e.T193912A2290545. http://dx.doi.org/10.2305/IUCN.UK.2014-RLTS. T193912A2290545.en.

Peterson, AT, Soberon, J., Sanchez-Cordero V. 1999. Conservatism of ecological niches in evolutionary time. *Science* 285: 1265–1267.

Phillips, SJ, Anderson, RP, Dudík M, Schapire RE, Blair ME. 2017. Opening the black box: an open source release of Maxent. *Ecography* 40(7): 887-893.

Phillips SJ, Anderson RP, Schapire RE. 2006. Maximum entropy modeling of species geographic distributions. *Ecological Modelling* 190(3): 231–259.

Roy PS, Roy A, Joshi PK, Kale MP, Srivastava VK, Srivastava SK, Sharma Y. 2015. Development of decadal (1985–1995–2005) land use and land cover database for India. *Remote Sensing* 7(3): 2401-2430.

Thuiller W, Lavorel S, Araújo MB, Sykes MT, Prentice IC. 2005. Climate change threats to plant diversity in Europe. *Proceedings of National Academy of Science USA* 102(23): 8245–8250

Title PO, Bemmels JB. 2017. ENVIREM: An expanded set of bioclimatic and topographic variables increases flexibility and improves performance of ecological niche modeling. *Ecography* 41: 291-307. Doi: 10.1111/ecog.02880.

Warren DL, Glor RE, Turelli M. 2010. ENMTools: a toolbox for comparative studies of environmental niche models. *Ecography* 33(3): 607-611.

5

Bunium persicum (Boiss.) Fedtsch. Distribution, Botany and Agro-technology

Sajan Thakur and Harish Chander Dutt*

Ecological Engineering Laboratory
Department of Botany, University of Jammu, Jammu - 180006, Jammu & Kashmir, INDIA
*Email: hcdutt@rediffmail.com

ABSTRACT

Bunium persicum (Boiss.) Fedtsch., (Family: Apiaceae) is an important plant of temperate regions and its seeds are used as carminative substance in various food recipes. The seeds of the species act as a good source of economy to the hilly and mountain people. As the species is restricted in its habitat, therefore, it requires mass cultivation practices. Many attempts are already made to cultivate species near to its natural habitat. In recent years some efforts have been done to domesticate this plant in certain areas of India like Shong area in Kinnaur district of Himachal Pradesh, India, but the cultivation remained very limited. Therefore, in addition to its collection from the wild sources the species needs more scientific intervention in agriculture sector. This species has typical Palaeo-Mediterranean distribution. In India the species is distributed in wild only in North West Himalaya (J&K, HP and UK) which makes its ocurrence rarity in the country. The species restricted its distribution between 1850-3100 meters above mean sea level. Phylogenetically the species is near to *Carum carvi* L., therefore, *C. carvi* is the potential adulterant of *B. persicum.*

Keywords: Medicinal plant, *Bunium persicum*, distribution, botany, agrotechnology India

INTRODUCTION

Medicinal and aromatic plants have been used for many centuries and are still popular in today's as an alternative therapies. Herbs often represent the original sources of most of the drugs. Parsley family (Apiaceae) is the 16th largest family of flowering plants with more than 3,700 species under 434 genera. It includes economically important aromatic medicinal plants such as *Bunium persicum* (Boiss.) Fedtsch., *Centella asiatica* (L.) Urban, *Hydrocotyle asiatica* L., *Pleurospermum candollei* (DC.) C.B.Clarke, *Angelica glauca* Edgew., *Trachyspermum ammi* L. and several others.

Bunium L. is an important aromatic plant genus under the family Apiaceae which has typical Palaeo-Mediterranean distribution. The systematics of *Bunium* has been elaborated insufficiently so far, and rests exclusively on morphological characters. *Carum* L. is the closest genus of *Bunium*, whose seeds and essential oils have been used in food and medicine throughout the world (Jassbi et al. 2005). About 56 species are described under the genus *Bunium* as evident from published literatures.

Data Collection

Data with respect to its taxonomy, botany and agriculture has been collected from the open source data bases like Wikipedia, Wikimedia common, http:// www. iucnredlist.org, http://www.catalogueoflife.org, http://www.theplantlist.org, http://www.ipni.org, http://www.efloras.org, science direct, NCBI etc. The data collected from various databases was thoroughly studied and the technical language of taxonomy or agrotechnology was transformed to easy English Script so that non technical persons can easily understand it. The geographical distribution of the species provided in the different databases was understood and depicted on the line maps using Adobe Photoshop.

Taxonomic characters

Bunium persicum (Boiss.) Fedtsch., Rastit. Turkestan. 612. 1915. Herbaceous plant, 15-70 cm tall, branched perennial herb with underground bulbs *ca*.1-2 cm in size. Lower leaves are petiolate and upper sessile having filiform shape with 2-3 pinnatisect condition. Involucre may have 1-5 bracts or may be absent whereas involucel compose 2-5 narrow linear bractlets. Pedicels are twice or thrice longer than flowers. Petals are pinkish white and *ca*. 1 mm in length. Fruit stalk is thin and *ca*. 5-10 mm long. Fruits black brown, oblong, slightly curved, 3-4 mm long with prominent ridges and flat stylopodium; styles are reflexed; furrows 1-vittate; commissure 2-vittate; vittae large (Khosravi 1993, Nasir and Ali 2005).

Synonyms

Carum heterophyllum Regel & Schmalh., *Carum persicum* Boiss., *Bunium bulbocastanum* L., *Elwendia persica* (Boiss.) Pimenov & Kljuykov.

Common Names

Kala-zirah, Kalajirah, Shiajira, Siah-zirah, Siyajira (Hindi); Krihun zeur (Kashmiri), Shime jeerige (Kannada), Jiraka, Kalajira, Karavi, Kasmirajiraka, Krsnajiraka (Sanskrit); Cimaiccirakam, Kavarcirakam, Kevalika, Pilappu Shiragam (Tamil); Seema jeelahara, Simajilakara (Telugu); Zireh Koohi, Zireh e Irani (Persian); Black cumin (English).

Geographical Distribution

Bunium persicum is considered to have originated from Central Asia to Northern India and grows naturally in temperate to alpine regions at 1850-3100 meters above msl (Azizi et al. 2009). At global level of distribution, it is found to grow at extreme high altitudes of Iran, Afghanistan, Pakistan, China, Kyrgyzstan, Tajikistan, Uzbekistan, Turkmenistan, Pakistan and India (Panwar 2000, Hanelt et al. 2001, Anonymous 2018).

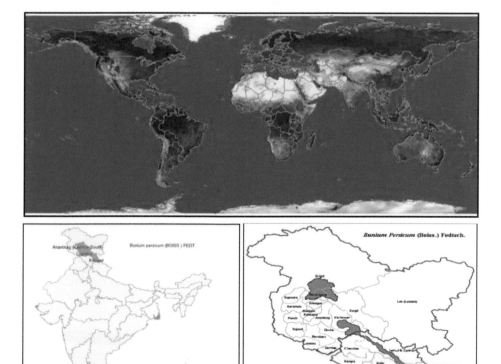

Fig. 1: (a) Worldwide distribution; (b) distribution in India; (c) distribution in NW Himalaya (Courtesy: www.Google.com)

The species is found in sparse patches confined to high altitudes of Gilgit in PoK (Pakistan Occupied Kashmir), North West India, Baltistan, Chitral, Swat, Hazara, Frontier Province of Baluchistan and Ziarat. In India the species is collected from (i) Jammu and Kashmir (Kishtwar High Altitude National Park, Kargil, Zanskar, Gurez and Padder valley), (ii) Himachal Pradesh (Kinnaur, Lahul-Spiti, Pangi and Bharmaur) and (iii) some ranges of Uttrakhand (Roscov et al., 2018). The plants growing in Baluchistan are similar to those growing in other parts of India but they show different range of occurrence near Juniper zone and a characteristic smell in their fruits.

AGROTECHNOLOGY

Land Preparation

First of all crop stubbles or weeds along with rocks and deep-rooted sedges are removed from the land and then it is properly ploughed 2-3 times to make a fine tilt. The soil is then mixed properly with farmyard manure or vermicompost at the rate of 3500 kg/ha. Proper drainage is provided by maintaining a sufficient gradient and slope. The proper level and slope is maintained in the field within beds and between beds to facilitate proper irrigation and drainage.

Propagation

Kalajira can be propagated through seeds as well as tubers. Firstly the propagation is done mainly through seeds, and then subsequent growth and production is maintained through tubers/bulbs. The tubers are formed between 10-15 cm deep in the soil from the first and second year seedlings. Each plant requires a total three to four years to produce a good quantity of healthy and viable seeds. The seed collection is done mainly in the month of August, and the sowing is done during October – November. The seeds are small in size, so they are first mixed with sand and then sown in line at 3-4 cm deep in soil. The seeds remain dormant in the winters and germinate after the melting of snow in March-April. They require a chilling pre-treatment (3-5°C) for 4-5 months to acquire maximum germination.

Nursery

Fully matured bold seeds are collected from healthy plants with robust vegetative growth. Seeds are sown at the rate of 1-1.5 kg per ha in the first sowing. The seed sowing is done in November. For nursery development, 5-10 deep furrows are prepared in line with 40-50 cm gap in between. The mixture of seed and soil is properly placed within the furrows and then immediately covered with a thin layer of soil. The seed germination takes place in April. However, nursery preparation is not required for crop produced through tubers, because tubers can be placed at the appropriate distance in the field.

Transplantation

The seedlings are grown in full sun and not in shade. Partial shady condition also leads to retarded growth and yield. When the size of the plantlet becomes large enough to handle, the seedlings are transferred to the greenhouse conditions for first winter. In late spring or early summer when there is no frost, the plantlets are transplanted out into the field. The plants in first year, grow about 20-30 cm and in second year up to 60-80 cm. The spacing in the field is maintained theoretically about 10-15 cm between the rows and 15 cm between the plants. But the distance about 45-50 cm between the individuals as well as between the rows is better to achieve the maximum production.

The other way of cultivation of the plant species is through tubers. One to two years old tubers has a tendency of producing only 1-2 buds where as, a three year old tubers produces maximum of 4-5 buds and thus plants. The buds of tubers also require long chilling period for their better sprouting and then initiation of flowering. Vegetatively propagated crop through tubers get healthy and good quantity of seeds in one year only whereas the seed cultivation require 3-4 years for the same.

Irrigation

As a plant of hilly region the water requirement is not very high. It requires only 2-4 irrigations in the form of sprinkling for good harvest. The water sprinkling is required only during the periods of sowing, flowering and fruiting.

Weeding

A weed free field is the pre-requisite for the better growth of plant. A minimum of 3-4 weeding operations are necessarily required after every 20-25 days period. In the first year as the plantlets are fragile so careful hand weeding is preferred for their better growth. Thorough weeding is done in the second year as the plants used to attain a good height. Upper ground foliage and tubers are attacked by white grubs, hairy caterpillars, armyworms and semi-loopers. Pesticides like, a mixture of BHC and HCH or Aldrin (5%) dust at the rate of 25 kg/ha and methyl parathion are some of the effective control measures for these pests. Instead of inorganic pesticides, the biological control of such pests can be done through *Beauveria bassiana*, an entomopathogenic fungus (Biopesticide) at the rate of 5 ml/L on fully grown plants. For juvenile plants less concentration may be applied.

Harvesting and Postharvest Management

The browning of seeds in July-August indicates their maturation and the crop becomes ready to harvest. Depending on the climate the maturation of seeds vary from low altitude to high altitude regions. Crop matures early in low altitude areas as compared to high altitude areas due to temperature variation. The total harvesting period lasts for only 10-15 days in mid August. The mature plants are harvested daily in the morning hours, and dried in sun light for 4-5 days for better maturity and storage. The shattering of umbels may lead to loss of seeds therefore much care is needed for the handling of plants. With the help of beating stick the seeds are separated from the plants and then threshed by hand. The threshed seeds are then cleared by winnowing. The seeds are then stored and packed in a paper bag or closed air tight container and then kept in a dark cool place.

Yield = 0.5 ton/ha. approximately

Market price: In general the price varies between Rs. 450-500 per kg of dried seeds.

Economic importance

The seeds of the plant are used traditionally as flavouring condiment in a number of ethnic dishes and food industry. These are very popular seasoning agent for non-vegetarian dishes in Central Asia (Karim et al. 1977). The seeds have stimulant, expectorant, antispasmodic, carminative, anti-obesity and diuretic properties (Chauhan 1999). Moreover seeds are frequently used in treating a lot of disorders like diarrhea, dyspepsia, fever, flatulence, stomach ache and many other diseases (Baser et al. 1997, Sardari et al. 1998, Panda 2004, Boskabady and Moghaddas 2004). The seeds of *B. persicum* contain essential oils like p-Mentha-1,4-dien-7-al, gamma-terpinene, beta-pinene and cuminaldehyde having antioxidant and antimicrobial activities (Shansavari et al. 2008, Moghtader et al. 2009, Talei and Mosavi 2009).

CONCLUSION

High commercial utility of seeds of *Bunium persicum* has increased its demand in national as well as international markets. Due to limited cultivation practice of this species, the major portion of the seeds is harvested extensively from its natural habitats. Because of unsustainable and unscientific exploitation of the species there has been increase in reduction of its natural populations. The special conservation efforts are needed to conserve this plant in North-western Himalayas. In recent past some important efforts were made to domesticate this plant in areas like Shong area in Kinnaur district of Himachal Pradesh, India, but the cultivation remained very limited. This may be because there is non-availability of suitable genotype for cultivation. The quantity of crop produced from both wild as well as cultivated resources at present is not enough to meet the requirements of its end users. Therefore, this crop needs much more attention by the farmers as well as agriculture scientists. The mass cultivation of the species can reduce the collection pressure on wild populations of the species.

REFERENCES CITED

Anonymous. 2018. http://www.catalogueoflife.org accessed on March 28, 2018

Azizi M, Davareenejad G, Bos R, Woerdenbag HJ, Kayser O. 2009. Essential Oil Content and Constituents of Black Zira (*Bunium persicum* [Boiss.] Burdenko Fedtsch.) from Iran during Field Cultivation (Domestication). *Journal of Essential Oil Research* 21: 78–82.

Baser KHC, Oezek T, Abduganiev BE, Abdullaev UA, Aripov KN. 1997.Composition of the essential oil of *Bunium persicum* (Boiss.) B. Fedtsch from Tajikistan. *Journal of Essential Oil Research* 9: 597-598.

Boskabady MH, Moghaddas A. 2004. Antihistaminic effect of *Bunium persicum* on Guinea Pig Tracheal Chains. *Iranian Biomedical Journal* 8: 149–155.

Chauhan NS. 1999. *Medicinal and Aromatic Plants of Himachal Pradesh.* Indus Publishing Co., New Delhi.

Hanelt P, Bu"ttner R, Mansfeld R, Kilian R. 2001. *Mansfield's Encyclopedia of Agricultural and Horticultural Crops.* Springer, Berlin.

Jassbi AR, Mehrdad M, Soleimani M, Mirzaeian M, Sonboli A. 2005. Chemical Composition of the essential oils of *Bunium elegans* and *Bunium caroides. Chemistry of Natural Compounds* 41: 415–417.

Karim A, Pervez M, Bhatty MK. 1977.Studies on The Essential of the Pakistan Species of the family Umbelliferae. Part 10. *Bunium persicum* Boiss. (Siah Zira) Seed Oil. *Pakistan Journal of Scientific and Industrial Research* 20: 106-108.

Khosravi M. 1993. Black zira (botanical, agricultural ecology, and study the possibility of production). Agronomy. Mashhad, College of Agriculture, Ferdowsi University of Mashhad. MSc thesis.

Moghtader M, Mansori AI, Salari H, Farahmand A. 2009. Chemical composition and antimicrobial activity of the essential oil of *Bunium percisum* Bioss. Seed. *Iranian Journal of Medicinal and Aromatic Plants.* 25:20–28

Nasir E and Ali SI. 2005. *Flora of Pakistan.* University of Karachi, Karachi.

Panda H. 2004. *Aromatic plants cultivation, processing and uses.* Asia Pacific Business Press Inc, New Delhi

Panwar KS. 2000. *Black caraway-Kala zira. In: Arya PS (ed) Spice crops of India.* Kalyani Publishers, New Delhi, pp 172–178.

Sardari S, Amin GH, Micetichm RG, Daneshtalab M. 1998. Antifungal activity of selected Iranian and Canadian plants. *Pharmaceutical Biology* 36:180–188.

Shahsavari N, Barzegar M, Sahari MA, Naghdibadi H. 2008. Antioxidant Activityand Chemical Characterization of Essential Oil of *Bunium persicum. Plant Foods for Human Nutrition.* 63: 183–188.

Talei GR, Mosavi Z. 2009. Chemical composition and antibacterial activity of Bunium persicum from west of Iran. *Asian Journal of Chemistry* 21:4749–4754.

6

Dysoxylum binectariferum (Roxb.) Hook., a Plant of High Medicinal Value and A Source of Leads for Modern Medicine

Vikas Kumar[1,3], Sonali S. Bharate[1], Ram A. Vishwakarma[2,3] and Sandip B. Bharate[2,3,]*

[1]*Preformulation Laboratory, PKPD Toxicology & Formulation Division, CSIR - Indian Institute of Integrative Medicine, Jammu - 180001, Jammu and Kashmir, INDIA*
[2]*Medicinal Chemistry Division, CSIR - Indian Institute of Integrative Medicine Jammu - 180001, Jammu and Kashmir, INDIA*
[3]*Academy of Scientific and Innovative Research, CSIR - Indian Institute of Integrative Medicine, Jammu - 180001, Jammu and Kashmir, INDIA*
*Email: sbharate@iiim.ac.in

ABSTRACT

Dysoxylum binectariferum (Roxb.) Hook. is an evergreen tree, indigenous to India and widely distributed in different parts of India. Primarily, this plant is used in the timber market for the preparation of furniture and other wooden goods. Apart from utility in timber market, it has tremendous medicinal properties because of its unique secondary metabolites. Its hydroalcoholic extracts possess a wide range of pharmacological activities including immunomodulatory, anti-inflammatory, anti-cancer, anti-leishmanial, etc. The chromone alkaloid rohitukine, the major constituent of the plant, possess anti-fertility, anti-leishmanial, anti-adipogenic, anti-ulcerogenic, anti-cancer and immunomodulatory activities. The most extensively studied and advanced therapeutic property of this chromone alkaloid is its cyclin-dependent kinase inhibition activity, which has a direct link with its anti-cancer effect. Inspired by this natural product, two clinical candidates (flavopiridol and riviciclib) are discovered and studied in human cancer patients for treating various types of cancer. Another preclinical lead IIIM-290, which is directly derived from rohitukine *via* synthetic modification, has demonstrated promising oral efficacy in animal models of various cancer types. The high-value medicinal applications of this plant have necessitated the need of bringing this tree under captive cultivation. The present chapter discusses the biology pharmacological and secondary metabolites present in Dysoxylum binectariferum. The preclinical and clinical leads derived or inspired from rohitukine are also discussed in this communication.

Keywords: Medicinal Plant, *Dysoxylum binectariferum,* IIIM-290, anti-cancer, flavopiridol, India

INTRODUCTION

Dysoxylum binectariferum (Roxb.) Hook. (Family Meliaceae) is a plant which mainly occurs in India, China, and Sri Lanka. *Dysoxylum* Blume is a large genus habit either trees or shrubs comprising more than 100 species distributed all over the world. In India, it occurs primarily in Bengal, Assam, South India, and Andaman Islands. Endemic Indian species include *D. binectariferum* (Roxb.) Hook., *D. malabaricum* Bedd. ex Hiern, *D. beddomei* Hiern, and *D. ficiforme* (Wt.) Gamble. This plant has tremendous economic importance because of its utility in timber industry. The wood of this tree is widely used for making furniture and cabinets. The wood of this plant can be sawn and machined well to a smoothed surface for building construction, boxes, canoes, and turnery; also suitable for matchboxes and splints, cigar- boxes, and plyboard. Because of the presence of unique secondary metabolites in *Dysoxylum*, the different species are used in traditional systems of medicine either as crude plant extracts or as an ingredient of several herbal preparations for the treatment of skin diseases, inflammation, cardiovascular disorder, neurological disorders, and tumor. The indigenous people are using many plants in the genus as traditional medicine. *Dysoxylum richii,* is such an example, which is used by indigenous Fijians as a medicine to treat many diseases, such as rigid limbs, facial distortion in children, lumps under the skin, skin irritations, and as a remedy for sexually transmitted diseases. It is also reported as a remedy for fish poisoning and convulsions (Aalbersberg and Singh, 1991).

D. binecteriferum species is cultivated as a sacred tree and is used instead of sandal in some parts of southern India. A decoction of wood is useful in arthritis, anorexia, cardiac debility, expelling intestinal worms, inflammation, leprosy, and rheumatism. Approximately 125 compounds are reported from the *Dysoxylum,* amongst which 20 compounds are from *D. binectariferum.* Rohitukine is a chromone alkaloid, which was first isolated from the leaves and stems of *Amoora rohituka* (Roxb.) and then from the stem barks of *D. binectariferum* (Mohanakumara et al., 2009) both belonging to the family Meliaceae (Yang et al., 2004). Rohitukine was isolated at Hoechst India Ltd. as the constituent responsible for anti-inflammatory and immunomodulatory activity (Naik et al., 1988a). Medicinal chemistry efforts around this nature-derived chromone alkaloid led to the discovery of two promising clinical candidates for the treatment of cancer viz. flavopiridol and P276-00 (Jain et al., 2012). Flavopiridol was discovered in the late 1980s from a research programme initiated at Hoechst India Limited. Flavopiridol (Alvocidib; L868275; HMR-1275; NSC 649890 of Sanofi-Aventis + NCI) and P276-00 (Piramal Enterprises India) has advanced to clinical trials for the treatment of cancer. Though these flavones are entirely synthetic molecules, the basis for their novel structure is a natural product rohitukine. The mechanism of their anticancer activity is inhibition of cyclin-dependent kinases (Cdk) (Galons et al., 2010; Huwe et al., 2003; Kelland, 2000; Senderowicz, 2001), which are key regulators of cell cycle. Recently, CSIR-IIIM has discovered an orally effective rohitukine-derivative displaying anticancer activity in human xenograft models of leukemia and pancreatic cancer (Bharate et al., 2018). Because of the interesting pharmacological activities of rohitukine and its derivatives/ analogs, this tree has gained tremendous medicinal importance. In the present chapter, we have discussed geographical distribution, chemical constituents, pharmacological activities, and preclinical or clinical candidates discovered based on this plant.

DIVERSITY WITHIN THE GENUS *DYSOXYLUM*

The genus *Dysoxylum* belongs to Meliaceae family and is distributed widely in the rainforest climate regions including New Guinea, Malaysia, Australia, Indian subcontinents, New Caledonia, Fiji, South-east Asia, Southern China, Philippines, Taiwan, Caroline Islands, New Zealand and Niue. It mainly contains big flowering plants, which are usually dioecious, pachycauls, and either tree or shrubs. Leaves of the plants are having pinnate venation and spiral arrangement around the axis. Inflorescences showed wide variation throughout different species of the genus, and it varies from thyrses to racemose or spicate and sometimes reduced to fascicles or solitary flowers (Mabberley et al., 1995). *Dysoxylum* has been grouped with more than 100 species, and medicinally explored species of the genus mainly includes *D. binectariferum, D. acutangulum, D. malabaricum, D. richii, D. alliaceum, D. cumingianum, D. hainanense, D. magnificum, D. lenticellare, D. roseum, D. spectabile, D. beddomei* and *D. ficiforme* (Shilpi et al., 2016). In India, 12 different species of *Dysoxylum* are present which are mainly distributed in Bengal, Assam, Southern India, and Andamans (Shilpi et al., 2016). *D. binectariferum* is indigenous to India, identified from a moist forest at an altitude of 2000-3000 ft (Hook. .. ex Bedd., Tr. L. S. 25: 212, 1866). *D. binectariferum* is well-explored species for various disease conditions (Quattrocchi, 2012).

SYSTEMATIC AND TAXONOMY

Systematic Classification

Kingdom   : Plantae Haeckel, 1866

Phylum   : Tracheophyta

Class   : Equisetopsida C. Agardh, 1825

Subclass   : Magnoliidae Novák ex Takht., 1967

Super order   : Rosanae Takht., 1967

Order   : Sapindales Juss. ex Bercht. & J. Presl, 1820

Family   : Meliaceae Juss., 1789

Genus   : *Dysoxylum* Blume, 1825.

Binomial name   : *Dysoxylum binectariferum* (Roxb.) Hook., 1866.

Taxonomic enumeration

Dysoxylum binectariferum tree grow up to 30 ft or higher with a thick trunk containing reddish wood and pallid leaves with 5-9 glabrous, alternate, acuminate leaflets. Leaves are large having 9-18 inches length, and flowers show racemiform pinnacle inflorescence with short (shorter than leaves) and mostly glabrous panicles. Flowers are small and pale green in color, calyx is generally half the length of flower, petals are in valvate position except in apex, style is short and hairy, short four-celled hairy ovary and a staminal tube are most of the time present on both sides. Fruits are 2.5 inches long, reddish, four-celled, four-seeded and obovoid or pyriform or subglobose nearly glabrate or puberulous in shape. Seeds are dark purple colored polished and shining (Hooker, 1875). The photographs of the fully grown tree, young tree, and its various parts are shown in Figure 1.

Fig. 1: Photographs of *Dysoxylum binectariferum* tree (A-B) and its various parts - fruits (C), seeds (D), leaves (E) and bark (F)

Synonyms

Dysoxylum gotadhora (Buch.-Ham.) Mabb., Fl. China. 2008; *Alliaria gotadhora* (Buch.-Ham.) Kuntze, Rev. Gen. 109. 1891; *Dysoxylum cupuliforme* Li, Journ. Arn. Arb. 25: 301. 1944; *Dysoxylum grandifolium* Li, Journ. Arn. Arb. 25: 302. 1944; *Dysoxylum macrocarpum* Thw., Enum. Pl. Zeyl. 60. 1859; *Dysoxylum reticulatum* King, Journ. As. Soc. Beng. 65: 114. 1897; *Epicharis exarillata* Arn. ex Wight & Arn., Prod. 120. 1834; *Epicharis gotadhora* (Buch.-Ham.) M. Roem., Syn. Hesper. 103. 1846; *Guarea binectarifera* Roxb., (Hort. Beng. 28 (1814),) Fl. Ind. ed. 2, 2: 240. 1832; *Guarea gotadhora* Buch.-Ham, Mém. Wern. Nat. Hist. S. 6: 307, 1832; *Dysoxylum ficiforme* (Wight) Gamble, Fl. Madras 178. 1915.

Common Names

Akil, agunni agil, agilu, akuni, akunivakil, akuniyakil, bandardima, banderdima, bol-narang, chembi agil, erand, kaadugandha, kadgandha, kadugadda, kadugandha, karakil, khrang, khrang-kelaukaraung, masispel, naupak-ban, yerandi (Quattrocchi, 2012).

Ecology and Geographical Distribution

Dysoxylum binectariferum is recorded growing in evergreen and moist lowland forests. The moist, rainy, humid environment is most suitable for the plant. It is indigenous to India and widely distributed in different parts of India including Tamil Nadu, Kerala, Western Ghats, Andaman Islands, Assam, Sikkim, Bengal and north-east region of India. In the Western Ghats, it is present in moist forest ranging from Coorg to Anamalais and Tinnevelly but absent in Travancore (Gamble, 1967). Apart from India, it is mainly distributed in Nepal, Vietnam, and Srilanka (Mishra et al., 2014; Varshney et al., 2014) [http://www.catalogueoflife.org/annual-checklist/2017].

MEDICINAL ASPECTS

Material used as crude drug

In traditional Sidha/ Siddha medicinal system, most commonly the seeds of this plant were taken with hot water for the treatment of leprosy along with topical dressing with seed powder (Quattrocchi, 2012). The uses of various plant parts are well described in different traditional medicine system such as Ayurveda and Siddha.

Pharmacological activities of extracts

The ethanolic extract from the stem bark of *D. binectariferum* showed potent *in vitro* and marginal *in vivo* leishmanicidal effect against the *Leishmania donovani*; the causative agent of visceral leishmaniasis (Lakshmi et al., 2007). The ethanolic extract of stem bark has also been reported to possess promising contraceptive and hormonal properties and anti-breast cancer activity (Keshri et al., 2007; Shareef et al., 2016). In another study for anticancer activity, the ethanolic extract of stem bark of plant showed significant *in vitro* anticancer activity against SKOV3 (ovarian carcinoma cells), MDAMB273 (breast cancer cells), NCI /ADR-RES (adriamycin-resistant cells) and MCF7 (breast carcinoma cells) with IC_{50} values of 15, 7, 2.8 and 10 µg/µL, respectively (Mohanakumara et al., 2010). In another study, the ethanolic extract of leaves showed potent anticancer (cytotoxic) activity against A549 (lung carcinoma), MCF-7 (breast adenocarcinoma) and HepG2 (hepatocellular carcinoma) human cancer cell lines (Yan et al., 2014). The hydroalcoholic extract of fruits and leaves of *D. binectariferum* showed significant inhibition of proinflammatory cytokines TNF-α and IL-6 in THP-1 cell line (Kumar et al., 2017). The promising biological activities associated with extracts has led to pursuing the phytochemical investigations on this plant.

PHYTOCHEMISTRY ASPECTS

Phytochemical investigations have resulted in the isolation of a wide range of chemical compounds including alkaloids, tannins, triterpenes, tetranortriterpenoids, and steroids from different parts of this plant. Most of the natural products were isolated either from alcoholic or hydro-alcoholic extracts. Rohitukine **(1)** was the first alkaloid isolated from the methanolic extract of stem bark of this plant (Naik et al., 1988b). Later on, different alkaloidal molecules **2-6** with anti-inflammatory, immunomodulatory and anticancer activities were isolated from the stem bark (Jain et al., 2014; Jain et al., 2013; Kumar et al., 2017; Yang et al., 2004).

Table 1: Chemical constituents isolated from different parts of *Dysoxylum binectariferum*

Sr.No	Chemical compound	Plant part (extract)	Pharmacological Activity	References
1.	Rohitukine **(1)**	Stem bark, fruits and leaves (methanolic)	Anti-inflammatory, immunomodulatory, anticancer	(Naik et al., 1988b; Kumar et al., 2017)
2.	Rohitukine-N-oxide **(2)**	Stem bark (methanolic)	none	(Yang et al., 2004)
3.	Dysoline **(3)**	Stem bark (ethanolic)	Antiproliferative	(Jain et al., 2013)
4.	Chrotacumine K **(4)**	Fruits (methanolic)	Anti-inflammatory	(Kumar et al., 2017)
5.	Camptothecin **(5)** and 9-methoxy camptothecin **(6)**	Stem bark (ethanolic)	Anticancer	(Jain et al., 2014)
6.	Schumaniofioside A **(7)**	Leaves (methanolic)	Anti-inflammatory	(Kumar et al., 2017)
7.	Steroidal compounds **8-13**	Leaves (hydro-alcoholic)	Antiproliferative	(Yan et al., 2014)
8.	Triterpenoids **14-19**	Stem bark (hydro-alcoholic)	Cytotoxic and anti-inflammatory	(Hu et al., 2014)
9.	Dysobinin **(20)**	Fruits (alcoholic)	CNS depressant	(Singh et al., 1976)

Schumaniofioside A **(7)**, a chromone glycoside was isolated from leaves of *D. binectariferum*, and was found to be present in abundance (1% w/w of dried leaves) (Kumar et al., 2017). Six steroidal molecules **8-13** with antiproliferative properties were isolated from the hydroalcoholic extract of leaves of this plant (Yan et al., 2014). Six triterpenoids **14-19** were isolated from fruit alcoholic and stem bark hydro-alcoholic extracts which were found to have CNS depressant, cytotoxic and anti-inflammatory activities (Hu et al., 2014). Singh et al. (1976) reported dysobinin **(20)**, a tetranortriterpene from the alcoholic extract of fruits of *D. binectariferum* (Singh et al., 1976). Table 1 and Figures 2-3 enlist name, extract type, pharmacological activity and structures of different molecules isolated from the *D. binectariferum*.

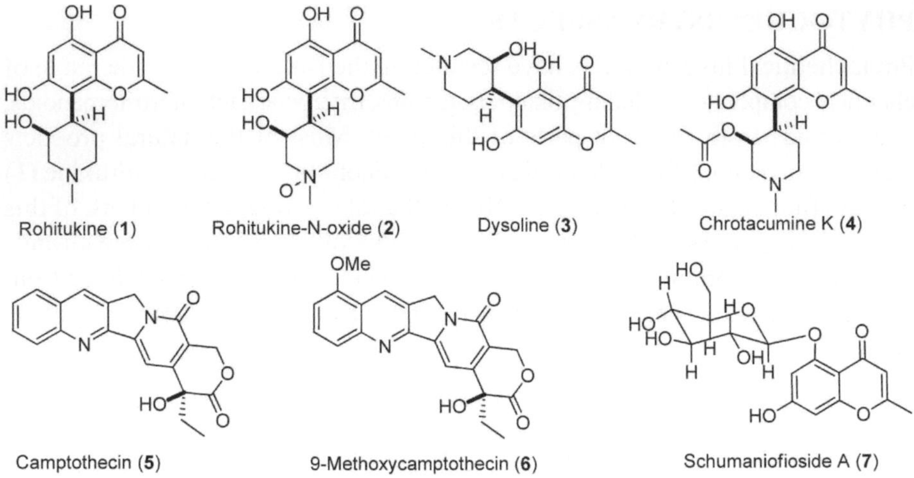

Fig. 2: Chemical structures of alkaloids **1-6** and chromone glycoside **7** from *D. binectariferum*

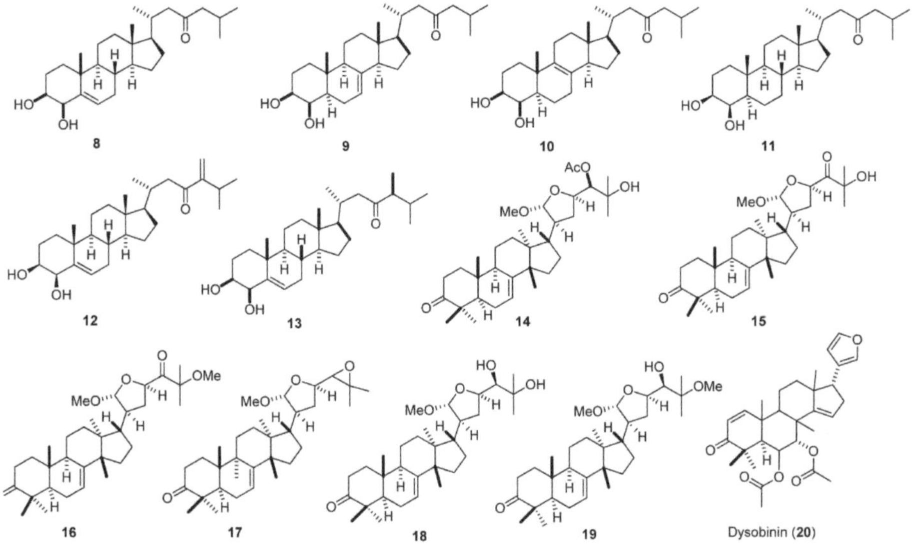

Fig. 3: Chemical structures of steroids/trierpenoid **8-20** from *Dysoxylum binectariferum*

PHARMACOLOGICAL PROPERTIES OF ISOLATED NATURAL PRODUCTS

Alkaloids

Total of six alkaloids **1-6** are isolated from the alcoholic extract of stem/stem bark/leaves of the plant. Alkaloids isolated from the *D. binectariferum* are known to have potent anticancer activity. Among different alkaloids, rohitukine possess anticancer, anti-inflammatory and immunomodulatory properties (Naik et al., 1988b). Dysoline (**3**), the regioisomer of rohitukine was tested for its anticancer potential against a panel of human cancer cell lines including Colo205 (colon), HCT116 (colon), HT1080 (fibrosarcoma), NCIH322 (lung), A549 (lung), Molt-4 (leukemia), and HL60 (leukemia). It showed potent anticancer activity against HT1080 cells with an IC_{50} of 0.21 µM. Further, dysoline also showed anti-inflammatory activity at 0.1 µM concentration with 47% and 83% inhibition of proinflammatory cytokines TNF-α and IL-6, respectively (Jain et al., 2013). Interestingly, bioassay-guided isolation of bark extract resulted in the isolation of well known anticancer alkaloids camptothecin (**4**) and 9-methoxy camptothecin (**5**) from the stem bark extract (Jain et al., 2014). Chrotacumine K (**6**), the acetyl ester of rohitukine was isolated from the methanolic extract of fruits, which possess anti-inflammatory activity. It showed 57% inhibition of TNF-α at 0.3 µM and was non-toxic to THP-1 cells up to 40 µM concentration (Kumar et al., 2017).

Steroids

Six different steroidal molecules **8-13** are isolated from the hydroalcoholic extract of leaves. These were screened for cytotoxicity determination against the human cancer cell lines A549, MCF-7, and HepG2. Compound **8**, **12** and **13** showed only moderate cytotoxicity with the IC_{50} >50 µM while remaining three compounds **9**, **10** and **11** showed potent anticancer activity with the IC_{50} between 1.5-9.6 µM (Yan et al., 2014).

Triterpenoids

Dysobinin (**20**) isolated from the alcoholic extract of fruits of *D. binectariferum* showed CNS-depressant and mild anti-inflammatory activity (Singh et al., 1976). Other triculane triterpenoids **14-19** were screened for anti-inflammatory and anticancer potential against eight different cell lines. Cytotoxicity studies were done against A549, BGC-823, HCT15, HeLa, HepG2, MCF-7, SGC-7901, and SK-MEL-2 cell lines and all six triterpenoids showed the IC_{50} <50 µM in most of cell lines except in BGC-823, HeLa and MCF-7 (moderate activity). In this study, selectively potent cytotoxic activity was found against HepG2 cell lines with the IC_{50} of 7.5-9.5 µM. Furthermore, the triterpenoid **14** showed selective COX-1 inhibition with 95% at 100 µM (Hu et al., 2014).

PRECLINICAL AND CLINICAL LEADS

The major secondary metabolite of this plant is a chromone alkaloid rohitukine, which is present in all parts of the plant. It has been studied for a wide range of pharmacological activities; and has shown promising activity in a series of animal models. Because of its interesting pharmacological profile, the total synthesis of this natural product is established. This natural product has also been studied in detail for its ADME properties including pharmacokinetics. Semi-synthetic modifications of this natural product have produced an anticancer lead compound IIIM-290 which was validated in animal models of various types of cancer. Apart from these two lead compounds (rohitukine, and IIIM-290), two clinical lead candidates (flavopiridol and riviciclib) were discovered which were not prepared using rohitukine as a starting material; however, they were inspired from the rohitukine scaffold. These four lead compounds are discussed in this section.

Rohitukine

Isolation of Rohitukine

Rohitukine is a chromone alkaloid which was isolated for the first time from *Amoora rohituka* and later on from the different parts of *D. binectariferum* (Harmon et al., 1979; Kumar et al., 2017; Naik et al., 1988b). Rohitukine content analysis showed that *A. rohituka* species contains only 0.083% w/w (dry weight) of rohitukine while *D. binectariferum* was found to have 0.9-7% w/w rohitukine. *D. binectariferum* species present in Western Ghats region of India was found to contain up to 7% rohitukine w/w of dry stem bark (Mohanakumara et al., 2010). Rohitukine was also isolated from different cultures of endophytic fungus *Fusarium proliferatum* (MTCC9690), *Fusarium oxysporum* (MTCC-11383), *Fusarium oxysporum* (MTCC-11384), and *Fusarium solani* (MTCC-11385) which were isolated from the inner bark tissue of *D. binectariferum* wherein the rohitukine content was found in the range of 186 µg to 359.55 µg 100 g^{-1} of dry mycelia weight (Mohanakumara et al., 2014; Mohanakumara et al., 2012). Rohitukine is successfully isolated from all plant parts of *D. binectariferum* including stem bark, leaves, and fruits. As a general procedure for isolation of alkaloids, the acid-base treatment provides enrichment of alkaloids. The earlier protocols adopted for rohitukine isolation comprises fractionation of the alcoholic extract into alkaloidal and non-alkaloidal portions by the step-wise acid-base treatment followed by partitioning. From the alkaloidal portion, rohitukine was either purified by crystallization (Naik et al., 1988b) or by column chromatography using either silica gel or alumina or HP-20 resin as a stationary phase (Jain et al., 2014). Few efforts were made for column-free isolation of rohitukine. In this direction, a solubility-dependent chromatography free method was developed for the isolation of rohitukine from the leaves of *D. binectariferum.* In this method, the hexane defatted leaves were extracted with a mixture of chloroform and methanol (85:15) and the dried extract was partitioned between water and ethyl acetate. From the water layer, the rohitukine was isolated by repetitive crystallization using water-acetone and methanol-acetone (Kumar et al., 2016b).

Total Synthesis of Rohitukine

Because of the numerous medicinal activities of this natural product, its total synthesis is accomplished. The first total synthesis of rohitukine (Naik et al., 1988b) was reported by Hoechst researchers in 1990s as depicted in Figure 4. Briefly, the synthesis comprises a total of 8 steps. The first step involves condensation of 1,3,5-trimethoxybenzene with 1-methyl-3-piperidinone in the presence of acetic acid saturated with HCl lead to the formation of 1,2,3,6-tetrahydro-4-(2,4,6-trimethoxyphenyl)-1-methylpyridine (**III**), which on hydroboration produced trans-arylpiperidinol (trans-**IV**). The inversion of trans-**IV** to cis-arylpiperidinol (cis-**VI**) was achieved via Swern oxidation followed by reduction with NaBH$_4$; yielding a mixture of cis/trans alcohols (cis-**VI** and trans-**VI**) in a 7:3 ratio. These isomers were separated via fractional crystallization or chromatography. Acylation of cis-**VI** followed by saponification provided *O*-hydroxyacetophenone intermediate **VIII**. Treatment of **VIII** with ethyl acetate and sodium metal yielded chromone **IX** which on demethylation provided rohitukine (Figure 4).

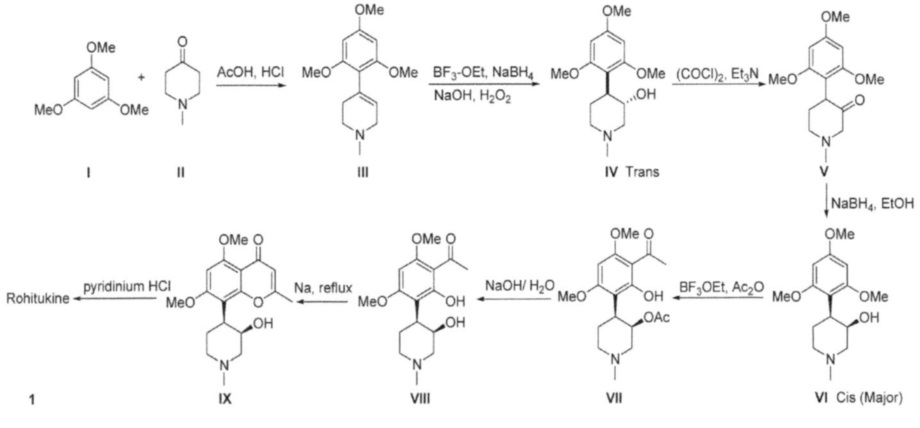

Fig. 4: Total synthesis of rohitukine

Pharmacological Activities

The first pharmacological activity reported for rohitukine was anti-inflammatory and immunomodulator activity by Naik *et. al.* in 1988. Rohitukine showed anti-inflammatory activity in carrageenan-induced rat paw edema assay with the ED$_{50}$ of 9 mg/kg, p.o. and also inhibited the reverse passive arthus reaction in the rats to 50.8 ± 5.9 % at a dose of 2.5 mg/kg, p.o (Naik et al., 1988b). Rohitukine showed anti-fertility activity in rats at 10 mg/kg daily dose and significantly reduced the implantation number in pregnant females (Keshri et al., 2007). In further studies, rohitukine was found to inhibit the gastric lesion formation in rats by inhibiting acid secretion and cytoprotective effect (Singh et al., 2012). Rohitukine showed weaker *in vitro* anti-leishmanial activity; however, it was ineffective in infected hamsters model (Lakshmi et al., 2007). Rohitukine was found to have *in vitro* anti-adipogenic and antioxidant activity and *in vivo* anti-dyslipidemic properties in Syrian golden hamsters in two different studies (Mishra et al., 2014; Varshney et al., 2014). Rohitukine has been studied extensively for its anticancer potential via screening it in a panel of cancer cell lines. Rohitukine possess significant anticancer activity against HL-60 and Molt-4 (leukemia) cell lines with GI_{50} of 10 and 12 µM, whereas it does not cause toxicity in normal cell lines fR2 and HEK-293 (GI_{50} > 50 µM). One of the mechanisms of its cytotoxicity in cancer cell lines is the inhibition of cyclin-dependent kinases. It inhibits Cdk2/A and Cdk9/T1 with IC_{50} values of 7.3 and 0.3 µM, respectively (Kumar et al., 2016b). Rohitukine also inhibits the growth of ovarian carcinoma cells SKOV3 (IC_{50} value of 20 µM) and breast cancer cells T47D, MDAMB273 and MCF7 (IC_{50} values of 50, 3, and 15 µM, respectively) (Kamil et al., 2015). The pharmacological activities of rohitukine are summarized in Figure 5.

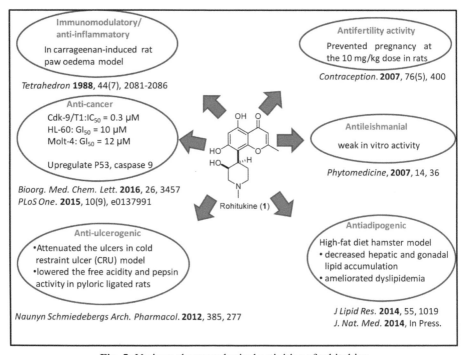

Fig. 5: Various pharmacological activities of rohitukine

Physicochemical Properties of Rohitukine

Different physicochemical properties *viz.* solubility [water, simulated gastric fluid (SGF), simulated intestinal fluid (SIF) and phosphate buffer saline pH 7.4 (PBS)], lipophilicity (Log P and Log D), stability and acid dissociation constant (pKa) were studied for this natural product. Rohitukine is a highly polar molecule with thermodynamic equilibrium solubility of 10.25, 16.94, 33.37 and 18.91 mg/ml in water, PBS, SGF, and SIF, respectively. It has Log P and Log $D_{(pH7.4)}$ of 0.73 and -0.55 while acid dissociation constant (pKa) was 5.83 (Kumar et al., 2016). Physicochemical properties of rohitukine indicated that it is prone to rapid elimination from the body by enhanced renal secretion and higher first pass metabolism. Lipophilicity values were also below the ideal value required for the oral delivery of a drug molecule which is 1-3 (Log P) (Kumar et al., 2016b). Efforts were made to improve the physicochemical properties of rohitukine via prodrug approach. The hexanoate ester prodrug of rohitukine showed optimum solubility and lipophilicity and was stable in GIT whereas it gets readily hydrolyzed to rohitukine in rat plasma and in the presence of esterases with hydrolysis half life of 58.85 and 36.70 min, respectively (Kumar et al., 2016a).

Pharmacokinetic and Metabolic Stability Studies

In vivo pharmacokinetic studies of rohitukine were done by both intravenous and oral route at 5 and 50 mg/kg dose, respectively. After oral administration, rohitukine showed rapid absorption from the gastrointestinal tract, and maximum plasma concentration was reached just within one hr. Rohitukine showed rapid elimination after intravenous administration compared to oral route, and the elimination half-life of rohitukine was found to be 0.81 h via intravenous route and 2.18 h via oral administration. The oral bioavailability of rohitukine was found to be only 34.25% in Sprague-Dawley rats. The *in vitro* metabolic stability performed in rat liver and intestinal microsomes proved that rohitukine is quite stable in both enzymatic conditions and is not susceptible to hepatic first-pass metabolism (Chhonker et al., 2014b). In another pharmacokinetic study, the similar oral pharmacokinetic profile was observed in hamsters at 50 mg/kg dose with elimination half-life of 2.62 h. Rohitukine showed ~ 60% plasma protein binding and was stable in the gastric environment. Tissue distribution studies in hamster at 50 mg/kg showed that rohitukine has significantly higher levels in liver and kidney, followed by spleen and lungs (Chhonker et al., 2014a). The complete preclinical data of rohitukine is summarized in Table 2.

Table 2: Pre-clinical data of rohitukine

Chemical structure:	

Source:	*Dysoxylum binectariferum* (bark, stem bark, leaves, fruits)
Target kinase :	Cdk-9/T_1, Cdk-2/A (Bharate et al., 2018)TNF-α, IL-6 (unpublished results)
In vivo activity:	Anti-inflammatory activity against carrageenan induced rat paw oedema, antifertility activity in rats at 10 mg/kg, inhibition of gastric acid secretion and cytoprotective effect in rats, antidyslipidemic activity in Syrian golden hamsters (Keshri et al., 2007; Mishra et al., 2014; Naik et al., 1988b; Varshney et al., 2014).
Physicochemical parameters:	Thermodynamic equilibrium solubility (mg/ml): water (10.25 ± 0.24), PBS (16.94 ± 0.25), SGF (33.37 ± 0.99), SIF (18.90 ± 0.25); Partition coefficient log P : -0.55 ± 0.01; Distribution coefficient logD : -0.55 ± 0.04, Ionization constant pKa : 5.83 ± 0.15 (Kumar et al., 2016b)
PK data:	(a) Pharmacokinetics data in male SD rats (Chhonker et al., 2014b). Dose: 5 / 50 mg/kg i.v. / p.o.; IV PK parameters: C_0 1653.83 ± 215.90 ng/ml, $t_{1/2}$ 0.81 ± 0.08 h, $AUC_{0-\infty}$ 1980.97 ± 262.09 ng.h/ml, V_d 4.53 ± 1.28 L/kg, Cl 3.97 ± 1.59 L/h/kg. Oral PK parameters: C_{max} 4883.33 ± 1843.15 ng/ml, t_{max} 1.00 ± 0.0 h, $t_{1/2}$ 2.18 ± 0.13h, $AUC_{0-\infty}$ 6786.61 ± 1653.18 ng.h/ml, V_d 8.26 ± 1.92 L/kg, Cl 2.63 ± 0.68 L/h/kg, F% 34.25 ± 2.23.(b) Pharmacokinetics data in Golden Syrian hamsters (Chhonker et al., 2014a). Dose: 50 mg/kg p.o.; Oral PK parameters: C_{max} 6.32 ± 0.32 µg/ml; t_{max} 0.5 ± 0.0 h, $t_{1/2}$ 2.90 ± 1.24 h; $AUC_{0-\infty}$ 10.66 ± 2.34 µg.h/ml; V_d 17.34 ± 11.34 L/kg; CL 3.95 ± 0.9 L/h/kg.
Current status:	Preclinical candidate
Patents:	US9932327, US9776989, US5292751(A), US07865247

Semisynthetic Derivatives of Rohitukine

Two studies are published on the synthetic modifications of naturally isolated rohitukine. The CSIR-CDRI have prepared sulphonyl derivatives of rohitukine and tested for cellular antiproliferative activity in MCF-7 and MDA-MB-231 breast cancer cell lines. These sulphonyl derivatives were found to be only moderately active, and the IC_{50} of 17-30 µM (Mishra et al., 2018). CSIR-IIIM have prepared four series of rohitukine derivatives (ether derivatives, Mannich reaction products, Baylis-Hillman reaction products, Claisen-Schmidt condensation products), which has led to the identification of potent inhibitor (IIIM-290) of cyclin-dependent kinase 9 with the IC_{50} of 1.9 nM. IIIM-290 showed potent antiproliferative activity against a panel of cancer cell lines and was found to be most active against SW-630 (colon), MOLT-4 (leukemia), HL-60 (leukemia), MIAPaCA-2 (pancreatic) human cancer cell lines with the IC_{50} of 0.3, 0.5, 0.9 and 1.0 µM, respectively. This lead compound has excellent metabolic stability profile in liver microsomes, hepatocytes, and S9 liver fractions, and does not have cytochrome P450 inhibition liability and efflux pump substrate liability. It showed excellent oral bioavailability (70%) in mice as well as in rats. It showed *in vivo* anticancer activity in pancreatic, colon, and leukemia xenografts at a dose of 50 mg/kg, po (Bharate et al., 2018). The preclinical data of IIIM-290 is summarized in Table 3.

Table 3: Pre-clinical data of IIIM-290

Chemical structure	
Other names	IIIM-290
Source	Rohitukine derivative (semisynthetic derivative)
Target kinase	Cdk-9/T_1, Cdk-2/A
In vivo activity	Human xenograft models of pancreatic, colon, and leukemia cancer at dose of 50 mg/kg, po (Bharate et al., 2018)
Physicochemical parameters	Thermodynamic equilibrium solubility (µg/ml) : water (20.31 ± 4.04), PBS (2.29 ± 0.07), SGF (8.81 ± 0.39), SIF (2.18 ± 0.16); Partition coefficient logP : 3.09; Distribution coefficient logD : 1.65, Ionization constant pKa : 5.4 (Bharate et al., 2018)
PK data	(a) Pharmacokinetics data in BALB/C mice [dose: 1 / 10 mg/kg i.v./p.o.] (Bharate et al., 2018): IV PK parameters: C_0 859 nM, $t_{1/2}$ 5.46 h, $AUC_{0-\infty}$ 653 nM.h, V_d 26.2 L/kg; Cl 55.4 ml/min/kg. Oral PK parameters: C_{max} 1286 nM, t_{max} 0.25h, $t_{1/2}$ 4.65 h, $AUC_{0-\infty}$ 4399 nM.h, F% 70.7.(b) Pharmacokinetics data in BALB/c mice [dose: 3 / 30 mg/kg i.v./ p.o.]: IV PK parameters: $t_{1/2}$ 2.55 h, $AUC_{0-\infty}$ 1862 nM.h, C_{max} 4743 nM.· Oral PK parameters: C_{max} 4108 nM, t_{max} 0.54 h, $t_{1/2}$ 3.62 h, $AUC_{0-\infty}$ 11266 nM.h, F% 61.(c) Pharmacokinetics data in SD rats (Bharate et al., 2018): Dose: 3 / 30 mg/kg i.v. / p.o.; IV PK parameters: $t_{1/2}$ 2.43 h, $AUC_{0-\infty}$ 2412 nM.h, C_{max} 3510 nM. Oral PK parameters: C_{max} 3623 nM, t_{max} 1 h, $t_{1/2}$ 3.23 h, $AUC_{0-\infty}$ 15876 nM.h, F% 66.
Current status	Preclinical candidate
Patents	WO2014170914A1, US20160052915, EP2986605, CA2908084, IN2013DE01142, US9932327B2

Rohitukine-inspired lead compounds - Flavopiridol and Riviciclib

The most extensively studied Cdk inhibitor flavopiridol has been discovered based on the structure of rohitukine. During the last two decades, numerous efforts were made to discover anticancer drugs based on kinases involved in cell cycle regulation. The chemical synthesis of rohitukine published in 1988 has opened the new ways to synthesize rohitukine derivatives of medicinal importance. Hoechst Marion Roussel (presently, Aventis Pharma) pharmaceutical company studied various synthetic analogs of rohitukine for enhancing the anti-inflammatory and immunomodulatory activities and they identified flavone acetic acid and flavone acetic acid ester with improved activities.

Further, in the late 1980s, the same company discovered "flavopiridol" via screening of rohitukine analogs for phosphorylation inhibition of epidermal growth factor receptor (EGFR-TK). Flavopiridol showed the IC_{50} of 21, 6 and 122 μM against EGFR-TK, protein kinase C and protein kinase A, respectively (Sedlacek et al., 1996). Further developmental studies of flavopiridol were carried out at Developmental Therapeutics branch of National Cancer Institute (Bethesda, MD), where flavopiridol was studied against 60 cancer cell lines and was found to inhibit the cell growth of several cell lines (Tan and Swain, 2002).

In cyclin-dependent kinase inhibition study in breast carcinoma cells, flavopiridol was found to inhibit Cdk-1, 2 and 4 with the IC_{50} between 100-400 nM (Carlson et al., 1996). Further x-ray crystallography studies have shown that deschloroflavopiridol (co-crystallized derivative of flavopiridol) showed interaction with the ATP binding site of Cdk-2, indicating that flavopiridol inhibits the kinase activity via restricting binding of ATP to its pocket (De Azevedo et al., 1996; Sedlacek, 2001; Tan and Swain, 2002; Zeidner and Karp, 2015).

Flavopiridol was found effective in numerous xenograft studies including head and neck squamous cell carcinoma (Patel et al., 1998), and leukemia (Arguello et al., 1998). Preclinical studies showed that for cytostatic action, the prolonged exposure of flavopiridol is required therefore in initial clinical trials; 72 h continues intravenous infusion was given. In such trial, at a maximum tolerated dose (MTD), the steady state plasma concentration was ~0.5 μM that was above / equal to the *in vitro* therapeutic concentration. But, the 72 h continuous infusion was associated with secretory diarrhea and thrombotic events, so prophylactic measures were taken in the phase-I trial (Sedlacek, 2001). In the phase-II study, 20 previously non treated patients of colorectal cancer were given 72 h continues i.v. infusion (50 mg/m^2/day) every 14 days. After treatment, only five patients had stable disease conditions and rest of them were showing no objective response (Aklilu et al., 2003). Similar results at the same dose, with no objective response, were observed in a prostate cancer patient in the phase-II studies (Liu et al., 2004). The 72 h continues i.v. infusion trials showed a significant improvement in mantle cell lymphoma patients in which 11% patient showed a partial response, 71% showed stable disease condition while in 18% the disease progression was observed (Kouroukis et al., 2003).

In further drug developmental studies, based on initial significant efficacy in tumor xenograft studies, flavopiridol was studied in various combinations with existing anticancer drugs. Flavopiridol enhanced the anticancer potential of irinotecan in colorectal cancer xenografts studies and phase-I trials were initiated to know the safe and effective dose for phase-II studies and the recommended phase II dose was found to be either irinotecan 100 mg/m^2 and flavopiridol 60 mg/m^2 or irinotecan 125 mg/m^2 and flavopiridol 50 mg/m^2 (Shah et al., 2005). Similar trials were done in combination with FOLFIRI (chemotherapy regimen of folinic acid, fluorouracil, and irinotecan) and maximum tolerated safe and effective dose for flavopiridol in combination was found to be either 80 mg/m^2 over 1 h or 35 mg/m^2 bolus + 35 mg/m^2 over 4 h (Dickson et al., 2010).

Recently, the phase-I clinical trial of flavopiridol in patients with chronic lymphocytic leukemia (CLL) has been carried out in 10 patients. The therapy was well tolerated and 22% patient showed improved response while 88% experienced reduced tumor burden (Awan et al., 2016). Presently, the flavopiridol has an orphan drug designation for the treatment of chronic lymphocytic leukemia (CLL) by FDA and the EMA (Wiernik, 2016). The preclinical data of flavopiridol is summarized in Table 4.

Table 4. Pre-clinical and clinical data of flavopiridol

Chemical structure	
Other names	Alvocidib
Source	Synthetic (rohitukine inspired synthetic analog)
Target kinase	EGFR-TK, protein kinase C, protein kinase A, various CDKs including 1, 2, 4, 7 & 9 (Zeidner and Karp, 2015)
In vivo activity	Xenograft models of head and neck squamous cell carcinomas, and colorectal cancer (Shah et al., 2005; Patel et al., 1998; Arguello et al., 1998)
Physicochemical parameters	Flavopiridol hydrochloride (Kumar et al., 2016b):Thermodynamic equilibrium solubility (µg/ml) : water (1926 ± 180), PBS (1141 ± 250), SGF (1291 ± 380), SIF (7378 ± 190); Partition coefficient logP : - 0.21 ± 0.03; Distribution coefficient logD : 1.24 ± 0.24, Ionization constant pKa : 5.56 ± 0.07.
	Flavopiridol free base (Dörwald, 2012; Li et al., 1999); Thermodynamic equilibrium solubility (µg/ml) : water (25); Partition coefficient logP : 1.96; Ionization constant pKa : 5.86

Contd.

PK data	Pharmacokinetics data in male Sprague Dawley rats (Xia et al., 2013) Dose : 3 mg/kg i.v. and 3 and 12 mg/kg p.o.IV PK parameters: $t_{1/2}$ 173.73 ± 31.97 min, $AUC_{0-\infty}$ 64.54 ± 10.38 mg.min/mL, Cl 0.012 ± 0.0025 ml/min, $MRT_{0-\infty}$ 250.69 ± 46.13. Oral PK parameters at 3 mg/kg : C_{max} 127.52 ± 45.54 ng/mL, T_{max} 18.00 ± 6.71 min, $t_{1/2}$ 354.68 ± 21.40 min, $AUC_{0-\infty}$ 48.98 ± 11.88 mg.min/mL, Cl 0.018 ± 0.0066 ml/min, $MRT_{0-\infty}$ 511.80 ± 30.88, % F 75.89.Oral PK parameters at 12 mg/kg : C_{max} 1317.35 ± 150.72 ng/mL, T_{max} 11.00 ± 5.48 min, $t_{1/2}$ 262.23 ± 45.41 min, $AUC_{0-\infty}$ 200.36 ± 60.51 mg.min/mL, CL 0.020 ± 0.0072 ml/min, $MRT_{0-\infty}$ 378.40 ± 65.53.
Clinical trials	Tested in numerous clinical trials as a stand-alone or in combination with other cytotoxic drugs (Aklilu et al., 2003; King et al., 2003; Sedlacek, 2001; Awan et al., 2016; Dickson et al., 2010; Kouroukis et al., 2003; Shah et al., 2005)
Current status	Presently the flavopiridol has an orphan drug designation for the treatment of chronic lymphocytic leukemia (CLL) by FDA and the EMA (Wiernik, 2016).
Patents	US7888341, US6660750, US6087366, US5849733, US5908934

Riviciclib (P276-00) is another rohitukine inspired anticancer agent with potent Cdk inhibition properties developed by the company Nicholas Piramal India Limited, Mumbai, India. This molecule was identified from the novel flavone analogs series of rohitukine by screening against the Cdk4/D1. Different molecules from the series showed potent *in vitro* kinase inhibition activity with the $IC_{50} < 250$ nM; amongst which P276-00 was the lead molecule showing 40-fold selectivity for Cdk4/D1 compared to Cdk2/E (Joshi et al., 2007a; Raje et al., 2009). P276-00 showed promising preclinical anticancer activity. It was tested against a panel of 16 cancer cell lines which were containing both cisplatin-sensitive and resistant cell lines.

The anti-proliferative potential of P276-00 was ~30-fold higher compared to cisplatin. The mean IC_{50} and IC_{70} against 16 cell lines for P276-00 was 0.55 µM and 1.1 µM, and for cisplatin, the values were 16 µM and 40 µM, respectively. Mechanistic analysis studies further showed that P276-00 resulted in G_1-G_2 cell cycle arrest. P276-00 showed good cytotoxicity against mantle cell lymphoma cells by causing an increase in G1 phase cells number and the increase was concentration and time-dependent. The mechanistic analysis resulted in down regulation of anti-apoptotic protein Mcl-1 which is known to be the major cause of mantle cell lymphoma (Shirsath et al., 2012). In human xenograft study using clonogenic assay, it was ~26-fold more potent compared to cisplatin. Interestingly, it was active against cisplatin-resistive tumors of the central nervous system, melanoma, renal, and prostate cancer. In xenograft studies against colon (HCT-116) and lung carcinoma (H-460), P276-00 showed a significant (P < 0.05 for both models) inhibition of tumor growth at 50 mg/kg i.p. once a day for 10 days and 30 mg/kg twice a day for 20 days, respectively (Joshi et al., 2007b). A combination study of P276-00 in two different approaches with gemcitabine and doxorubicin against pancreatic cancer and human non-small cell lung cancer (cell line and human xenograft study) resulted in improved anticancer profile compared to individual molecules (Rathos et al., 2012; Rathos et al., 2013). Presently this molecule is in phase-2 clinical trials for the treatment of mantle cell lymphoma. The initial phase-II single-arm, open-label, multicenter study to evaluate the efficacy of P276-00 is conducted in 13 patients. 11 out of 13 patients of relapsed or refractory mantle cell lymphoma showed disease progression after intra-venous administration of P276-00 at 185 mg/m²/day dose from days 1-5 of a 21-day cycle. These results showed that clinical trials either during early disease state or in combination therapy are required (Cassaday et al., 2015). The pre-clinical data of riviciclib is summarized in Table 5.

Table 5: Pre-clinical and clinical data of P276-00 (riviciclib)

Chemical structure	
Other names	P276-00
Source	Synthetic (rohitukine inspired synthetic analog)
Target kinase	CDK-4/D$_1$, CDK-1/B and CDK-9/T$_1$ (Joshi et al., 2007a)
In vivo activity	In : cisplatin-resistive central nervous system tumor, melanoma, renal and prostate cancer xenograft models; human colon (HCT-116) and lung carcinoma (H-460) xenograft at 50 mg/kg i.p. once a day for 10 days and 30 mg/kg twice a day for 20 days respectively; multiple myeloma tumor xenograft model (Joshi et al., 2007b; Raje et al., 2009).
Clinical trials	Phase-II clinical trials for the treatment of mantle cell lymphoma (Cassaday et al., 2015)
Current status	Phase-II clinical trial
Patents	US20170112809, US20160136132

Patents

- Vishwakarma RA; Bharate SB; Bhushan S; Mondhe DM; Jain SK; Meena S; Guru SK; Pathania AS; Kumar S; Behl A; Mintoo MJ; Bharate SS; Joshi P. Cyclin-dependent kinase inhibition by 5,7-dihydroxy-8-(3-hydroxy-1-methylpiperidin-4-yl)-2-methyl-4H-chromen-4-one analogs. WO2014170914A1, US20160052915, EP2986605, CA2908084, IN2013DE01142, US9932327B2 [Granted on 3 April 2018], GB /EP 2986605 [Appl no 14734915.3, granted on 15 November 2017].

- Vishwakarma RA; Jain SK; Bharate SB; Dar AH; Khajuria A; Meena S; Bhola SK; Qazi AK; Hussain A; Sidiq T; Uma Shaanker R; Ravikanth G; Vasudeva R; Patel MK; Ganeshaiah KN. New chromone alkaloid dysoline for the treatment of cancer and inflammatory disorders. IN 2013DE01077A; WO2014167580A1 US9776989 B2 [granted on 3 October 2017].

- Bharate SS; Kumar V; Gupta M; Gandhi S; Kumar A; Bharate SB; Vishwakarma RA. Sustained release formulations of *Dysoxylum binectariferum* – IN201811014818.

- Baosong C; Donghui IY. Method for extracting rohitukine. CN101139344(A) [granted on 12 March 2008].

- Houghton PJ; Woldermarian TZ; Mahmood N. Alkaloids and their antiviral agents. WO1996007409A1 [granted on 14 March 1996].

- Bhat SV; Shah V; Dohaswalla AN; Mandrekar SS; De SNJ; Dickneite G; Kurrle R; Schorlemmer HU; Sedlacek HH. Chromane alkaloid, process for its isolation from *Dysoxylum binectariferum*, and its use as a medicament. EP0137193(A2) [granted on 17 April 1985].

- Bhat SV; Shah V; Dohadwalla AN; Mandrekar SM; De SNJ; Dickneite G; Kurrle R; Schorlemmer HU; Sedlacek HH. Immunosuppressive chromone alkaloid. US4603137(A) [granted on 29 July 1986].

- Naik RG; Lal B; Rupp RH; Sedlacek HH; Dickneite G; Czech J. The use of 4H-1-benzopyran-4-one derivatives, new 4H-1-benzopyran-4-one derivatives and pharmaceutical compositions containing them. US5292751(A), US07865247 [granted on 8 March 1994].

FUTURE PERSPECTIVES

Dysoxylum binectariferum has a huge economic and medicinal value; and therefore, it should be brought under captive cultivation. The unique secondary metabolite of this plant has already delivered three preclinical/ clinical candidates, and thus it holds further promise to deliver more of such superior candidates in future for treatment of cancer and other diseases.

Abbreviations Used

Cdk, cyclin-dependent kinase; CLL, chronic lymphocytic leukemia; MTD, maximum tolerated dose; FDA, food and drug administration; EMA, european medicines agency; PK, pharmacokinetics; EGFR, epidermal growth factor receptor; ATP, adenosine triphosphate; NCI, national cancer institute; CNS, central nervous system; TNF-α, Tumor necrosis factor alpha; IL-6, Interleukin

6; MTCC, microbial type culture collection and gene bank; SGF, simulated gastric fluid; SIF, simulated intestinal fluid; PBS, phosphate buffer saline; pKa, dissociation constant; GIT, gastrointestinal tract.

REFERENCES CITED

Aalbersberg W, Singh Y, 1991. Dammarane triterpenoids from *Dysoxylum richii*. *Phytochemistry* 30: 921-926.

Aklilu M, Kindler H, Donehower R, Mani S, Vokes E. 2003. Phase II study of flavopiridol in patients with advanced colorectal cancer. *Ann. Oncol.* 14: 1270-1273.

Arguello F, Alexander M, Sterry JA, Tudor G, Smith EM, Kalavar NT, Greene JF, Koss W, Morgan CD, Stinson SF. 1998. Flavopiridol induces apoptosis of normal lymphoid cells, causes immunosuppression, and has potent antitumor activity in vivo against human leukemia and lymphoma xenografts. *Blood* 91: 2482-2490.

Awan FT, Jones JA, Maddocks K, Poi M, Grever MR, Johnson A, Byrd JC, Andritsos LA. 2016. A phase 1 clinical trial of flavopiridol consolidation in chronic lymphocytic leukemia patients following chemoimmunotherapy. *Ann. Hematol.* 95: 1137-1143.

Bharate SB, Kumar V, Jain SK, Mintoo MJ, Guru SK, Nuthakki VK, Sharma M, Bharate SS, Gandhi SG, Mondhe DM. 2018. Discovery and Preclinical Development of IIIM-290, an Orally Active Potent Cyclin-Dependent Kinase Inhibitor. *J. Med. Chem.* 61: 1664-1687.

Carlson BA, Dubay MM, Sausville EA, Brizuela L, Worland PJ. 1996. Flavopiridol induces G1 arrest with inhibition of cyclin-dependent kinase (CDK) 2 and CDK4 in human breast carcinoma cells. *Cancer Res.* 56: 2973-2978.

Cassaday RD, Goy A, Advani S, Chawla P, Nachankar R, Gandhi M, Gopal AK. 2015. A phase II, single-arm, open-label, multicenter study to evaluate the efficacy and safety of P276-00, a cyclin-dependent kinase inhibitor, in patients with relapsed or refractory mantle cell lymphoma. *Clin. Lymphoma Myeloma Leuk.* 15: 392-397.

Chhonker YS, Chandasana H, Kumar A, Kumar D, Laxman T, Mishra S, Balaramnavar V, Srivastava S, Saxena A, Bhatta R. 2014a. Pharmacokinetics, tissue distribution and plasma protein binding studies of rohitukine: a potent anti-hyperlipidemic agent. *Drug Res. (Stuttg).* 64: 1-8.

Chhonker YS, Chandasana H, Kumar D, Mishra SK, Srivastava S, Balaramnavar VM, Gaikwad AN, Kanojiya S, Saxena AK, Bhatta RS. 2014b. Pharmacokinetic and metabolism studies of rohitukine in rats by high performance liquid-chromatography with tandem mass spectrometry. *Fitoterapia* 97: 34-42.

De Azevedo WF, Mueller-Dieckmann H-J, Schulze-Gahmen U, Worland PJ, Sausville E, Kim S-H. 1996. Structural basis for specificity and potency of a flavonoid inhibitor of human CDK2, a cell cycle kinase. *Proc. Natl. Acad. Sci.* 93: 2735-2740.

Dickson MA, Shah MA, Rathkopf D, Tse A, Carvajal RD, Wu N, Lefkowitz RA, Gonen M, Cane LM, Dials HJ. 2010. A phase I clinical trial of FOLFIRI in combination with the pan-cyclin-dependent kinase (CDK) inhibitor flavopiridol. *Cancer Chemother. Pharmacol.* 66: 1113-1121.

Dörwald FZ. 2012. Lead optimization for medicinal chemists: pharmacokinetic properties of functional groups and organic compounds. John Wiley & Sons, pp 252.

Galons H, Oumata N, Meijer L. 2010. Cyclin-dependent kinase inhibitors: a survey of recent patent literature. *Expert Opin. Ther. Patents* 20: 377-404.

Gamble JS. 1967. Flora Of The Presidency Of Madras. Botanical Survey Of India Calcutta, Vol-2, pp 177.

Harmon AD, Weiss U, Silverton JV. 1979. The structure of rohitukine, the main alkaloid of *Amoora rohituka* (Syn. *Aphanamixis polystachya*) (Meliaceae). *Tetrahedron Lett.* 20: 721-724.

Hooker JD. 1875. The Flora of British India. Vol-1, L. Reeve & Co. London, pp 546.

Hu J, Song Y, Li H, Yang B, Mao X, Zhao Y, Shi X. 2014. Cytotoxic and anti–inflammatory tirucallane triterpenoids from *Dysoxylum binectariferum*. *Fitoterapia* 99: 86-91.

Huwe A, Mazitschek R, Giannis A. 2003. Small molecules as inhibitors of cyclin-dependent kinases. *Angew. Chem. Int. Ed.* 42: 2122-2138.

Jain SK, Bharate SB, Vishwakarma RA. 2012. Cyclin-dependent kinase inhibition by flavoalkaloids. *Mini-Rev. Med. Chem.* 12: 632-649.

Jain SK, Meena S, Gupta AP, Kushwaha M, Shaanker RU, Jaglan S, Bharate SB, Vishwakarma RA. 2014. *Dysoxylum binectariferum* bark as a new source of anticancer drug camptothecin: Bioactivity-guided isolation and LCMS-based quantification. *Bioorg Med Chem Lett.* 24: 3146-3149.

Jain SK, Meena S, Qazi AK, Hussain A, Bhola SK, Kshirsagar R, Pari K, Khajuria A, Hamid A, Shaanker RU, Bharate SB, Vishwakarma RA. 2013. Isolation and biological evaluation of chromone alkaloid dysoline, a new regioisomer of rohitukine from *Dysoxylum binectariferum*. *Tetrahedron Lett.* 54: 7140-7143.

Joshi KS, Rathos MJ, Joshi RD, Sivakumar M, Mascarenhas M, Kamble S, Lal B, Sharma S. 2007a. *In vitro* antitumor properties of a novel cyclin-dependent kinase inhibitor, P276-00. *Mol. Cancer Ther.* 6: 918-925.

Joshi KS, Rathos MJ, Mahajan P, Wagh V, Shenoy S, Bhatia D, Sivakumar M, Maier A, Fiebig H-H, Sharma S. 2007b. P276-00, a novel cyclin-dependent inhibitor induces G1-G2 arrest, shows antitumor activity on cisplatin-resistant cells and significant in vivo efficacy in tumor models. *Mol. Cancer Ther.* 6: 926-934.

Kamil M, Jadiya P, Sheikh S, Haque E, Nazir A, Lakshmi V, Mir SS. 2015. The chromone alkaloid, rohitukine, affords anti-cancer activity via modulating apoptosis pathways in A549 cell line and yeast mitogen activated protein kinase (MAPK) pathway. *PloS one* 10: e0137991.

Kelland LR, 2000. Flavopiridol, the first cyclin-dependent kinase inhibitor to enter the clinic: current status. *Expert Opin. Investig. Drugs* 9: 2903-2911.

Keshri G, Oberoi RM, Lakshmi V, Pandey K, Singh MM. 2007. Contraceptive and hormonal properties of the stem bark of *Dysoxylum binectariferum* in rat and docking analysis of rohitukine, the alkaloid isolated from active chloroform soluble fraction. *Contraception* 76: 400-407.

Kouroukis CT, Belch A, Crump M, Eisenhauer E, Gascoyne RD, Meyer R, Lohmann R, Lopez P, Powers J, Turner R. 2003. Flavopiridol in untreated or relapsed mantle-cell lymphoma: results of a phase II study of the National Cancer Institute of Canada Clinical Trials Group. *J. Clin. Oncol.* 21: 1740-1745.

Kumar V, Bharate SS, Vishwakarma RA. 2016a. Modulating lipophilicity of rohitukine via prodrug approach: Preparation, characterization, and in vitro enzymatic hydrolysis in biorelevant media. *Eur. J. Pharm. Sci.* 92: 203-211.

Kumar V, Gupta M, Gandhi SG, Bharate SS, Kumar A, Vishwakarma RA, Bharate SB. 2017. Anti-inflammatory chromone alkaloids and glycoside from *Dysoxylum binectariferum*. *Tetrahedron Lett.* 58: 3974-3978.

Kumar V, Guru SK, Jain SK, Joshi P, Gandhi SG, Bharate SB, Bhushan S, Bharate SS, Vishwakarma RA. 2016b. A chromatography-free isolation of rohitukine from leaves of *Dysoxylum binectariferum*: Evaluation for in vitro cytotoxicity, CDK inhibition and physicochemical properties. *Bioorg. Med. Chem. Lett.* 26: 3457-3463.

Lakshmi V, Pandey K, Kapil A, Singh N, Samant M, Dube A. 2007. In vitro and in vivo leishmanicidal activity of *Dysoxylum binectariferum* and its fractions against *Leishmania donovani*. *Phytomedicine* 14: 36-42.

Li P, Tabibi SE, Yalkowsky SH. 1999. Solubilization of flavopiridol by pH control combined with cosolvents, surfactants, or complexants. *J. Pharm. Sci.* 88: 945-947.

Liu G, Gandara DR, Lara PN, Raghavan D, Doroshow JH, Twardowski P, Kantoff P, Oh W, Kim K, Wilding G. 2004. A Phase II trial of flavopiridol (NSC #649890) in patients with previously untreated metastatic androgen-independent prostate cancer. *Clin Cancer Res.* 10:924-928.

Mabberley DJ, Pannell CM, Sing AM. 1995. Flora Malesiana: Series I. Spermatophyta Volume 12, part 1. Meliaceae. Rijksherbarium, Foundation Flora Malesiana.

Mishra SK, Srivastava P, Rath SK, Mahdi AA, Agarwal SK, Lakshmi V. 2018. Synthesis and Biological Evaluation of Sulphonyl Derivatives of Naturally Occurring Chromone Alkaloid of Rohitukine as Anticancer Agents. *Der. Pharma Chemica* 10: 174-180.

Mishra SK, Tiwari S, Shrivastava S, Sonkar R, Mishra V, Nigam SK, Saxena AK, Bhatia G, Mir SS, 2014. Pharmacological evaluation of the efficacy of *Dysoxylum binectariferum* stem bark and its active constituent rohitukine in regulation of dyslipidemia in rats. *J. Nat. Med.*: https://doi.org/10.1007/s11418-11014-10830-11413.

Mohanakumara P, Sreejayan N, Priti V, Ramesha BT, Ravikanth G, Ganeshaiah KN, Vasudeva R, Mohan J, Santhoshkumar TR, Mishra PD. 2010. *Dysoxylum binectariferum* Hook. f (Meliaceae), a rich source of rohitukine. *Fitoterapia* 81: 145-148.

Mohanakumara P, Soujanya K, Ravikanth G, Vasudeva R, Ganeshaiah K, Shaanker RU, 2014. Rohitukine, a chromone alkaloid and a precursor of flavopiridol, is produced by endophytic fungi isolated from *Dysoxylum binectariferum* Hook.f and *Amoora rohituka* (Roxb). Wight & Arn. *Phytomedicine* 21: 541-546.

Mohanakumara P, Zuehlke S, Priti V, Ramesha BT, Shweta S, Ravikanth G, Vasudeva R, Santhoshkumar TR, Spiteller M, Shaanker RU. 2012. *Fusarium proliferatum*, an endophytic fungus from *Dysoxylum binectariferum* Hook. f, produces rohitukine, a chromane alkaloid possessing anti-cancer activity. *Antonie Van Leeuwenhoek* 101: 323-329.

Naik RG, Kattige SL, Bhat SV, Alreja B, de Souza NJ, Rupp RH. 1988a. An antiinflammatory cum immunomodulatory piperidinylbenzopyranone from *Dysoxylum binectariferum* : Isolation, structure and total synthesis. *Tetrahedron* 44: 2081-2086.

Patel V, Senderowicz AM, Pinto D, Igishi T, Raffeld M, Quintanilla-Martinez L, Ensley JF, Sausville EA, Gutkind JS. 1998. Flavopiridol, a novel cyclin-dependent kinase inhibitor, suppresses the growth of head and neck squamous cell carcinomas by inducing apoptosis. *J. Clin. Invest.* 102: 1674-1681.

Quattrocchi U. 2012. CRC world dictionary of medicinal and poisonous plants: common names, scientific names, eponyms, synonyms, and etymology. CRC Press, Boca Raton Florida, pp 1502-1503.

Raje N, Hideshima T, Mukherjee S, Raab M, Vallet S, Chhetri S, Cirstea D, Pozzi S, Mitsiades C, Rooney M. 2009. Preclinical activity of P276-00, a novel small-molecule cyclin-dependent kinase inhibitor in the therapy of multiple myeloma. *Leukemia* 23: 961.

Rathos MJ, Joshi K, Khanwalkar H, Manohar SM, Joshi KS. 2012. Molecular evidence for increased antitumor activity of gemcitabine in combination with a cyclin-dependent kinase inhibitor, P276-00 in pancreatic cancers. *J. Transl. Med.* 10: 161.

Rathos MJ, Khanwalkar H, Joshi K, Manohar SM, Joshi KS, 2013. Potentiation of in vitro and in vivo antitumor efficacy of doxorubicin by cyclin-dependent kinase inhibitor P276-00 in human non-small cell lung cancer cells. *BMC Cancer* 13: 29.

Sedlacek H. 2001. Mechanisms of action of flavopiridol. *Crit. Rev. Oncol. Hematol.* 38: 139-170.

Sedlacek H, Czech J, Naik R, Kaur G, Worland P, Losiewicz M, Parker B, Carlson B, Smith A, Senderowicz A. 1996. Flavopiridol (L86 8275; NSC 649890), a new kinase inhibitor for tumor therapy. *Int. J. Oncol.* 9: 1143-1168.

Senderowicz AM. 2001. Development of cyclin-dependent kinase modulators as novel therapeutic approaches for hematological malignancies. *Leukemia* 15: 1-9.

Shah MA, Kortmansky J, Motwani M, Drobnjak M, Gonen M, Yi S, Weyerbacher A, Cordon-Cardo C, Lefkowitz R, Brenner B. 2005. A phase I clinical trial of the sequential combination of irinotecan followed by flavopiridol. *Clin. Cancer Res*. 11: 3836-3845.

Shareef M, Ashraf MA, Sarfraz M. 2016. Natural cures for breast cancer treatment. *Saudi Pharm. J*. 24: 233-240.

Shilpi JA, Saha S, Chong SL, Nahar L, Sarker SD, Awang K. 2016. Advances in Chemistry and Bioactivity of the Genus *Chisocheton* Blume. *Chem. Biodivers*. 13: 483-503.

Shirsath NP, Manohar SM, Joshi KS. 2012. P276-00, a cyclin-dependent kinase inhibitor, modulates cell cycle and induces apoptosis *in vitro* and *in vivo* in mantle cell lymphoma cell lines. *Mol. Cancer* 11: 77.

Singh N, Singh P, Shrivastva S, Mishra SK, Lakshmi V, Sharma R, Palit G. 2012. Gastroprotective effect of anti-cancer compound rohitukine: possible role of gastrin antagonism and H+ K+-ATPase inhibition. *Naunyn Schmiedebergs Arch Pharmacol*. 385: 277-286.

Singh S, Garg HS, Khanna NM. 1976. Dysobinin, a new tetranortriterpene from *Dysoxylum binectariferum*. *Phytochemistry* 15: 2001-2002.

Tan AR, Swain SM. 2002. Review of flavopiridol, a cyclin-dependent kinase inhibitor, as breast cancer therapy, *Semin. Oncol*. 29: 77-85.

Varshney S, Shankar K, Beg M, Balaramnavar VM, Mishra SK, Jagdale P, Srivastava S, Chhonker YS, Lakshmi V, Chaudhari BP. 2014. Rohitukine inhibits in vitro adipogenesis arresting mitotic clonal expansion and improves dyslipidemia in vivo. *J. Lipid Res.* 55: 1019-1032.

Wiernik PH. 2016. Alvocidib (flavopiridol) for the treatment of chronic lymphocytic leukemia. *Expert Opin. Investig. Drugs* 25: 729-734.

Xia B, Liu X, Zhou Q, Feng Q, Li Y, Liu W, Liu Z. 2013. Disposition of orally administered a promising chemotherapeutic agent flavopiridol in the intestine. *Drug. Dev. Ind. Pharm.* 39: 845-853.

Yan HJ, Wang JS, Kong LY. 2014. Cytotoxic steroids from the leaves of *Dysoxylum binectariferum*. *Steroids* 86: 26-31.

Yang DH, Cai SQ, Zhao YY, Liang H. 2004. A new alkaloid from *Dysoxylum binectariferum*. *J. Asian. Nat. Prod. Res*. 6: 233-236.

Zeidner JF, Karp JE. 2015. Clinical activity of alvocidib (flavopiridol) in acute myeloid leukemia. *Leuk. Res.* 39: 1312-1318.

7

Biochemical Analysis of three Species of Hemiparasitic Taxa in the Polluted Area: Significant Role in Mitigators of Pollutants

*Shibdas Maity, *Souradut Ray and *Amal Kumar Mondal*

Plant Taxonomy, Biosystematics and Molecular Taxonomy Laboratory
UGC-DRS-SAP & DBT-BOOST, Department of Botany and Forestry, Vidyasagar University
Midnapore-721102, West Bengal, INDIA
*Email: shibdasmaity345@gmail.com, amalcaebotvu@gmail.com

ABSTRACT

In the present investigation on three different plant species from pollution and control area were carried act for biochemical analysis such as total chlorophyll content, leaf extract pH, relative water content and ascorbic acid test. The results indicate accordance to order of tolerance index of plant species in air pollution are *Macrosolen cochinchinensis, Loranthus parasiticus* and *Viscum album*. The changes in pollution tolerance index are biochemically induced due to the surrounding environmental conditions. Thus, it is observed that *Macrosolen cochinchinensis* is highly tolerant species among the selected three species. Thus, in future this plant species may play an important role in pollution indicator. APTI analysis show the pollution tolerating decreasing order for there species can be *Macrosolen cochinchinensis > Loranthus parasiticus >Viscum album* and as ascorbic acid result indicate that *Macrosolen cochinchinensis* show mostly defence mechanism and highly tolerance level against different kind of pollutant. Besides these three hemiparasite plant species have some valuable ethnomedicinal important. In India tribal people used leaf and stem bark in curing abortion, miscarriage, vaginal bleeding during pregnancy, treat circulatory and respiratory system problems.

Keywords: Parasitic plant, Biochemical analysis, Pollution Tolerance Index (PTI), Pollution Indicator, Ethnomedicinal aspects.

INTRODUCTION

It is the time to think about our environment which is not our rather than global concern. As a citizen of India, we must care about the environment where we are living to be aware of the upcoming threats. Heavy metals are natural constituent of the lithosphere, whose geochemical cycles and biochemical balances have been drastically altered by human activity (Sebastiani et al. 2004). Pollution due to heavy metals placed human health at risk and it is responsible for several environmental problems, including the decrease of microbial activity, soil fertility and crop yields (Yang et al. 2005). The annual toxicity of all the mobilized metals exceeds the total combined toxicity of all radioactive and organic residues generated in the same period (Nriagu and Pacyna 1988). Pollution by heavy metals normally coincides with the increase in industrialization of a given region and becomes more severe when there are neither controls nor adequate environmental norms (Pilon-Smits 2005).

The accumulation of heavy metals in vascular plants provokes significant biochemical and physiological responses, modifying several metabolic processes (Macfarlane et al. 2003). Reduction in the net photosynthetic rate due to toxicity of these metals causes decreases in growth and productivity (Van Assche and Clijsters 1983). Several studies have demonstrated that pH is one of the most important factors in the control of the concentration of these metals in the soil solution (King 1988). Elliott et al. (1986) proposed that under acidic conditions the phenomenon of adsorption is more important in the control of metal bioavailability, while precipitation reactions and complexation have greater influence under neutral and alkaline conditions. According to Patra et al. (2004), plants that grows in environments contaminated with traces of metals show strategies of escape or tolerance to metal toxicity that have been selected during evolution. Several plant species have developed tolerance to metals in a relatively short period of about thirty years (Hall 2002). The exposure of these pollutants to the leaves indirectly cause a reduction in the concentration of their photosynthetic pigments, viz., chlorophyll which affects the plant productivity, germination of seeds, length of pedicles, and number of flowers inflorescence (Nithamathi et al. 2005).

Chlorophyll is the principal photoreceptor in photosynthesis, the light-driven process by which carbon dioxide is "fixed" to yield carbohydrates, oxygen and also protect chlorophyll from photoxidative destruction (Siefermann-Harms 1987). When plants are exposed to the environmental pollution above the normal physiologically acceptable range, photosynthesis gets inactivated (Miszalski et al. 1990). Since, the plants leaf samples used for this experiment were constantly exposed to soil pollutants (polluted area- industrial area, automobile area and less polluted area – jungle, they had absorbed, accumulated and integrated pollutants on their surface and showed specific response. Hence, plants can be used as bioindicators in various field of research (Joshi 1997). Besides chlorophyll, ascorbic acid, water consumption and leaf pH were also affected. Air Pollution Tolerance Index of plant species also studies in West Midnapore District (Maity and Mondal 2015).

STUDY AREA

The present study was carried out within the different intersection in West Bengal Midnapore and Kharagpur zone, one of the important cities of India and is extended from latitude $22^{\circ}41'61''$ in the north to longitude $87^{0}38'44''$ in the east with a tropical climate during the time period of January 2015. The sites selected for the present study includes forest area and industrial area. This industry in the city increases the density of pollutants and also the selected plant species are present in this polluted area which remains in direct contact with these types of pollutants. Therefore, this particular plant species were selected for the study purposes.

PHYTOCHEMICAL ANALYSIS

Tree species were selected from Midnapore and Kharagpur zone. Leaf samples were collected in triplicates of fully matured leaves and were immediately transferred to the laboratory for analysis. Leaf samples were preserved in a refrigerator for further examination. The following are the different methods used to determine APTI.

Estimation of Total Chlorophyll Content (TCh)

3g of fresh leaves were blended and then extracted with 10 ml of 80% acetone and left for 15 minutes for thorough extraction. Then the liquid portion was poured into another text-tube and centrifuged at 2,500rpm for 3 minutes. The supernatant was then collected and the absorbance was then taken at 645nm and 663nm using Systronics UV spectrophotometer (Arnon 1949).

Chlorophyll a = 12.7DX663 – 2.69 DX645 x V/1000W mg/g

Chlorophyll b = 22.9Dx645 – 4.68 Dx663 x V/1000W mg/

TCh = Chlorophyll a + b mg/g

where, Dx = Absorbance of the extract at the wavelength in nm, V = total volume of the chlorophyll solution (ml), and W = weight of the tissue extract (g).

Leaf Extract pH

5g of fresh leaves was homogenized in 10ml deionised water. This was then filtered and the pH of leaf extract was determined after calibrating pH meter-HI 98130 with buffer solution of pH 4, pH 7 and pH 9 (Agbaire and Esiefarienrhe 2009).

Relative Water Content of Leaf (RWC)

Fresh leaves were weighed and then immersed in water over night, blotted dry and then weighed to get the turgid weight. Then, the leaves were dried overnight in an hot air oven at 70°C and reweighed to obtain the dry weight (Singh 1977). Calculations were made using the formula:

RWC = [(FW − DW)/(TW − DW)] x 100

where, FW = Fresh weight, DW = dry weight, and TW = turgid weight.

Ascorbic Acid (AA) Content

1g of the leaf sample was measured into a test tube, 4ml of oxalic acid – EDTA extracting solution was added. Then 1ml of orthophosphoric acid followed by 1ml 5% tetraoxosulphate (vi) acid, 2ml of ammonium molybdate and then 3ml of water was added. The solution was then allowed to stand for 15 minutes, after which the absorbance at 760nm was measured with Systronics UV-Vis spectrophotometer 118. The concentration of ascorbic acid in the leaf samples were then extrapolated from a standard ascorbic acid curve (Bajaj and Kaur, 1981). Ascorbic acid content was measured by Titrimetric method of (Sadasivam, 1987) using 2,6, Dichlorophenol indo phenol dye. 500mg of leaf sample was extracted with 4% oxalic acid and then titrated against the dye until pink colour develops. Similarly a blank is also developed.

APTI

The air pollution tolerance indices of twelve common plants were determined by the following standard method (Singh and Rao 1983). The formula of APTI is given as

APTI = [A (T+P) + R]/10

Where, A = Ascorbic acid content (mg/g), T = Total chlorophyll content

(mg/g), P = pH of leaf extract, and R = Relative water content of leaf (%).

Total Chlorophyll Content Estimation with Respect to Air pollution

Chlorophyll estimation is one of the important criteria for pollution measurement, Loss in total chlorophyll content of plant depends on the degree of pollution but it is observed less effective in case of air pollution.

The experiment which we had done for pollution measurement through chlorophyll estimation in different plant samples show a dramatic results which summarize in follows:

In the present result, the chlorophyll content level reduced from 36.93 to 35.16 in case of *Solanum nigrum* whereas in case of *Macrosolen cochinchinensis* it is reduced from 11.08 to 9.45, *Loranthus parasiticus* it is reduced from 15.39 to 7.87, *Viscum album* is reduced from 18.23 to 18.20, and therefore, it is more sensitive but *Macrosolen cochinchinensis* is highly tolerance according to reduction result of chlorophyll estimation (Fig. 1).

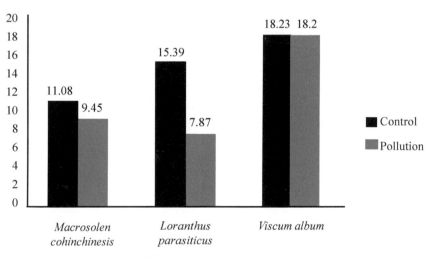

Fig. 1: Chlorophyll Content

Analysis of pH Value of Leaf Extracts of Air Pollution Tolerating Plant

Chlorophyll aiding in starch synthesis can indicate environmental pollution likewise pH helps in physiological responses caused by stress. pH change influence stomatal sensitivity and leaves with low pH are more susceptible to pollution but those having neutral pH are more tolerant.

pH value which we have observed for different plant sample are summarize as follows:

In our result, pH value of leaf extract of control plant and pollution area plant of different plant species sample i.e *Macrosolen cochinchinensis* is 6.3 (C) and 6.0 (P), *Loranthus parasiticus* 15.39 (C) and 5.6 (P) and *Viscum album* 6.2

(C) and 6.1 (P) are given respectively. It can be concluded that *Loranthus parasiticus* shows lower pH value, indicates as a susceptible to polluting condition; but in case of *Macrosolen cochinchinensis* there are little change of pH value so, it should be treated as a pollution tolerating plant (Table 1).

Table 1: Leaf extract pH of three different Plant species.

Sl.No	Plant species	Leaf extract pH	
		C	P
1.	*Macrosolen cochinchinensis*	6.3	6.0
2.	*Loranthus parasiticus*	5.9	5.6
3.	*Viscum album*	6.2	6.1

Relative Water Content Estimation of Air Pollution Tolerating Plant

Water is an essential factor for the transportation of food, minerals. Plants with relatively high water content are highly resistant to pollution (Tanaka et.al. 1982). The amount of relative water content which we have observed for different plant sample are summarize as follows: In our result amount of relative water content of control plant and pollutant plant sample i.e *Macrosolen cochinchinensis* is 3.29 (C) and 3.10 (P), *Loranthus parasiticus* 2.48 (C) and 2.19 (P) and *Viscum album* 3.19 (C) and 3.13 (P). From this result it can be concluded that *Macrosolen cochinchinensis* is highly resistant to pollution because it shows very minute change of relative water content than control plant (Fig. 2).

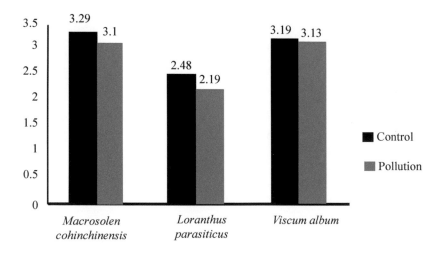

Fig. 2: Relative Water Content

Ascorbic Acid content Estimation of Air Pollution Tolerating Plant

Ascorbic acid content estimation is one of the important criteria for pollution measurement. Ascorbic acid being a strong reductant protects chloroplast against sulphur dioxide induced H_2O_2, $O^{2\text{É}}$ and OH accumulation. Similarly, it protects the enzymes of CO_2 fixation cycle and chlorophyll from inactivation. Defence mechanism in plants cause increased level of ascorbic acid.

The experiment which we had done for pollution measurement through ascorbic acid content estimation in different plant samples show a dramatic result which summarized is as follows. In this result, the ascorbic acid content of control plant and pollutant plant sample i.e *Macrosolen cochinchinensis* is 0.186 (C) and 0.156 (P), *Loranthus parasiticus* 0.091 (C) and 0.049 (P) and *Viscum album* 0.165 (C) and 0.159 (P) are given respectively. It shows that in *Macrosolen cochinchinensis* the difference of ascorbic acid content is very low between control plant and pollutant plant, therefore, this plant has high defensive mechanism for tolerating pollution. All biochemical test result analysed under APTI formula.

So observed that *Macrosolen cochinchinensis* is highly tolerant species respectively among the selected 3 species (Table 2 and Fig.3).

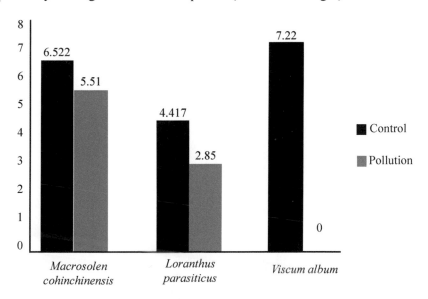

Fig. 3: Graphical Representation of APTI Analysis

Table 2: Comparison of different biochemical parameters of air pollution tolerating plants.

Sl.No.	Plant species	Total chlorophyll		Leaf extract pH		Ascorbic acid		Relative water content (gm)		APTI	
		C	P	C	P	C	P	C	P	C	P
1.	*Macrosolen cochinchinensis*	11.08	9.45	6.3	6.0	0.186	0.156	3.29	3.10	6.522	5.510
2.	*Loranthus parasiticus*	15.39	7.87	5.9	5.6	0.091	0.049	2.48	2.19	4.417	2.850
3.	*Viscum album*	18.23	18.20	6.2	6.1	0.165	0.159	3.19	3.13	7.220	0.699

Ethnobotanical Uses

Viscum album

- Mistletoe leaves and young twigs are used by herbalists.
- For attempting to treat circulatory and respiratory system problems.
- Use of mistletoe extract in treatment of cancer.

Loranthus parasiticus

- prevention and treatment of treatened abortion, miscarriage, and vaginal bleeding during pregnancy.
- for treating arthralgia, rheumatic sciatica, and aches and pain in the loins and legs.

Macrosolen cochinchinensis

- Leaf paste used as folk medicine for jaundice.
- Tea from leaves used for headaches.

Fig. 4: a- *Viscum album, b- Loranthus parasiticus* and
c- *Macrosolen cochinchinensis*

CONCLUSION

These species has well known economic and aesthetic values and thus recommended for extensive planting. Thus, pollution tolerance index might be very useful in the selection of appropriate species. The results of the study concluded that each of biochemical parameter plays a significant role in the determination of susceptibility level of the plant species with reference to their tolerance and performance index. Estimation of these indicates will be a reliable method for the selection of appropriate species which can be used as bioindicators and mitigators of pollutants in an urban and industrial region.

Abbreviation

APTI   : Air pollution Tolerance Index

RWC   : Relative Water Content

AA   : Ascorbic Acid

FW   : Fresh Weight

DW   : Dry Weight

TW   : Turgid Weight

TCH   : Total Chlorophyll Content

Conflict of Interest

We have no conflict of interest.

ACKNOWLEDGEMENTS

We are thankful to Plant Taxonomy, Biosystematics & Molecular Taxonomy Laboratory and also thankful to UGC DRS-SAP Laboratory, UGC-DRS-SAP and DBT-BOOST-WB Funded Department, Department of Botany & Forestry for infrastructural facility.

REFERENCES CITED

Agbaire PO, Esiefarienrhe E. 2009. Air pollution tolerant indices (APTI) of some plants around Otorogun grass plant in Delta State, Nigeria. *Journal of Applied Sciences and Environmental Management* 13(1):11-14.

Arnon DI 1949. Copper enzymes in isolated chloroplast polyphenol oxidase in *Beta vulgaris*. *Plant Physiology* 24: 1-15.

Bajaj K L, Kaur, G. 1981. Spectrophotometric determination of L-ascorbic acid in vegetables and fruits. *Analyst* 106 (1258): 117-120.

Elliott HA, Liberati MR, Huang CP 1986. Competitive adsorption of heavy metals by soils. *Journal of Environmental Quality.* 15:214-217.

Hall JL 2002. Cellular mechanisms for heavy metal detoxification and tolerance. *Journal of Experimental Botany.* 53:1-11.

Joshi OP, Wagola DK, Pawar, K. 1997. Urban air pollution effect on two species of *Cassia.Pollution Research* 16(1): 1-3.

King LD 1988. Retention of metals by several soils of the southeastern United States. *Journal of Environmental Quality* 17:239-246.

MacFarlane GR, Pulkownik A, Burchett MD. 2003. Accumulation and distribution of heavy metals in the grey mangrove, *Avicennia marina* (Forsk.) Vierh.: *Biological indication potential. Enviromental Pollution*123:139-151.

Maity S, Mondal, AK 2015. Air Pollution Tolerance Index Of Some Plant Species Of West Midnapore Distric, West Bengal, India. *Advancement in Plant Sciences,* Lambert Academic Publishing 1: 51-54.

Nithamathi, C.P. and Indira, V. 2005. Impact of air pollution on *Ceasalpinia sepiariaL.* in Tuticorin city. *Indian Journal of Environmental and Ecoplanning* 10(2): 449-452.

Nriagu JO, Pacyna JM. 1988. Quantitative assessment of worldwide contamination of air, water and soils by trace metals. *Nature* 333:134-139.

Patra M, Bhowmik N, Bandopadhyay B, Sharma A 2004. Comparison of mercury, lead and arsenic with respect to genotoxic effects on plant systems and the development of genetic tolerance. *Journal of Experimental Botany* 52:199-223.

Pilon-Smits E. 2005. Phytoremediation. *Annual Review of Plant Biology* 56:15-39.

Sebastiani L, Scebba F, Tognetti R. 2004. Heavy metal accumulation and growth responses in poplar clones Eridano (*Populus deltoides* x *maximowiczii*) and I-214 (*P.* x *euramericana*) exposed to industrial waste. *Environmental and Experimental Botany* 52:79-88.

Siefermann- Harms, D. 1987. The light- harvesting and protective functions of carotenoids in photosynthetic membranes. *Physiology Plantarum* 69: 561-568.

Singh SK and Rao DN 1983. Evaluation of plants for their tolerance to air pollution. *Indian Proceedings Symposium on Air Pollution Control* 218-224.

Van Assche F, Clijsters H 1983. Multiple effects of heavy metals on photosynthesis. In: Marcelle R (ed), *Effects of Stress on Photosynthesis* 371-382.

Yang X, Feng Y, He Z, Stoffella PJ. 2005. Molecular mechanisms of heavy metal hyperaccumulation and phytoremediation. *Journal of Trace Elements in Medicine and Biology* 18: 339-353.

8

Exploring Commercial Cultivation of Tissue Cultured Raised Banana in Himalayan Shivalik Range

Sabha Jeet[1], VP Rahul[1], Rajendra Bhanwaria[1], Chandra Pal Singh[1], Rajendra Gochar[1], Amit Kumar[1], Kaushal Kumar[1], Jagannath Pal[1] and Bikarma Singh[2]

[1]*Genetic Resources and Agrotechnology Division*
CSIR-Indian Institute of Integrative Medicine, Jammu 180001
Jammu and Kashmir, INDIA
[2]*Plant Sciences (Biodiversity and Applied Botany Division)*
CSIR-Indian Institute of Integrative Medicine, Jammu 180001
Jammu and Kashmir, INDIA
Email: sabhajeet@iiim.ac.in

ABSTRACT

Agrotechnology of banana has been introduced for the first time by CSIR-Indian Institute of Integrative Medicine Jammu for commercial cultivation of banana in Himalayan Shivalik range. The aim of the present investigation is to study the potential of Tissue Culture (TC) banana production in a diverse environment of Shivalik range, to make the state self-sufficient in banana production, employment generation and generate revenue for the farmers and to studies on the post harvest handling and marketing of TC banana. Initially, the sapling of this high quality tissue culture variety {BHIM Grand naine (G-9)} Banana was brought from Agro-division of Cadilla Pharmaceuticals Limited Ahmedabad, Gujarat. The field experimental trial was conducted during 20016-2017 at CSIR-IIIM Research Station, Chatha Farm, Jammu in which four treatment combination arranged in Randomized Complete Block Design (RCBD). Significant differences were observed among the treatments for all the parameters studied. Among the date of planting, banana planted at 10th August reported significantly highest pseudostem height, pseudostem girth, more average number of leaves/plant, first flower emergence (days), first finger ripening (days), more number of finger per hand, more number of hand/ branch, higher weight of finger/hand (kg), higher weight of single finger (g), more finger diameter (cm), higher length of finger (cm), more weight of finger/ bunch (kg), more percentage of plant harvested at a time, higher yield (64.22 tonnes/ha), gross return (Rs. 12,84,400/ha), net return (Rs. 9,58,730.5/ha) and B:C ratio (3.94) as compared to other treatments. On an average the yield of per plant was 20-30 kg and 50- 64 tonnes per hectare. In terms of economy as per market analysis, price of banana in Jammu is approximately Rs. 20 per kg. Thus, on an average it gave Rs. 250-300 per banana plant. On the basis of market demand, approximately Rs 6.17- 9.50 lakh net return can be obtained by cultivation of one hectare of land which is alternative business for the farmers of the Jammu and Kashmir. This may leads to revolution of banana cultivation in Shivalik range of Himalaya.

Keywords: Food plant, Tissue culture raised banana, Shivalik range, Commercial value

INTRODUCTION

Jammu and Kashmir referred as 'Paradise of Earth' is mainly home to temperate fruit like apple for which the state is famous across the globe. The main factor which influences temperate fruit bearing trees is soil, climate and environment which are highly favorable and unparalleled in the province of Kashmir. Apple is commercially the most important temperate fruit and is fourth among the most widely produced finger in the world after banana, orange and grapes. Banana is the 4th most important food crop in world. It is considered as "Poor Man's Apple". It is a staple food and exported commodity. It contributes to the food security of millions of people in the developing world and, when traded in local markets, provides income and employment to the rural population.

Banana is basically a tropical crop, grows well in temperature range of 13°C – 38⁰C with RH=Relative humidity regime of 75-85% and thrives from 0- 1800 m above sea level, with rainfall requirements of 1000 mm annually. In India, this crop is being cultivated in climate ranging from humid tropical to dry mild subtropics through selection of appropriate varieties like Grand Naine. Chilling injury occurs at temperatures below 12°C. The normal growth of the banana begins at 18°C, reaches optimum at 27°C, then declines and comes to a halt at 38°C. Higher temperature causes Sun scorching. High velocity wind which exceeds 80 km per hrs damages the crop. The crop grows well in deep, well drained loams of high fertility and high content of organic matter.

In India, planting season varies from one geographic location to another and in most parts, very cold or hot seasons are unsuitable. Planting during winter leads to initial exposure to unfavorable conditions of hot summer and heavy winds during critical stages of growth. In general, planting banana before the commencement of monsoon stands to help the plants build up rapid growth and establishment before onset of the cold weather. But, in view of the divergent climatic and soil conditions prevalent in India, banana are grown all through out the year. In Israel, which experiences severe cold winter, planting of banana is generally done during March, i.e., spring planting. In the sub-tropics, planting is done during the dry season with pre-irrigation to facilitate better establishment. In the sub-tropics of South Africa, summer planting helps to avoid winter flowering. In North-Western Australia, where the summer is very hot, planting during winter (June-July) is practiced. In Puerto Rico, planting during December facilitates harvesting during February - April, which fetches a very high price. Under Nigerian conditions, planting during January to May is often prone to cyclone damage; August to December planting is found ideal. In Bangladesh, September-October and February-March are the two main seasons for successful banana cultivation (Mustaffa and Kumar 2012).

The propagation of plant by plant part or group of cells in a test tube under very controlled and aseptic conditions is called tissue culture. Tissue culture banana production technology is a superior technology over traditional method (sucker propagated) of banana production with respect to optimal yield, uniformity, disease free planting material and true to type plants (Hanumantharaya et al. 2009). Tissue culture is a technique that allows mass multiplication of planting materials in short duration of time. It is a form of biotechnology that refers to production of plants from very small plant parts, tissues or cells grown aseptically under laboratory conditions where the environment and nutrition are rigidly controlled. The basis of the TC technology lies on the ability of many plant species to regenerate a whole plant from a plant part. Using tissue culture, it is possible to produce 1000 plantlets from one sucker in one year in comparison to ten suckers produced per banana in the same period.

The demand for tissue culture banana is expected to increase by 25 to 30 percent estimated to the extent of about 5.0 million every year. The present global biotech business is estimated at around 150 billion US dollars. Around 50-60% of this constitutes agri-business. The annual demand of tissue-cultured products constitutes nearly 10% of the total, amounting to 15 million US dollars. The estimated annual growth rate is approximately 15%. Among the developing nations, however, India is positioned advantageously to exploit this market due to availability of cheap skilled labour, low input cost and low energy cost when compared to the developed nations. The Govt. of India identified micro-propagation of plants as an industrial activity and several subsidies and incentives were offered. Many state governments, prominent among them the state of Karnataka, have given the plant tissue culture ventures as thrust sector industry (Anonymous 2007). Grand Naine is a popular variety grown mostly in all export oriented country of Asia , Africa ,South America , is superior selection of Giant Cavendish which is introduced in India (1990) due to many desirable traits like excellent fruit quality, immunity to fusarium wilt , it has proved better variety (Singh & Chundawat 2002).

NATIONAL AND INTERNATIONAL STATUS

Banana (Musa spp.) growing areas of the world are mainly situated between the equator and latitudes 20°N and 20°S. Climatic conditions in these areas are mainly tropical, with relatively small temperature fluctuations from day to night and from summer to winter. Since banana plants show wide adaptability to a range of environments, they can also be grown in subtropical areas. In the areas of New South Wales, Western Australia, South Queensland, South Africa, Israel, Taiwan, Spain (the Canary Islands), Egypt, Morocco, and parts of Brazil, banana plantations are situated between latitudes 20°-30°. In Turkey, banana plantations are situated at 36° latitude.

Banana plants reproduce asexually by shooting suckers from a subterranean stem and then shoots have a vigorous growth and can produce a ready-for-harvest bunch in less than one year. Suckers continue to emerge from a single mat year after year, making banana a perennial crop. It grows quickly and can be harvested all year round. During 2012-13, the world acreage of banana was 50,07,520 hectares, while the world production 10,36,32,349 Metric Tones and productivity was 20.7 Metric Tone/Hectare. In case of production of Banana, it was also maximum in India i.e., 2,65,09,096 Metric Tones (25.58%), followed by China 1,05,50,000 Metric Tones (MT) (10.18%), Philippines 92,25,998 Metric Tones (8.90) and Ecuador 70,12,244 Metric Tones (6.77). These four countries produce more than 51 percent of world production of Banana. The productivity of Banana is maximum in Indonesia i.e., 58.9 MT/Ha., followed by Guatemala 40.9 MT/ha., India 34.2 MT/ha, China 26.4 MT/ha (Table 1)

Table 1: Global status of production and productivity of banana on commercial scale

Sr.No.	Country	Area (ha)	Production (MT)	Productivity (MT/ha)
1	India	775995(15.50%)	26509096 (25.58)	34.20
2	China	400000 (7.99)	10550000 (10.18)	26.80
3	Philippines	454179 (9.07)	9225998 (8.90)	20.30
4	Ecuador	210894 (4.21)	7012244 (6.76)	33.33
5	Brazil	481116 (9.61)	6902184 (6.66)	14.30
6	Indonesia	105000 (2.10)	6189052 (5.97)	58.90
7	Angola	115749 (2.31)	2991454 (2.89)	25.80
8	Guatemala	66000 (1.32)	2700000 (2.61)	40.90
9	Tanzania	442190 (8.83)	2524740 (2.44)	5.70
10	Mexico	72617 (1.45)	2203961 (2.13)	30.30
11	Others	1883780 (37.61)	26823720 (25.87)	14.20
	Total (World)	50007520 (100)	103632349 (100)	20.77

(*Source:* Anonymous 2015).

India leads the world in banana production with an annual output of about 26509.0 MT from an area of 776.0 thousand hectare. The edible banana is believed to have originated in the hot, tropical regions of South-East Asia. India is believed to be one of the centres of origin of banana. In India, major banana producing states are Tamil Nadu, Maharashtra, Karnataka, Gujarat, Andhra Pradesh, Assam and Madhya Pradesh (Karuna and Rao 2013). It may be seen that the production of Banana was the highest in Tamil Nadu i.e., 5136.20 thousand tones (19.38%), followed by Gujarat 4523.49 thousand tones (12.23%), Karnataka 2529.60 thousand tones (9.54%), Bihar 1702.41 thousand tones (6.42%) and Madhya Pradesh 1701.00 thousand tones (6.42%). Total production of Banana in these five states was more than 84 percent (Table 2) (Anonymous 2015).

Table 2: Major banana producing states of India

Sr.No.	States	Area (000, ha)	Production (000, MT)	Productivity (MT/ha)
1.	Tamil Nadu	111.36 (14.35%)	5136.20 (19.38)	46.10
2.	Gujarat	97.40 (12.55)	4523.49 (17.06)	64.10
3.	Maharashtra	92.65 (11.94)	3600 (13.58)	43.90
4.	Andhra Pradesh	82.00 (10.57)	3242.80 (12.23)	35.00
5.	Karnataka	70.58 (9.10)	2529.60 (9.54)	26.00
6.	Bihar	51.51 (6.65)	1702.41 (6.42)	51.50
7.	M.P.	33.06 (4.26)	1701.00 (6.42)	66.00
8.	West Bengal	44.70 (5.76)	1077.80 (4.07)	24.10
9.	Assam	27.49 (3.54)	837.02 (3.16)	16.20
10.	Odisha	25.76 (3.32)	521.31 (1.97)	19.00
11.	Others	139.49 (17.97)	1637.47 (6.17)	11.70
	Total	776.00 (100)	26509.10 (100)	34.20

Experimental Trials

Field experiment on tissue culture raised banana was carried out at CSIR-IIIM research station, Chatha farm, Jammu during of 2016-2017. Jammu is located at 32.73°N and 74.87°E and has an average elevation of 350 m above mean sea level. The city lies at uneven ridges of low heights at the Shivalik hills. It is surrounded by Shivalik range to the north, east and southeast while the Trikuta range surrounds in north-west. The city spreads around Tawi river with the old city overlooking from the north (right bank), while the new neighbourhoods spread around the southern side (left bank) of river. Jammu, like the rest of north-western India, features a humid subtropical climate with extreme summer highs reaching 46 °C, and temperatures in the winter months occasionally falling below 4 °C. June is the hottest month with an average temperature of 40.6 °C, while January is the coldest month with average lows reaching 7 °C temperature.

Average yearly precipitation is about 42 inches (1,100 mm) with the bulk of the rainfall in the months from June to September, although the winters can also be rather wet. In winter dense smog causes much inconvenience and temperature even drops to 2 °C. In summer, particularly in May and June, extreme intense sunlight or hot winds can raise the temperature to 46 °C. Following the hot season, the monsoon lashes the city with heavy downpours along with thunderstorms: rainfall may total up to 669 mm in the wettest months.

The experimental field was clay loam with low organic carbon (0.25 %), low available N (195 kg ha^{-1}), medium available P (15.02 kg ha^{-1}), medium available K (228 kg ha^{-1}) and 14.32 kg ha^{-1} available S with pH 7.6. Experiments were conducted in randomized complete block design (RCBD) replicated fifth times with four number of treatments *viz.*, date of planting T_1 (10th August), T_2 (10th September), T_2 (10th October) and T_2 (10th November). FYM were applied 10 tonnes T_2 before one month of planting. The major fertilizer basically nitrogen (287 kg ha^{-1}) phosphorus (203.75 kg ha^{-1}) and potassium (427.50 kg ha^{-1}) was applied as basal and top dressing and mixed with soil at 50 days intervals. The source of fertilizer was from Urea, DAP, SSP and MOP. The others nutrients like *Aspergillus niger* (12.5 kg ha^{-1}), Mychorrhiza (2.5 kg ha^{-1} mixed with FYM), Magnesium sulphate (62.50 kg ha^{-1}) and Zinc sulphate (12.5 kg ha^{-1}) were applied as basal and mixed with soil. The sapling of this high quality tissue culture variety known as BHIM Grand naine (G-9) Banana was brought from Agro- division of Cadilla Pharmaceuticals Limited Ahmadabad, Gujarat and planted with the narrow spacing 2x2 m.

Transplanting of banana was done with manual digging pit of 1x1x1 feet, add 2.50 kg well decomposed farmyard manure with 5 gram Kalisena (*Aspergillus nizer*) along with organic manure mixed thoroughly with soil. after 5 g of phorate, 75 g DAP 75 g of murate of potash were added in each pit. Pit Preparation was done 8-10 days before planting. At the time planting add Five gram of 'Josh' powder (*Mychorrhiza*) in pit which increase root growth. Polythene cover from the plants were removed carefully and place it firmly in the pit. Safety were followed ensure that root ball must be inside the pit. Plants were fided well and irrigate it gently. Manually weeding was done at 50- 75 days interval to control weeds. Banana being a succulent, evergreen and shallow rooted crop requires large quantity of water for increasing productivity. Water requirement of banana has been worked out to be 1,800 – 2,000 mm per annum. In winter, irrigation was given at an interval of 20-25 days while in summer it at an interval of 10-15 days. However, during rainy season irrigation was provided if required as excess irrigation will lead to root zone congestion due to removal of air from soil pores, thereby affecting plant establishment and growth. In all, total 20-25 irrigations were provided to the crop. Other agricultural practices were applied as per recommendation of Banana. Plant data basically pseudostem height (cm), pseudostem girth (cm), average number of leaves/plant were recorded at 30 days intervals while first flower emergence (days), first finger ripening (days), no. of finger per hand, no. of hand/ branch, weight of finger/ hand (kg), weight of single finger (g), finger diameter (cm), length of finger (cm), weight of finger/ bunch (kg) percentage of plant harvested at a time and yield was recorded at different stages and at harvest.

Growth and Yield Attributes of TC Banana

The means of growth and yield contributing characters are shown in Table 3. Pseudostem height, pseudostem girth, average no of leaf/ plant, first flower emergence (days), first finger ripening (days), no. of finger per hand, number of hand/ branch, weight of finger/hand (kg), weight of single finger (g), finger diameter (cm), length of finger (cm), weight of finger/ bunch (kg), percentage of plant harvested at a time and mortality (%)/ chilling injury were significantly influenced by different date of planting. Among four date of planting, taller plants, higher stem girth with maximum no of leaves, first flower emergence (days), first finger ripening (days), higher no. of finger per hand, higher no. of hand/ branch, high weight of finger/hand (kg), high weight of single finger (g), more finger diameter (cm), length of finger (cm), weight of finger/ bunch (kg), more percentage of plant harvested at a time and low mortality (%)/ chilling injury was observed under Banana planted at 10th August as compared to 10th September, 10th October and 10th November. This was might be due to good environmental condition and favorable temperature during the said month for good plant growth. The sapling of banana planted at 10th October and 10th November having insufficient time to pick up the growth and development due to low temperature in December–January (temperature goes down below 5 °C) and check the growth and yield attributing characters ultimately plants suffered from chilling injury.

The climatic characteristic features in the subtropics are wide temperature fluctuations between day and night, low/high temperature extremes in winter and summer and low and poorly distributed rainfall respectively. In cold subtropical climates, there are typical physiological problems such as choke throat, November dump, winter leaf sunburn, under-peel discoloration, and growth cessation (Robinson 1996). In Egypt, banana under plastic covered greenhouses failed to give positive results due to hot and dry summer (average maximum temperature about 39°C during July and August) which converted the greenhouse air to excessive hot condition especially at the leaf level. The high temperatures causes marginal leaf desiccation, leaf deformations beside bunch and finger malformations. On the other hand, dust accumulations on the plastic cover during the summer reduce light transmission and limit photosynthetic activity.

The main purpose of this investigation were to protect the banana against winter low temperature and wind damage which in turn will usually give higher yields and better finger quality. This process improves microclimatic conditions in terms of radiation, air temperature, relative humidity and effectively reduces water consumption Rafaie et al. (2012). Gubbuk, (2008) reported that, dwarf Cavendish, a commonly grown banana cultivar in Turkey. Nainwad et al. (2005) reported that variability in yield contributing characters of banana varieties propagated by tissue culture and conventional sucker clearly indicates that all yield contributing characters viz. weight of bunch, number of hands per bunch number of finger per bunch, length of finger, girth of finger and length of bunch, showed significant variation within the plants of tissue culture and conventional suckers. It showed maximum variability among tissue cultured varieties (Grand Nain) in respect of weight of bunch (12.67%), number of hands per bunch (8.46%), number of fingers per bunch (10.46%), length of finger (10.91 %), girth of finger (8.37%), and length of bunch (8.68%).

Grand Nain is very popular cultivar under protected cultivation due to the higher yield and quality presently. Stem circumference, stem height, total leaf number, bunch stalk circumference, days from shooting to harvest, number of hands per bunch, number of fingers per bunch, finger circumference, finger length and bunch weight were measured in the protected cultivation. Cultivars 'Grand Nain' was found to be better than 'Dwarf Cavendish' in terms of total production, expressed as the number of hands and fingers per bunch and bunch weight under protected cultivation. Gubbuk and Pekmezci (2004) study to determine the yield and quality of 'Dwarf Cavendish' banana (Musa spp. AAA), cultivated in open fields and also in protected (plastic greenhouse) cultivation. The site is located in the central south coastal region (altitude 50 m, latitude 36°33'N) of Turkey. In both cultivation systems they determined the following: pseudostem circumference, pseudostem height, total leaf number, bunch stalk circumference, days from shooting to harvest, number of hands, number of fingers, finger circumference, finger length, and bunch weight.

Table 3: Growth and yield attributes of Tissue Culture raised Banana

Treatments (Date of Planting)	Pseudo-stem height (cm)	Pseudo-stem girth (cm)	Average no.of leaves/plant	First flower emergence (days)	First finger ripening (days)	No. of finger per hand	No. of hand/branch	Weight of finger/hand (kg)	Weight of single finger (g)	Finger diameter (cm)	Length of finger (cm)	Weigth of finger/bunch (kg)	Percen-plant of plant harvested at a time	Mortality (%)/Chillin ginjury
10th August	185.50	78.50	34.00	320.00	415.00	17.50	7.00	3.50	220.00	7.00	11	25.60	85.00	2.00
10th September	169.60	70.00	32.00	335.00	435.00	15.00	6.00	3.25	185.00	6.50	10	22.70	75.00	7.00
10th October	150.40	60.39	28.00	350.00	460.00	13.00	4.00	2.60	150.00	4.50	6	16.00	60.00	29.00
10th November	120.66	56.00	25.00	380.00	475.00	10.80	4.00	2.20	100.00	4.50	6	14.50	60.00	32.00
SEm±	4.89	2.90	1.11	5.50	4.50	0.60	0.35	0.29	5.89	0.66	0.83	2.11	3.55	2.33
CD (P=0.05)	15.85	10.54	3.56	20.00	15.40	2.50	1.40	1.10	18.00	2.20	2.78	6.90	12.00	7.80

Growth parameters and yield attributes

Fig. 1: Commercial cultivation of tissue culture raised banana (BHIM) at IIIM.

Yield and Economics of TC Banana

Data pertaining to bunch yield of banana by different date of planting is presented in Table 4. It was observed that there was significant variation in bunch yield. It was quite evident from the data at all the date of transplanting produced significant variation on bunch yield performance. Banana planted on 10th August reported highest yield (64.22 tonnes/ha), highest gross return (Rs. 12,84,400/ha), highest net return (Rs. 958730.5/ha) and B: ratio (3.94) as compared to all other treatments, except yield obtained (56.81 tonnes/ha) planted on 10th September was found to be non significant. This might be due to taller plants height, higher stem girth with more no of leaves causing timely flower emergence (days), early finger set and ripening (days), higher number of finger per hand, higher no. of hand/ branch, high weight of finger/hand (kg), more weight of single finger (g), more finger diameter (cm), higher length of finger (cm), more weight of finger/ bunch (kg), more percentage of plant harvested at a time and low mortality /chilling injury (%) causes highest yield. Banana planted at 10th October and 10th November causes lowest yield 44.46 tonnes/ha and 42.00 tonnes/ha, respectively. This was might be due to the sapling planted in these month could not found suitable temperature for growth and development because in the month of December and January on average minimum temperature of

Jammu division goes down 7°C which is below the base temperature of 10°C. Therefore, sapling suffers chilling injury ultimately check yield attributing characters and bunch yield. Mustaffa & Kumar (2012) reported that planting during winter leads to initial exposure to unfavorable conditions of hot summer and heavy winds during critical stages of growth. In Jalgaon and Burhanpur district of Maharashtra, and Madhya Pradesh during 2011-12. The per hactare yield of TC banana was 60.50 metric tons which was higher than the yield of sucker banana 45-50 metric tons. The gross return obtained from sucker banana were Rs. 4,58,991 which was higher than the gross returns obtained from sucker banana were Rs. 3,01,605 (Bairwa et al. 2015).

Table 4: Yield and economics of tissue culture banana

Treatments (Planting times)	Yield and economics				
	Yield (ton/ha)	Cost of cultivation (Rs/ha)	Gross return (Rs/ha)	Net return (Rs/ha)	B:C ratio
10ᵗʰ August	64.22	3,25,669.50	12,84,400	958730.50	3.94
10ᵗʰ September	56.81	3,26,608.10	11,36,200	809591.90	3.48
10ᵗʰ October	44.46	3,34,067.50	8,89,200	555132.50	2.66
10ᵗʰ November	42.00	3,35,426.00	8,40,000	504574.00	2.50
SEm±	2.50	-	-	-	-
CD (P=0.05)	7.88	-	-	-	-

Tissue culture techniques generates homogenous population of plant endowed with totipotency of elite mother plants, which are not only agro-climatically adopted, but also attributed with vibrant growth, pest resistance and consequently higher productivity (Vasane and the Kothari 2006). Gubbuk and Pekmezci (2004) study to determine the yield and the quality of 'Dwarf Cavendish' banana (Musa spp. AAA), cultivated in open fields and also in protected (plastic greenhouse) cultivation. The site is located at the central south Coastal Region (altitude 50 m, latitude 36°33'N) of Turkey. In both cultivation systems they determined average annual yield under plastic greenhouse was 53% higher (65.5 t/ha) than in the open field (42.8 t/ha). Furthermore, in protected cultivation, chilling injury and low temperature differences do not negatively affect the plants and fingers. The data in respect to gross return (Rs/ha) at different date of transplanting have been presented in Table 4. It is evident from finding that planting of banana at 10ᵗʰ August gave highest gross return (Rs 12,84,400/ha), net return (Rs 09,58,730/ha) and B:C ratio as compared to other treatments. Hanumantharaya et al. 2009 also reported and an study was conducted in Tungabhadra and Malaprabha areas of Northern Karnataka, where larger area was concentrated under both sucker propagated and tissue culture methods of banana cultivation. Study was based on data collected from 80 farmers in 12 villages of two taluks in Tungabhadra and Malaprabha command areas of Karnataka. Results of this investigation revealed that, in cost of sucker banana was Rs. 82,298 and tissue culture banana was Rs. 1,17,563. The gross returns were obtained Rs. 1,60,113.81 and Rs. 1,97,295.94, respectively. The net returns obtained were Rs. 77,815.81 and Rs. 79,732.94, respectively. Shakila and Manivamnan (2002) found that split application of N and K fertilizers (200 and 300 g/plant in seven split) significantly increased bunch weight and finger characters resulting in higher yield in tissue culture banana cv. Robusta. Naresh Babu *et al*. (2004) observed that application of 240g N/ plant in four doses at 2,4,6 and 8 month after planting recorded higher number of finger/bunch, yield in banana. Gogoi (2004) observed that combined application of biofertilizers & half of the recommended dose of inorganic fertilizers increased the yield of banana & soil NPK availability. Sabarad (2004) recommended that inoculation of VAM fungus, *Trichoderma* in combination with 180:108:225 g NPK /plant produced better growth bunch and yield.

CONCLUSION

It is quite evident from the data that date of transplanting of sapling and on the basis of net returns and B:C ratio it is recommended that planting of tissue culture banana at 10th August is best time of planting for getting highest bunch yield and net returns in Shivalik range.

ACKNOWLEDGEMENT

Authors are thankful to the Director, Dr Ram A. Vishwakarma, CSIR-Indian Institute of Integrative Medicine, Jammu for providing the necessary financial support for the research. This represents institutional publications number IIIM/2242/2018.

REFERENCES CITED

Anonymous. 2007. Report of the working group on horticulture, plantation crops and organic farming for the XI five year plan 2007-12, Govt of India, Planning commission.

Anonymous. 2015. Post harvest profile of Banana. Govt. of India, Ministry of Agriculture& Farmer Welfare, DAC, Directorate of marketing and inspection, Nagpur, Maharashtra.

Bairwa KC, Singh A, Jhajaria A, Singh H, Goyam BK, Lata M, Singh N. 2015. Conventional suckers with tissue culture Banana production in central India. A case study. *African Journa of Agricultural Research* 10 (14): 1751-1755.

Beyza B, Temirkaynak M and Mehmet O. 2009. Production of Banana in Turkey, West Mediterranean Agricultural Research Institute, Antalya-Turkey.

Gogoi D. 2004. Combined application of biofertilizer & inorganic fertilizer increased soil NPK availability. *Banana Nutrition-A Review, Agriculture Review* 30(1): 24-31

Gubbuk H, Pekmezci M. 2004. Comparison of open field and protected cultivation of banana (Musa spp. AAA) in the coastal area of Turkey. *New Zealand Journal of Crop and Horticultural Science* 32: 375-378.

Gubbuk H. 2008. Comparison of 'dwarf cavendish' and 'grand nain' banana cultivars under protected cultivation. *Bulletin UASVM, Horticulture* 65 (1): 513.

Hanumantharaya MR, Kerutagi MG, Patil BI, Kanamadi VC, Bankar B. 2009. Comparative economic analysis of TC banana and sucker propagated banana production in Karnataka, Karnataka. *Journal of Agricultural Sciences* 22(4) : 810-815.

Karuna Y, Kameswara R. 2013. Studies on phenological characters of different banana cultivars (Musa) in Visakhapatnam, Andhra Pradesh. *International Journal of Science and Research* 1689-1693.

Mustaffa MM, Kumar V. 2012. Banana production and productivity enhancement through spatial, water and nutrient management. *Journal of Horticultural Science* 7(1):1-28.

Nainwad RV, Kulkarni RM, Kulkarni NH, Kalalbandi BM. 2005. Extent of variation in growth and yield attributes of some tissue culture vs conventional sucker planted banana varieties *Karnataka Journal of Agricultiural Sciences* 18(1): 221-222.

Naresh B, Sharma A, Singh S. 2004. Effect of different nitrogen doses and their split application on growth, yield & quality of Jahajee banana. *South Indian Horticulture* 52: 35-40.

Refaie KM, Esmail AAM, Medany. 2012. The response of Banana production and finger quality to shading nets. *Journal of Applied Science Research* 8 (12): 5758-5764.

Sabarad AI 2004. Effect of Trichoderma on growth & yield of banana. *Karnataka Journal of Horticulture* 36:1-4.

Shakila A and Manimnan K 2002. Response of tissue cultured banana Robusta to split application of fertilizer. *Global Conference on Banana & Plantain.* 28-31 Oct, (Abstract P-129)

Singh HP, Chundawat BS. 2002. Improved Technology of banana. Ministry of Agriculture, Government of India. pp 1-46.

Vasane SR, Kothari RM. 2006. Optimization of secondry hardening process of Banana plantlets (*Musa paradisiaca* L. var. *grand nain*). *Indian Journal of Biotechnology* 5: 394-399.

9

Advent of Isabgol (*Plantago ovata* Forsk.) Husk as a Cost Effective and Promising Gelling Agent in Plant Tissue Culture Experiment

Sougata Sarkar*, Sana Khan², Janhvi Pandey³ and RK Lal⁴

¹Genetic Resources and Agrotechnology Division, CSIR-Indian Institute of Integrative Medicine, Canal Road, Jammu – 180001, Jammu and Kashmir, INDIA ²Plant Biotechnology Department, ³Soil Science Department, ⁴Plant Breeding Department, CSIR-Central Institute of Medicinal and Aromatic Plants (CIMAP) Near Kukrail Picnic Spot Road, Lucknow-226015, Uttar Pradesh, INDIA Email: *recalling.de.parted@gmail.com

---

ABSTRACT

Due to the exorbitant price of the popular gelling agent agar, Plant Tissue Culture (PTC) has gradually become a costly technique in recent times. So, alternative sources of agar needs to be popularized which at one hand should be able to meet all the positive properties of agar and on the other hand be cost effective as well. The seed husk obtained from *Plantago ovata* Forsk. is such an alternative which was successfully used as a gelling agent for media preparation in PTC experiments for *in-vitro* shoot and root development in *Nicotiana tabacum* L. and *Ocimum gratissimum* L. The nodal explants were grown in MS basal medium (MSo) supplemented with 3% sucrose and were gelled with progressive concentrations of isabgol husk to ascertain the optimum concentration for better survival and growth response, while 0.8% agar acted as a negative control. The response of the explants on media gelled with husk was compared to that of media gelled with 0.8% agar. It was inferred that 0.3% husk in the medium was optimum for best growth response which was comparable to that of agar. Consequently, the estimated cost of a liter of MSo reduced by almost Rs.61.15/- (compared to Himedia's Agar) to Rs.177.57/- (compared to Sigma's Agar) after using isabgol seed husk as a gelling agent in media preparation.

Keywords: Agar, Gelling Agent, Media, *Plantago ovata*, Plant Tissue Culture.

INTRODUCTION

Commercially, agar is extracted from species of red algae genera *Gelidium*, *Gracillaria* and *Pterocladia* (McLachlan 1985). Although it is inert in nature, some investigations have, raised doubts about its biological inertness and non-toxic nature (Singha 1980, Debergh 1983, Kohlenbach and Wernicke 1983, Arnold and Ericksson 1989). The wide spread use of agar in plant tissue culture (PTC) technique is resulting in overexploitation of its sources. Because of these reasons and the exorbitant price of PTC grade agar Rs.3925/500g (PCT0901-500G-HIMEDIA 2014-15) and Rs.22403.09/1kg (A9799-1kg-SIGMA 2014-15), attempts are being made to identify suitable alternative gelling agents. Sorvari (1986ab,c) reported the use of starch from barley, corn, wheat, potato and rice as an alternative gelling agents. Starch-gelled media weakly solidified and polyester nets were used to prevent the sinking of tissues into the medium. Henderson and Kinnersley (1988) compared the responses of tobacco and carrot cell cultures on media gelled either with corn starch or agar singly or with a combination of these two.

From the above investigations, it is evident that the problems which come in the way of the universal acceptance of substitutes of agar as an alternative gelling agent are their (a) inferior gelling quality, (b) lower clarity than agar, and (c) metabolizable nature which leads to softening of the media during the culture period. Moreover, autoclaving of starch leads to the release of sugar which in turn has their own effect (osmotic or metabolic) on the overall growth response of cultures. As all the substitutes of agar and their hydrolytic products may not be biologically inert, they are expected to have limited use and that too only for explants whose response is not adversely affected by the presence of starch in the medium. In the recent past, barley, corn, potato, rice, wheat and tapioca have been tried and tested to be used as gelling agents with mixed responses of success (Sorvari 1986a,b,c; Henderson and Kinnersley 1988, Tiwari and Rahimbaev 1992, Zimmerman et al. 1995, Nene et al. 1996). Nevertheless, the favorable properties of isabgol seed husk like (a) polysaccharide and colloidal nature, (b) gelling ability even in cold water, (c) comparable inertness like agar, (d) non-metabolizable nature during the culture period and (e) practical clarity of the media in its gelled form are surely positive indications in the direction of its potential to become a universal gelling agent in PTC media preparation techniques. Compared to opaque agar, husk jelled medium offered no serious problems with respect to photography or observations of roots penetrating in the medium. Thus, it can be used effectively as a substitute to agar in PTC experiments.

The husk derived from seeds of *Plantago ovata* Forsk. (Family -
Plantaginaceae) is used as emollient, demulcent, laxative and in the treatment
of dysentery and diarrhoea (Chopra et al. 1958). Dried seeds of the plant contain
over 30% mucilage. The husk, which contains all of the mucilaginous matter, is
separated from the seeds by crushing and winnowing manually. Crushing is
done with flat-stone grinding mills. After thorough cleaning, the seeds are allowed
to pass through these mills six to seven times for total husk removal. The crushed
material consisting of husk and kernel is sieved to remove the kernels, after
which it is passed through screens of different mesh size to separate it out into
products of different fineness (Anonymous 1969). The efficacy of isabgol is
entirely due to the mucilage forming tendency of the husk. This action has been
found to be purely mechanical as mucilage, which swells into a jelly-like mass,
stimulates intestine peristalsis and remains practically unaffected by the digestive
enzymes and bacteria (Chopra et al. 1958).

Mucilage of isabgol is colloidal and polysaccharidic in nature. It is mainly
composed of xylose, arabinose, galactouronic acid, rhamnose and galactose.
Two polysaccharide fractions are separable from its mucilage. One fraction
(eq. wt. 700; uronic acid 20%) is soluble in cold water and upon hydrolysis
yields D-xylose (46%), aldobiouronic acid (40%), L-arabinose (7%) and an
insoluble residue (2%). The other fraction (eq. wt. 4000; uronic acid 3%) is
soluble in hot water, forming a highly viscous solution which sets to a gel when
cooled and yields upon hydrolysis D-xylose (80%), L-arabinose (14%),
aldobiouronic acid (0.3%) and traces of D-galactose (Laidlaw and Percival
1949, 1950). The unique gel forming ability of this easily available, very cost
effective Rs.55/100g (Dabur, dated 12/2014) and purely herbal plant material
needs to be exploited for media solidification in PTC experiments as an alternative
to conventional agar. In this regard, the inertness of agar is an advantage over
husk when nutrient availability by the plantlets from the medium is concerned.

To diagnose whether husk can stand a chance with agar in the context of
inertness, ICP-MS was performed in three replicates for both husk and agar to
realize the actual inertness in them. Thereafter, the following experiment was
designed to ascertain the usefulness of isabgol husk over agar as a PTC gelling
agent in view of metabolism and consistency in the medium during the course
of culture period, thereby perceiving a suitable alternative to conventional gelling
agent like agar.

TECHNIQUES

Nitrogen Analysis

P. ovata seed husk was washed with 0.1 N HCl, deionised water and then with distilled water. They were then dried in oven at 70°C and finely grinded. 0.25 g of the sample was acid (1.2: 1 H_2SO_4 / H_2O_2 mixture) digested at 250°C (Jones and Case, 1990) in three replicates. Total nitrogen was then estimated by Flow inject analyser. 0.25 g PTC grade agar was acid (1.2: 1 H_2SO_4 / H_2O_2 mixture) digested at 250°C (Jones and Case, 1990) in three replicates. Total nitrogen was then estimated by Flow Inject Analyser.

Potassium Analysis

P. ovata seed husk was washed with 0.1 N HCl, deionised water and then with distilled water. They were then oven dried at 70°C and finely grinded. 0.25 gm of the sample was acid (HNO_3: $HClO_4$ = 10:4) digested (AOAC, 1990) in three replicates. Potassium was then measured by flame photometer.0.25 g PTC grade agar was acid (HNO_3: $HClO_4$ = 10:4) digested (AOAC, 1990) in three replicates. Potassium was then measured by Flame Photometer.

Phosphorous Analysis

P. ovata seed husk was washed with 0.1 N HCl, deionised water and then with distilled water. They were then oven dried at 70°C and finely grinded. 0.25 gm of the sample was acid (1.2:1 H_2SO_4 / H_2O_2 mixture) digested at 250°C (Jones and Case, 1990) in three replicates. Phosphorous was then analysed by Flow Inject Analyser. 0.25 g PTC grade agar was acid (1.2:1 H_2SO_4 / H_2O_2 mixture) digested at 250°C (Jones and Case, 1990) in three replicates. Phosphorous was then analysed by Flow Inject Analyser.

Heavy Metals and Micronutrient Analysis

P. ovata seed husk was washed with 0.1 N HCl, deionised water and then with distilled water. They were then oven dried at 70°C and finely grinded. 0.25 gm of the sample was acid (HNO_3: $HClO_4$ = 10:4) digested (AOAC 1990) in three replicates. Heavy metal and micronutrients analysis was done via ICP-OES, model 53000V. 0.25 g PTC grade agar was acid (HNO_3: $HClO_4$ = 10:4) digested at (AOAC 1990) in three replicates. Heavy metal and micronutrients analysis was done via ICP-OES, model 53000V.

Establishment of Sterile Culture

The nodal explants of *N. tabacum* and *O. gratissimum* were collected from CSIR-CIMAP experimental field. These nodal explants were initially rinsed under running tap water for 10-15 min. Further these explants were washed with 70% ethanol for 1 minute, followed by surface sterilization using 0.1% of

HgCl$_2$ for 2 min for *N. tabacum* and 1 min for *O. gratissimum* respectively. The treated explants were finally rinsed with double distilled water for four to five times and placed on MS$_0$ medium (Murashige and Skoog, 1962). The nodal explants were allowed to grow into full plantlets.

Experimentation Design: Isabgol seed husk and PTC Agar

The *in-vitro* established plants were used as donor plants to check the potential of different gradients of husk to standardized concentration of agar in PTC studies. The nodal explants from the already established *in-vitro* cultures were inoculated on MS$_0$ supplemented with 3% (w/v) sucrose as a common component which were gelled with 0.75, 1.8, 2.2, 2.4, 2.6, 2.8, 3, 3.2, 3.4, 3.6, 4, 4.2, 4.4, 4.6, 4.8, 5, 5.2, 5.4 and 5.6% isabgol seed husk (w/v) and 0.8% (w/v) agar respectively. The pH of the medium was adjusted to 5.8±0.2 prior to autoclave at 121 psi for 15 minutes. The tissue culture medium with different gradations of isabgol seed husk was treated as experiments and 0.8% agar acted as a negative control. The experiments were replicated thrice.

Shoot Induction and Proliferation

The potency/efficiency of isabgol seed husk and agar was examined on the basis of induction, proliferation and growth of the nodal segment along with the number of days taken for shoot initiation. The data for shoot initiation, proliferation rate (percent of explants showing shoot proliferation) and number of shoots/ explant was recorded.

Growth Condition and Maintenance

All *in-vitro* established plantlets/cultures were maintained under cool fluorescent light (3,000 lux) with 16L: 8D-h photoperiods at 25 ± 2 ÚC. The sterile explants were allowed to grow and proliferate in their own accord as no PGRs were used in the medium to enhance their regeneration potency.

ICP RESULTS

The feasibility of using isabgol seed husk as an alternative to PTC grade agar or using agar all alone like always, largely depends on the inertness of the two candidates in use. ICP results gave us a clear picture on this aspect.

Presence of all the macro nutrients in detectable levels were found in both the competing gelling agents (agar and husk) as evident from the ICP results (Table 1, Figure 1). Agar contained more nitrogen than husk while husk contained much more potassium than agar. Whereas, agar and husk had almost equal amounts of phosphorus in them. Presence of ten micronutrients was detected in both the competing gelling agents (agar and husk) as evident from the ICP results (Table 1, Figure 1). Agar contained much more aluminium, iron, magnesium and sodium than husk while husk contained more calcium than agar.

Table 1: Estimation of macro, micro and heavy metal elements in PTC agar and isabgol husk

Components		Elements	PTC agar (%)	Isabgol husk (%)
Macro elements	1.	Nitrogen	0.157±0.009	0.124±0.004
	2.	Potassium	0.001±0.001	0.67±0.001
	3.	Phosphorus	0.015±0.009	0.028±0.004
Micro elements	1.	Aluminium	210±23.98	35.867±1.81
	2.	Barium	0.6±0.23	3.13±0.35
	3.	Calcium	622.67±8.35	690.66±25.78
	4.	Cobalt	-0.133±0.52	-0.467±0.27
	5.	Copper	0.267±0.07	0.4±0.004
	6.	Iron	143.53±11.39	43.2±4.89
	7.	Magnesium	592.73±1.33	37.67±3.27
	8.	Manganese	19±0.12	4.53±0.13
	9.	Sodium	952.13±53.62	378.13±26.62
	10.	Zinc	4.67±1.94	0.667±0.41
Heavy metal elements	1.	Cadmium	0.533±0.07	0.267±0.07
	2.	Chromium	8.267±1.29	2.733±0.07
	3.	Nickel	3.33±0.58	1.20±0.20
	4.	Lead	1±1.80	1.47±1.14

From the overall study, regarding the inertness of the two competing gelling agents and taking the results of ICP into account, it is evident that isabgol husk is far more a suitable candidate than agar when inertness is concerned. It is because of the fact that the former considerably surpasses the later only in three occasions (in nutrient composition of potassium, lead and calcium) while the later surpasses the former considerably in nine occasions (in nutrient composition of nitrogen, cadmium, chromium, nickel, aluminium, calcium, iron, manganese and sodium). Therefore, isabgol husk fairs better than PTC agar in this regard.

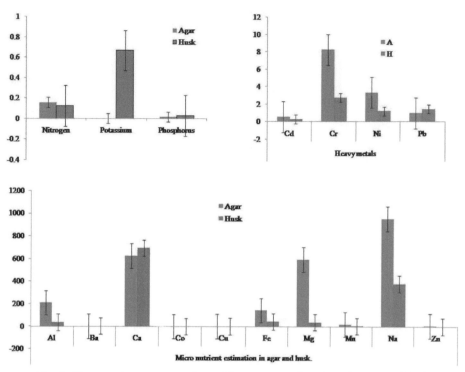

Fig. 1: Comparative representation of the presence of nutrients in agar and husk

PTC RESULTS

The suitability of isabgol seed husk as a gelling agent was investigated in relation to morphogenic processes (for shoot and root development) for which semi-solid media were commonly employed.

Selection of an optimum concentration of husk for proper gelling of the medium

Response to axillary shoot proliferation in the cultures with varying gradients of seed husk as a gelling agent in the medium was evaluated against conventional agar which was used as a negative control. It is clear from the Table 2 and Table 3 that isabgol seed husk exhibited different response in relation to shoot and root development when grown in different concentrations of husk. It was found that at lower concentrations (0.75, 1.8, 2.2, 2.4 g/l) of isabgol husk, the medium was too soft for the nodal explants to stand erectly. This condition improved gradually with increase in the concentration of husk (2.6, 2.8, 3.0, 3.2, 3.4, 3.6, 4.0, 4.2, 4.4, 4.6, 4.8 g/l) but after a particular concentration, the medium became increasingly dense (5.0, 5.2, 5.4 and 5.6 g/l) and so it became difficult to insert explants in the medium. Thus, a considerably broad range of gelling capacity of husk in the medium could be selected as 2.6, 2.8, 3, 3.2, 3.4, 3.6, 4.0,

4.2, 4.4, 4.6 g/l, which in turn may be narrowed down to an effective gelling capacity having a range of 2.8, 3, 3.2, 3.4 g/l. Out of this effective range, 3 g/l concentration of the husk was found to be most suitable in a majority of experiments. About three weeks old nodal explants were found to be proliferating and achieving proper growth and development in the control medium (MS_0 + 30g Sucrose + 8g agar).

Simultaneously, the experimental explants containing isabgol seed husk as a gelling agent in place of agar in MS_0 was also observed for growth and development. The observations recorded during the third week of culture are presented in the Table 2. Lower concentration of husk (0.75, 1.8, 2.2, 2.4 g/l) exhibited improper solidification of the medium due to improper gelling, primary root development was poor, primary axis did not give rise to secondary proliferations and overall growth was very slow. Higher concentration of husk (5.0, 5.2, 5.4 and 5.6 g/l) exhibited dense gelling of the medium due to overdose of husk, slow root and shoot development, increase in the internodal distance, slow and less axillary branch development, leaf area increase although leaves become pale and distribution of roots is prevalent all along the media.

Considerable range (2.6, 2.8, 3.0, 3.2, 3.4, 3.6, 4.0, 4.2, 4.4, 4.6, 4.8 g/l) exhibited better growth response as gelling ability of the husk became proper gradually, initiation of primary root development began, leaves appeared although yellowish-green in colour, branches also appeared and growth rate was satisfactory. Effective range (2.8, 3, 3.2, 3.4 g/l) exhibited the best growth response as gelling of the medium became better, primary and secondary root development became proper, growth of the nodal explants became proper, leaves turned green, axillary shoots development became proper along with proper branching and overall growth was proper. The best growth response of explants was observed in the medium containing 3 g/l (i.e. 0.3%) husk. The explants demonstrated best primary and secondary root development, best growth of the nodal explants, green leaves, best axillary shoots development along with proper branching and overall growth was better than any composition of husk in this effective range.

Effect in the growth response of explants with the variation in husk concentration in medium

The different concentration of husks exhibited different effects in growth, proliferation and shoot induction in the explants (Table 2). The density of the medium seems to play a key role in the uptake of nutrients from the medium by the explants, as time taken for root initiation and development is affected greatly in this regard. Therefore, some explants could recover from atrophy after some time (perhaps after proper rhizogenesis) and some could not at all. In the later

Table 2: Response of *Nicotiana tabacum* explants to agar and different composition of husk in MS_0.

Composition for 1 liter media	Days to shoot initiation	Number of shoots/explants	Feature of medium	Features of explants
MS_0 + 30 g Sucrose + 8 g/l agar = Control	6 ± 0.72	4.6 ± 0.03	Negative control	Proper gelling of the medium Primary and secondary rhizogenesis proper Axillary shoot bud development proper Branching proper Leaves green
MS_0 + 3% Sucrose + 0.75 (g/l) husk	12 ± 0.65	2.0 ± 2.33	Lower concentration of husk/liter media	Proper growth of the explant Improper solidification of the medium due to improper gelling. Primary root development is poor Primary axis did not give rise to secondary proliferations. Very slow growth
MS_0 + 3% Sucrose + 1.80 (g/l) husk	12 ± 0.45	2.0 ± 1.20		Gelling properInitiation of primary root development. Leaves yellowish-green in colour. Branching proper. Slow growth rate.
MS_0 + 3% Sucrose + 2.20 (g/l) husk	10 ± 0.32	2.0 ± 1.02		
MS_0 + 3% Sucrose + 2.40 (g/l) husk	15 ± 1.46	2.5 ± 0.3		
MS_0 + 3% Sucrose + 2.60 (g/l) husk	7 ± 0.34	3.0 ± 0.32		
MS_0 + 3% Sucrose + 2.80 (g/l) husk	7 ± 0.88	2.5 ± 0.98		Gelling better Primary roots gives rise to secondary roots Growth of the nodal explants proper

Contd.

Composition for 1 liter media	Days to shoot initiation	Number of shoots/explants	Feature of medium	Features of explants
MS_0 + 3% Sucrose + 3.00 (g/l) husk	5 ± 0.09	4.5 ± 0.15		Leaves greenish Axillary shoots development proper but slow. 1. Gelling better. 2. Primary and secondary root development proper. 3. Growth of the nodal explants proper. 4. Leaves green. 5. Axillary shoots development proper. 6. Proper branching.
MS_0 + 3% Sucrose + 3.20 (g/l) husk	5 ± 0.22	4.5 ± 0.03		1. Gelling proper. 2. Growth from the nodal explants proper 3. Leaves green 4. Maximum axillary shoots buds development
MS_0 + 3% Sucrose + 3.40 (g/l) husk	6 ± 0.71	2.5 ± 0.65		1. Proper primary and secondary root development 2. Growth of the nodal explants proper 3. Leaves green 4. Axillary shoots development
MS_0 + 3% Sucrose + 3.60 (g/l) husk	5 ± 0.33	1.5 ± 0.5		1. Primary and secondary root development proper. 2. Growth of the nodal explants proper 3. Leaves green

Contd.

Composition for 1 liter media	Days to shoot initiation	Number of shoots/explants	Feature of medium	Features of explants
MS_0 + 3% Sucrose + 4.00 (g/l) husk	8 ± 1.28	2.5 ± 1.2		4. Axillary shoot development proper 1. Primary and secondary root development proper 2. Growth of the nodal explants proper 3. Leaves green 4. Axillary shoots development 5. Overall growth of the plant is inferior to 3.6% husk
MS_0 + 3% Sucrose + 4.20 (g/l) husk	7 ± 0.42	2.5 ± 0.11	Higher concentration of husk/liter media	1. Dense gel due to overdose of husk 2. Slow root and shoot development 3. Internodal distance increases 4. Leaf laminar area increases 5. Well distribution of roots is prevalent all along the media.
MS_0 + 3% Sucrose + 4.40 (g/l) husk	10 ± 0.14	2.0 ± 0.55		
MS_0 + 3% Sucrose + 4.60 (g/l) husk	10 ± 0.10	1.5 ± 0.23		
MS_0 + 3% Sucrose + 4.80 (g/l) husk	10 ± 0.11	1.0 ± 0.15		
MS_0 + 3% Sucrose + 5.00 (g/l) husk	15 ± 0.32	1.5 ± 0.09		
MS_0 + 3% Sucrose + 5.20 (g/l) husk	15 ± 0.45	1.5 ± 0.07		1. Highly dense medium. 2. Root development is dense and all along the media. 3. Slow shoot development 4. Overall growth is slow 5. Leaves pale 6. Slow and less axillary branches development
MS_0 + 3% Sucrose + 5.40 (g/l) husk	16 ± 0.33	2 ± 0.25		
MS_0 + 3% Sucrose + 5.60 (g/l) husk	16 ± 0.37	2 ± 0.25		

Note: (i) Considerable range = MSo+3% Sucrose + [2.60(g/l) husk 4.80 (g/l) husk]
(ii) Effective range = MSo + 3% Socrose ± [2.80 (g/l husk to 3.40 (g/l) husk]

case, higher density of the medium may have contained the process of root induction, leading to the death of the explant due to atrophy. In other cases, where the medium was a little dense than optimum, root initiation, its growth and development occurred. This helped the explants to receive proper nourishment on one hand, moreover foliar activity (photosynthesis) also added up to their growth and development on the other hand.

Selection of an optimum concentration of husk in media for growth of *Nicotiana tabacum* explants

Maximum number of shoot initiation per explant (4.5 ± 0.15) was achieved in MS medium containing 3.0 g/l husk as a gelling agent in minimum number of days (5 ± 0.09). This result was comparable with the negative control that produced 4.6 ± 0.03 shoot initiations per explant in 6 ± 0.72 days. The minimum number of shoot initiations per explant (1.5 ± 0.07) was produced in the medium containing 5.20 g/l husk in 15 ± 0.45 days. Medium gelled with more husk (> 4.6 g/l), has led to the gradual decline in rate of shoot proliferation in explants. That is why even after 16 days of culture, merely 2 shoot buds could be observed to proliferate Table 2. The growth responses exhibited by negative control, lower, higher and considerable/effective concentration of husk in MS_o medium is represented by Figures 2 and 3.

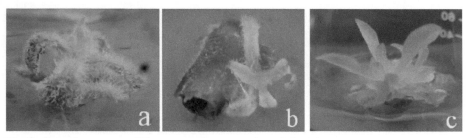

Fig. 2: Growth response of *Nicotiana tabacum* explants (a,b, c) in lower concentration of husk in MSo media.

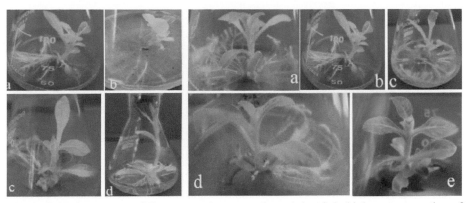

Fig. 3: Growth response of *Nicotiana tabacum* explants (a,b,c,d) in higher concentration of husk in MS0 media (Left side) and response (a,b,c,d,e) in considerable and effective concentration of husk in MS0 media (Right side).

Selection of an optimum concentration of husk in media for proper growth of *Ocimum gratissimum* explants

Maximum number of shoot initiation per explant (5.5±0.25) was achieved in MS medium containing 3.0 g/l husk as a gelling agent in minimum number of days (8 ± 0.19). This result was comparable with the negative control that produced 5.6 ± 0.93 shoot initiations per explant in 7 ± 0.88 days. The minimum number of shoot initiations per explant (1.0 ± 2.10) was produced in the medium containing 1.80 g/l husk in 13 ± 0.45 days. Medium gelled with more husk (>5.0 g/l), has led to the gradual decline in rate of shoot proliferation in explants. That is why even after 13 days of culture, merely 2 shoot buds could be observed to proliferate Table 3. The growth responses exhibited by negative control, lower, higher and considerable/effective concentration of husk in MS_0 medium is represented by Figures 4 and 5.

Fig. 4: Growth response of *Ocimum gratissimum* explants (a, b and c) in lower concentration of husk in MSo media.

Table 3: Response of *Ocimum gratissimum* explants to agar and different composition of husk in MSo

Composition for 1 liter media	Days to shoot initiation	Number of shoots/ explants	Feature of medium	Features of explants
MSo + 30 g Sucrose + 8 g/l agar = Control	7 ± 0.88	5.6 ± 0.93	Negative control	Proper gelling of the medium Primary and secondary rhizogenesis proper Axillary shoot bud development proper Branching proper Leaves green Proper growth of the explant
MSo + 3% Sucrose + 0.75 (g/l) husk	14 ± 0.56	1.0 ± 3.33	Lower concentration of husk/liter media	Improper solidification of the medium due to improper gelling. Primary root development is poor Primary axis did not give rise to secondary proliferations. Very slow growth
MSo + 3% Sucrose + 1.80 (g/l) husk	13 ± 0.45	1.0 ± 2.10		
MSo + 3% Sucrose + 2.20 (g/l) husk	13 ± 0.42	2.0 ± 1.22		
MSo + 3% Sucrose + 2.40 (g/l) husk	11 ± 2.46	3.5 ± 4.3		
MSo + 3% Sucrose + 2.60 (g/l) husk	10± 0.44	4.0 ± 0.32		Gelling proper Initiation of primary root development. Leaves yello wish-green in colour. Branching proper. Slow growth rate.
MSo + 3% Sucrose + 2.80 (g/l) husk	9 ± 0.92	4.5 ± 0.98		Gelling better Primary roots gives rise to secondary roots Growth of the nodal explants proper

Contd.

Composition for 1 liter media	Days to shoot initiation	Number of shoots/explants	Feature of medium	Features of explants
MSo + 3% Sucrose + 3.00 (g/l) husk	8 ± 0.19	5.5 ± 0.25		Leaves greenish Axillary shoots development proper but slow. Gelling better. Primary and secondary root development proper. Growth of the nodal explants proper.
MSo + 3% Sucrose + 3.20 (g/l) husk	8 ± 0.22	5.5 ± 0.03		Leaves green. Axillary shoots development proper. Proper branching. Gelling proper. Growth from the nodal explants proper
MSo + 3% Sucrose + 3.40 (g/l) husk	8± 0.71	5.0 ± 0.65		Leaves green Maximum axillary shoots buds development Proper primary and secondary root development Growth of the nodal explants proper
MSo + 3% Sucrose + 3.60 (g/l) husk	8± 0.35	4.8 ± 0.8		Leaves green Axillary shoots development Primary and secondary root development proper. Growth of the nodal explants proper

Contd.

Composition for 1 liter media	Days to shoot initiation	Number of shoots/ explants	Feature of medium	Features of explants
MSo + 3% Sucrose + 4.00 (g/l) husk	9 ± 0.28	4.0 ± 1.2		Leaves green Axillary shoot development proper Primary and secondary root development proper Growth of the nodal explants properLeaves greenAxillary shoots development Overall growth of the plant is inferior to 3.6% husk
MSo + 3% Sucrose + 4.20 (g/l) husk	9 ± 0.42	4.0 ± 0.21		
MSo + 3% Sucrose + 4.40 (g/l) husk	10 ± 0.14	3.0 ± 0.55		
MSo + 3% Sucrose + 4.60 (g/l) husk	10 ± 0.81	3.5 ± 0.33		
MSo + 3% Sucrose + 4.80 (g/l) husk	10 ± 0.91	3.2 ± 0.17		
MSo + 3% Sucrose + 5.00 (g/l) husk	11 ± 0.32	3.5 ±0.90	Higher concentration ofhusk/liter media	Dense gel due to overdose of husk Slow root and shoot development Internodal distance increases Leaf laminar area increases Well distribution of roots is prevalent all along the media.
MSo + 3% Sucrose + 5.20 (g/l) husk	12 ± 0.45	3.0 ± 0.09		Highly dense medium. Root development is dense and all along the media. Slow shoot development Overall growth is slow
MSo + 3% Sucrose + 5.40 (g/l) husk	12 ± 0.88	2 ± 0.85		Leaves pale Slow and less axillary branches development
MSo + 3% Sucrose + 5.60 (g/l) husk	13 ± 0.92	2 ± 0.88		

Where (i) Considerable range = MSo + 3% Sucorse + [2.60 (g/l) husk + 4.80 (g/l) husk + 4.80(g/l husk]
(ii) Effective range = MSo + 3% Sucrose + [8.80 (g/l) husk + 3.40 (g/l) husk

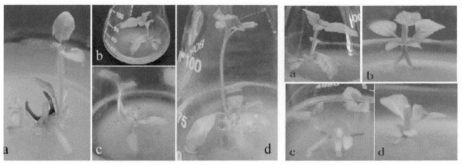

Fig. 5: Growth response of *Ocimum gratissimum* explants (a,b,c,d) in higher concentration of husk in MS0 media (Left) and response (a,b,c,d) in considerable and effective concentration of husk in MS0 media.

CONCLUSION

In all the experiments, there was no softening of the husk-gelled medium during the entire course of culture, indicating that it is not metabolized during culture. It performed at par with the agar-gelled media in all experiments statistically as evident from the results of Table 1, 2 and 3. From the point of view of adaptability as a PTC gelling agent, 3.0 g/l of isabgol husk has shown maximum number of shoot initiations in minimum time in both the experimental explants of *N. tabacum* and *O. gratissimum* which is almost equivalent to the negative control (when used 8 gm/l medium) in all aspects. Therefore, it is apparent from the result that using an optimum concentration (3.0 g/l) of isabgol husk as an alternative to PTC grade agar is in no way unequivalent to the later. From the point of view of economical benefit, the amount of agar used for 1 liter of PTC media is 8g (0.8%). In this relation, the cost of agar/liter of medium is estimated to be Rs. 62.80 and Rs. 179.22 for HIMEDIA and SIGMA respectively whereas, it is Rs 1.65 only (0.3% per liter medium) if we use husk instead of agar.

This wide difference in the cost price of PTC grade agar and isabgol seed husk, can bring a drastic reduction in the overall cost of PTC technique if husk be used as an alternative to agar. It has been calculated that the cost of preparation of 1 liter of MS0 may be reduced to Rs. 61.15/- (with respect to the agar of Himedia) and Rs. 177.57/- (with respect to the agar of Sigma) by using isabgol seed husk as a gelling agent as an alternative to the conventionally used agar Figure 6. In fact, from the point of view of cost effectiveness, husk is far more a suitable candidate than agar –both in the quantity and cost.

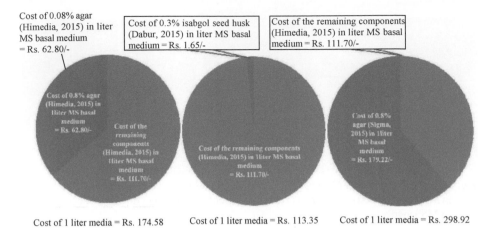

Cost of 0.08% agar (Himedia, 2015) in liter MS basal medium = Rs. 62.80/-

Cost of 0.3% isabgol seed husk (Dabur, 2015) in liter MS basal medium = Rs. 1.65/-

Cost of the remaining components (Himedia, 2015) in liter MS basal medium = Rs. 111.70/-

Cost of 1 liter media = Rs. 174.58   Cost of 1 liter media = Rs. 113.35   Cost of 1 liter media = Rs. 298.92

Fig. 6: Comparison of the cost of 1 liter MS media using agar and husk as a gelling agent

Abbreviations

PTC : Plant Tissue Culture

ICP-MS : Inductively Coupled Plasma– Mass Spectrometry

PGR : Plant Growth Regulator

ACKNOWLEDGEMENT

The first author acknowledges the DST, Government of India for awarding the INSPIRE Fellowship for Ph.D. program at CSIR-CIMAP, Lucknow, India and is also thankful to Dr. Laiq Ur Rahman, Scientist, Plant Biotechnology Department, CSIR-CIMAP, Lucknow (U.P.) for providing necessary facility and support in PTC studies. The authors are also thankful to the Director, CSIR-CIMAP for providing the required facilities and encouragement.

REFERENCES CITED

Anonymous. 1969. The wealth of India, Vol 8. Council of Scientific and Industrial Research, New Delhi, India.

AOAC 1990. AOAC official methods of analysis. 15th ed. Association of Official Analytical Chemists, Arlington, Virginia, pp. 84–85.

Arnold SV, Ericksson T. 1984. Effect of agar concentration on growth and anatomy of adventitious shoots of Picea abies (L.) Karst. *Plant Cell Tissue Organ Culture* 3:257–264.

Chopra RN, Chopra IC, Handa KL, Kapur LD. 1958. Chopra's indigenous drugs of India. UN Dhur & Sons Pvt, Calcutta, India.

Debergh PC. 1983. Effect of agar brand and concentration on the tissue culture media. *Physiologia Plantarum* 59:270–276.

Henderson WE, Kinnersley AM. 1988. Corn starch as an alternative gelling agent for plant tissue culture. *Plant Cell Tissue Organ Culture* 15:17–22.

Jones JB, Case VW. 1990. Sampling, handling and analyzing plant tissue samples. In: Westerman, R.L. (Ed.), Soil Testing and Plant Analysis. third ed., Soil Science Society of America, Book Series No. 3, Madison, Wisconsin, 389–427.

Kohlenbach HW, Wernicke W. 1978. Investigations on the inhibitory effect of agar and the function of active carbon in anther culture. *Z Pflanzenphysiol* 86:463–472.

Laidlaw RA, Percival EGV. 1949. Studies on seed mucilages. Part III. Examination of polysaccharide extracted from the seeds of *Plantago ovata* Forsk. *J Chem Soc* 1600–1607.

Laidlaw RA, Percival EGV. 1950. Studies of seed mucilages. Part V. Examination of polysaccharide extracted from the seeds of *Plantago ovata* Forsk. by hot water. *J Chem Soc* 528–537.

McLachlan J. 1985. Macroalgae (sea weeds): industrial sources and their utilizations. *Plant Soil* 89:137–157.

Murashige T, Skoog F. 1962. A revised medium for rapid growth and bioassays with tobacco tissue cultures. *Physiol Plant* 15: 473–497.

Nene YL, Shiela VK, Moss JP. 1996. Tapioca – a potential substitute for agar in tissue culture media. *Curr Sci* 70:493–494.

Singha S. 1980 Influence of two commercial agars on in vitro shoot proliferation of 'Almey' crab apple and 'Seckel' pear. *Hortic Sci* 19:227–228.

Sorvari S. 1986a. The effect of starch gelatinized nutrient media in barley anther cultures. *Ann Agric Fenn* 25:127–133.

Sorvari S. 1986b. Differentiation of potato tuber discs in barley starch gelatinized nutrient media. *Ann Agric Fenn* 25:135–138.

Sorvari S. 1986c. Comparison of anther culture of barley cultivars in barley starch and agar gelatinized media. *Ann Agric Fenn* 25:249–254.

Tiwari S, Rahimbaev I. 1992. Effect of barley starch in comparison and in combination with agar and agarose on anther culture of Hordeum vulgare L. *Current Science* 62:430–432.

Zimmerman RH, Bhardwaj SV, Fordham I. 1995. Use of starch gelled medium for tissue culture of some fruit crops. *Plant Cell Tissue Organ Culture* 43:207–213.

10

Medicinal Importance of *Artemisia annua* L. and Discovery of Artemisinin

Praveen Kumar Verma,* Aliya Tabassum and Sanghapal D. Sawant*

*Medicinal Chemistry Division, CSIR-Indian Institute of Integrative Medicine
Canal Road, Jammu - 180001, Jammu and Kashmir, INDIA
Email: praveen.ihbt@gmail.com, sdsawant@iiim.ac.in

ABSTRACT

Artemisia annua L. is known for the content of anti-malarial drug, artemisinin. It is widely distributed in subtropical, temperate and subtropical zones worldwide and traditionally used for the treatment of various ailments. Artemisinin has been the frontline treatment since the late 1990s and saved countless lives, especially among the world's poorest children. The major chemical constituents of *Artemisia annua* L. are sesquiterpenoids and some phenolics. The essential oil of *Artemisia* contains artemisia ketone, camphor, caryophyllene oxide, 1,8-cineole and α-pinene, and these essential oils constituents are known for their medicinal properties. Artemisinin was discovered in 1971 by a Chinese Medicinal Chemist Youyou Tu (Nobel Prize in 2015). As the demand for artemisinin remains high worldwide, exploring the chemical and the genetic variation especially knowledge, about its mechanisms for high-yielding would be more important. The plant extract is also known for other biological activities such as anti-hypertensive, anti-microbial, immuniosuppresive, and antiparasitic activities. To increase the production of artemisinin and to investigate the effect of climate change on artemisinin content will be the major research area in future from *Artemisia* genus. Exploring the alternative resources such as microbial production and semisynthetic approaches from other easily available natural molecules would be of high demand. Some other molecules from *Artemisia* should also be focused for the future development of man kind and in drug discovery science. An overview of present status of medicinal applications, phytochemistry, and future perspectives for the research possibilities on the active ingredient Artemisinin from *A. annua* is discussed in this chapter.

Keywords: *Artemisia annua*, Artemisinin, Phytoconstituents, Semisynthesis, Natural Resources, India

INTRODUCTION

The genus *Artemisia* L.(Asteraceae) includes well-known plants such as *A. dracunculus* L. (tarragon), *A. absinthium* L. (absinthe) and *A. vulgaris* L. (mugwort) used in medicine, perfumery food and drink industry. The word 'Artemisia' comes from the ancient Greek word: 'Artemis'=The Goddess (the Greek Queen artemisia) and 'absinthium'=Unenjoyable or without sweetness. The word 'wormwood' is influenced by the traditional use as a cure for intestinal worms and sea-dragon bites. *Artemisia annua* L. (Fig. 1), commonly known as "sweet wormwood" is an annual, aromatic herb growing in Asia, and has been used in China to treat fevers for more than 2,000 years. This plant naturally grows 30 to 100 cm tall, although in cultivation it is possible for plants to reach upto 200 cm tall. It is a potential crop in the United States for the production of anti-malarial drug artemisinin, a sesquiterpenoid lactone peroxide (Charles et al. 1990). It is the only natural botanical source for artemisinin (Jain et al. 1996, Klayman 1985) and a potential source for essential oils for the perfume industry (Simon et al. 1990). The growing period of *A. annua* from seeding till harvest is 190–240 days, depending on the climate and altitude of the production area. The plant should be harvested at the beginning of flowering because of the higher content of artemisinin at that time (World Health Organization 2006).

The artemisinin content depends on plant ecotypes, ecological interactions, seasonal and geographic variations (Delabays et al. 1993, Liersch et al. 1986, Singh et al. 1988, Charles et al. 1990, Woerdenbag et al. 1994). In fact, artemisinin is found absent in some *A. annua*, and some Chinese germplasm has relatively higher artemisinin levels than those of Europe, North America, East Africa and Australia (Delabays et al. 1993, Charles et al. 1990, Woerdenbag et al. 1994, Jain et al. 1996, Trigg 1989, Klayman et al. 1984). In Youyang (Chongqing), China, the hometown of *A. annua*, the plants have high (0.9%) levels of artemisinin. In 2006 the county became a national protected geographic area recognized by the General Administration of Quality Supervision, Inspection and Quarantine of China. As the demand for artemisinin remains high around the world, finding suitable geographic regions for *A. annua* is a critical research area (World Health Organization 2005).

Fig. 1: *Artemisia annua* L. (Photo from CSIR-IIIM, Jammu farm)

HISTORY OF ARTEMISININ: FROM DISCOVERY TO NOBEL PRIZE

Scientists worldwide had screened over 240,000 compounds for the treatment of Malaria, however, they were not successful in discovering potent compounds for treatment of malaria. In 1967, a drug discovery project (Project 523) lead by Professor Youyou Tu was set up in China at the Chinese Academy of Medical Sciences in Beijing (Tu 2011). They screened over 2000 traditional Chinese recipes and made 380 herbal extracts which were tested on malaria-infected mice (Miller and Su 2011). A herbal extract used for over 1600 years in traditional Chinese therapy for "intermittent fever" the symptom of malaria, was found to be more effective (Quinghaoshu Antimalaria Coordinating Research Group 1979). The *A. annua* was extracted at low temperature with ethyl ether and named as qinghaosu and chemically characterized in 1971. The active anti-malarial moieties and the physico-chemical properties were determined *in vitro* and *in vivo* in both animal models and in human. The drug was distributed to rest of the world in 1979 (Quinghaoshu Antimalaria Coordinating Research Group, 1979). The artemisinin was discovered in 1970s by a Chinese scientist Youyou Tu, who later was awarded with a Nobel Prize in Physiology and Medicine in 2015 for her discovery of qinghaosu/artemisinin and the more potent derivative dihydroartemisinin (Nobel Media, 2015). It saved the lives of millions of peoples worldwide affected with malaria (Krungkrai et al. 2016, Nobel Media 2015).

DISTRIBUTION, HABITAT AND DIVERSITY

Artemisia annua is widely distributed in subtropical, temperate and cool temperate zones worldwide (Fig. 2). This species has originated from China and extends slowly mainly in middle, eastern and southern parts of Europe and in northern, middle and eastern parts of Asia. However, it also grows in the Mediterranean region and North Africa, as well as in South-West Asia. It grows widely in Canada and United States. It is mainly collected from the wild for industrial use, however, few countries are currently cultivating on a large scale, such as China, Kenya, United Republic of Tanzania and Vietnam. Small-scale cultivation has been undertaken in India and other countries in Africa, south Europe and South America (World Health Organization 2006). There are about 500 species of *Artemisia* reported in the world and out of which 45 species are found in India. Huang and Ling (1996) reported 200 species of *Artemisia* growing in China.

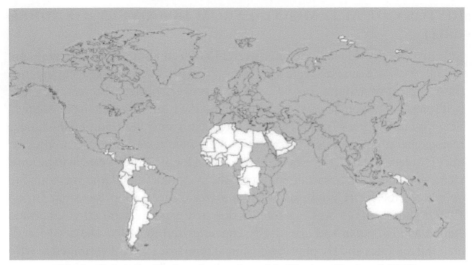

Fig. 2: Global distribution of genus Artemisia (*Courtsy*: Koul et al. 2018)

In India *Artemisia* species are distributed in the Northern plains, eastern Himalaya and in the Peninsular region including the Western Ghats. There are only a few species, *viz.*, *A. capillaries* Thunb. (syn. *A. scoparia*) found in Gujrat, Panjab, Upper Gangetic plains. *A. indica* Willd. and *A. nilagarica* (C.B.Clarke) Pamp. var. *nilagarica*, (earlier reported as *A.vulgaris* L.) is distributed in Mt Abu, Western Ghats, Nilgiries, Karnataka, Deccan peninsula, Maharashtra, (Chopra et al. 1956, Hazra et al. 1995) and *A. capillaries* Thunb. var. *scoparia* in Panjab and Indo-Gangetic Plains and Bihar, (Chopra et al. 1956) and Maharashtra. Further, *A. carvifolia* Buch.Ham. is distributed in the Gangetic Plains. *A. myriantha* Wall. ex Besser var. *pleocephala* distributed in Karnataka (Hazra et al. 1995).

Artemisia species found in Indian Himalayas are enumerated by Stewart (1972) from Kashmir Himalayas and Ladakh, and, from the high altitude region in the Western Himalayas, by Rau (1975), who enumerated total 13 species. Kaul and Bakshi (1984) studied 20 species from Kashmir Himalayas. Later, Hazra et al. (1995), described 32 species from India. Karthikeyan et al. (2009) has reported 45 species of *Artemisia* with their varieties. Further, two new species were reported by Kaul and Bakshi (1984), *viz.*, *A. banihalensis* and *A. cashemirica* from Kashmir Himalaya.

TRADITIONAL USE

Ninteen species of *Artemisia* are recognized as medicinal herbs in Himalayan regions (*A. absinthium* L., *A. biennis* Willd., *A. brevifolia* Wall. ex DC., *A. desertorum* Spreng., *A. dracunculus* L., *A. dubia* Wall., *A. gmelinii* Webb ex Stechmann, *A. indica* Willd., *A. japonica* Thunb., *A. laciniata* Willd., *A. macrocephala* Jacq. ex Besser, *A. maratima* L., *A. moorcroftiana* Wall. ex

DC., *A. nilagarica* (C.B.Clarke) Pamp., *A. parviflora* Buch. -Ham. ex Roxb., *A. roxburghiana* Wall. ex Besser, *A. scoparia* Waldst. & Kit., *A. sieversiana* Willd., and *A. vulgaris* L.) (Sah et al. 2010, Semwal et al. 2015). *A. dracunculus* (tarragon) is used worldwide, in cluding the Himalayan region, as a flavoring agent for food. Native peoples of Nubra Valley (Kashmir) (Kumar et al. 2009), Kibber Wildlife Sanctuary (Himachal Pradesh) and the Lahaul Valley (Himachal Pradesh) use a paste from the leaves to treat wounds on the legs of donkeys and yaks (Kumar et al. 2009, Deve et al. 2014). Plant extract is also used to relieve toothache, reduce fever, and as a treatment for dysentery, intestinal worms, and stomachache.

In the Garhwal Himalaya, Uttarakhand, the leaves of *A. japonica* is used as an incense and insecticide (Bhat et al. 2013). In veterinary medicine, this plant is used for the treatment of internal parasites (e.g., round worm) (Pande et al. 2007). *A. maritima* is used to treat stomach problems and for expelling intestinal worms in Himalayas (Ashraf et al. 2010). *A. parviflora* leaves are used in the Kumaun Himalaya to treat skin diseases, burns, cuts, and wounds, while the volatiles from the plant are used to repel insects (Mehra et al. 2014). The indigenous peoples of Jammu and Kashmir use *A. parviflora* as a diuretic and to treat gynecological disorders (Semwal et al. 2015). This species is used in veterinary medicine as an anthelmintic. Decoction of leaves and buds is used to stock animals (e.g., horses, mules, sheep and buffaloes) for round worm (Kumari et al. 2009); whole as this species used as a fodder plant in mid-altitude range of Uttarakhand (Singh et al. 2008). People living in the Kedarnath Wildlife Sanctuary in Western Himalaya of Chamoli-Rudraprayag, (Uttarakhand) use the plant extract to relieve fever (Bhat et al. 2013), and is rubbed on the skin to treat allergic reactions. *A. roxburghiana* is used to treat skin allergies in Jammu and Kashimir (Gairola et al. 2014) and in Uttarakhand is used to treat eye diseases, wounds, cuts, and external parasites in veterinary medicine. Inhabitants of the Nanda Devi National Park, (Uttarakhanda) used a paste of leaves of *A. scoparia* to heal cuts and wounds (Rana et al. 2010). The leaf powder is used for the treatment of diabetes, abdominal complaints, colic, cough, and cold and as a blood purifier.

PHYTOCHEMISTRY

Isolated Secondary Metabolites

The main chemical constituents isolated from *Artemisia annua* by different research groups are sesquiterpenoids (artemisinin, artemisinic acid, artemisinol, arteannuic acid methyl ester, artemisitene, arteannuin A, B, C, cadinanolide, deoxyartemisinin. Apart from these some flavonoids, coumarins, triterpenoids, steroids, phenolics, purines, lipids, aliphatic compounds, and mono terpenes (Fig. 3) have also been isolated from different parts of *A. annua* (Bhakuni et al. 2001).

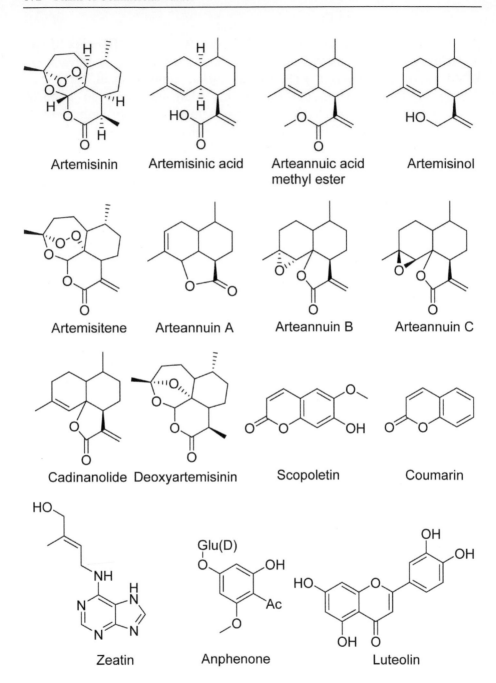

Fig. 3: Structure of major compounds of *Artemisia annua*

Essential Oil Composition

Essential oils are volatile and complex compounds characterized by a strong odour. They are formed by aromatic plants as secondary metabolites mainly terpenes. They are usually obtained by steam or hydro-distillation, although there are several methods for extracting them. The strong and aromatic smell of some species of *Artemisia* genus is due to high concentrations of volatile terpenes. The chemical composition of essential oils of *Artemisia* genus has been extensively studied in several species around the world. The major constituents of different *Artemisia* species are well compiled by Abad et al. (2012). In India, the major chemical constituent present in the essential oil of *A. annua* (Jain et al. 2002) is outlined in Table 1.

Table 1: Major (> 1%) essential oil components of *Artemisia annua* in India.

Compound	Amount (%)
artemisia ketone	52.9
camphor	6.0
camphene	1.5
artemisia alcohol	3.5
cis-p-menth-2-en-l -o	1.1
caryophyllene oxide	4.3
1,8-cineole	8.4
α-pinene	5.2
α-copaene	1.1

a(*Courtesy*: Jain et al. 2002)

Semisynthetic Approach for the Production of Artemisinin

Artemisinin is extracted on an industrial scale from *Artemisia annua*. Unfortunately, artemisinin is currently too expensive to meet the distribution needs of the world. Moreover, crop disruptions caused by natural disasters, poor planning, and geopolitical events have led to shortages and price fluctuations. Therefore, the synthetic biology approach has been adopted now a days for the production of artemisinin in which a chemical precursor of artemisinin (artemisinic acid) is produced by microbes which thereafter converted to artemisinin by semisynthetic approaches (Fig. 4). Some extensively available naturally occurring molecules (Fig. 5) have also been utilized as the starting material for the synthesis of artemisinin (Zhu and Cook 2012).

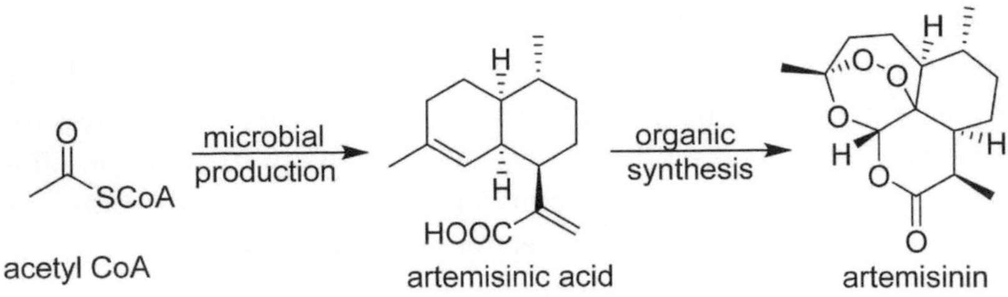

Fig. 4: Synthetic biology approach for the production of artemisinin
(*Courtesy*: Zhu and Cook 2012).

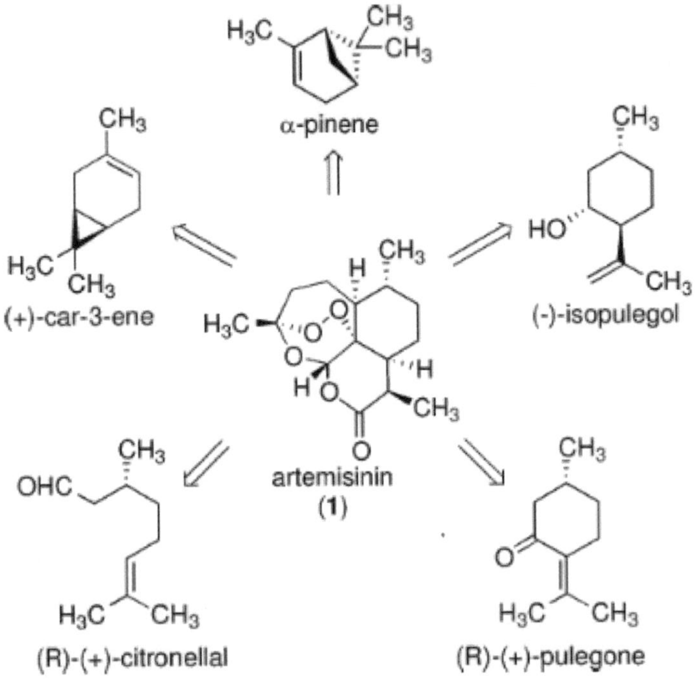

Fig. 5: Naturally occurring molecules used for the synthesis of artemisinin
(*Courtesy*: Zhu and Cook 2012).

Chemistry: Mode of Action of Artemisinin and its Derivatives

Different derivatives of artemisinin are under use for the antimalarial activity
(Fig. 6). Artemisinin is a 15-carbon sesquiterpene lactone bearing an
endoperoxide group, which is essential for anti-malarial activity (White 2008).
Dihydroartemisinin is an active metabolite. To increase solubility of artemisinin,
arteether and artemether were synthesized as lipid soluble structures. Artesunate
and Artelinic acid were synthesized as water soluble derivatives. Artemisinin
have multiple sites of action (Fig. 7) for its rapid killing effect (Krungkrai and
Yuthavong 1987, White 2008, Krungkrai et al. 2008).

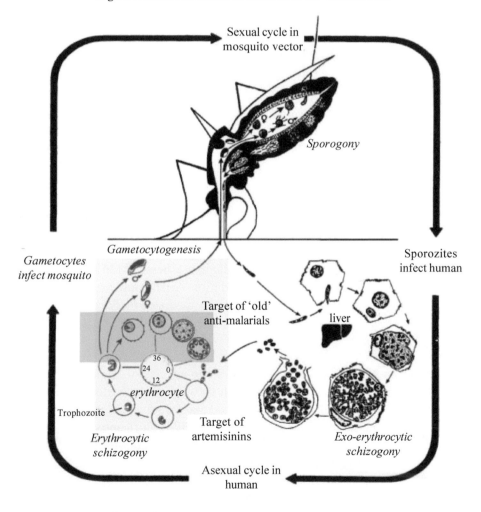

Dihydroartemisinin; R = H
Arteether; R = Et
Artemether; R = Me
Artesunic acid; R = (OC)CH$_2$CH$_2$COOH
Artelinic acid; R = CH$_2$Ph(COOH)-P

Artemisinin

Fig. 6: Chemical structures of artemisinin and its derivatives

Sexual cycle in
mosquito vector

Sporogony

Gametocytes
infect mosquito

Gametocytogenesis

Sporozites
infect human

Target of 'old'
anti-malarials

liver

36
24   0
12

erythrocyte

Trophozoite

*Erythrocytic
schizogony*

Target of
artemisinins

*Exo-erythrocytic
schizogony*

Asexual cycle in
human

Fig. 7: Malaria cycle and target of artemisinin's action
(*Courtesy*: Hommel 2008).

One of the well recognized and accepted mechanisms is that free or heme-bound iron (Fe) catalyzes the conversion of drug to free radicals, i.e., the reduction of the endoperoxide bridge by an electron from Fe^{2+} to a free radical and the ferrous iron to Fe^{3+}. The free radicals alkylate and oxidize proteins as well as lipids, results in the rapid killing of the parasite. However, this phenomenon should occur in the food vacuole of the parasite, especially during hemoglobin digestion to release amino acids for survival. Other targets for artemisinin action include: (i) activation of mitochondrial electron transport system resulting in reactive oxygen species production (Li et al. 2005, Wang et al. 2010); (ii) inhibition of mitochondrial oxygen utilization through cytochrome c oxidase complex (Krungkrai et al. 1999); (iii) inhibition of sarcoplasmic reticulum calcium adenosine triphosphatase (Eckstein-Ludwig et al 2003).

PHARMACOLOGY ASPECTS

According to Das (2017), *Artemisia annua* has been investigated scientifically to validate the potential in cure of variety of ailments. Some major pharmacological activities reported by different researchers are given below in sub-heads:

Antimalarial activity

Artemisinin and its derivatives showed anti-malarial effect by inhibition of heme polymerization. *In-vitro* experiment shows inhibition of digestive vacuole proteolytic activity of malarial parasite by artemisinin. *Ex vivo* treatment of artemisinin to parasites is also shown accumulation of hemoglobin in the parasites suggesting inhibition of hemoglobin degradation. Artemisinin has found to be a potent inhibitor of heme polymerization activity mediated by *Plasmodium yoeli* lysates as well as *Plasmodium falciparum* (Penissi et al. 2006).

Antihypertensive activity

The aquous extract of the aerial parts of some species of *Artemisia* for 2-4 weeks feeding diabetic rats and rabbits with 100-390 mg kg^{-1} causes significant reduction in blood level. It has been determined that the subchronic administration of 100 and 200 mg kg^{-1} aquuous extract of *A. annua* significantly inhibited the phenylephrine-induced contraction and potentiated the endothelium-dependent relaxation of rat aortic rings in Krebs solution (Allen et al. 1997).

Antimicrobial activity

The essential oil of *A. annua* showed anti-microbial activity against microorganisms. The maximum activity against fungal microorganisms *Saccharomyces cerevisiae* (MIC = 2 mg/ml) and *Candida albicans* (MIC = 2mg/ml). Moderate inhibitory activity against *S. aureus* and *E. coli* was found with MIC value of 32mg/ml and 64 mg/ml, respectively (Meier-zu-Biesen 2011).

Immunosuppressive activity

A. annua has been widely used for the treatment of autoimmune diseases such as systemic lupus erythematosus and rheumatoid arthritis in traditional Chinese medicine. Ethanolic extract significantly suppressed concanavalin A (Con A) and lipopolysaccharide (LPS)- stimulated splenocyte proliferation *in vitro* in a concentration-dependent manner (Verdian-rizi 2009).

Anti-parasitic activity

The *Neospora canum*, a protozoal parasite infected cells were treated with artemisinin at 10 or 20 µg/ml for 11 days showed that all microscopic foci of *N. caninum* were completely eliminated (Verdian-rizi 2009).

Artemisinin showed anti-leishmanial activity both in promastigotes and amastigotes with IC_{50} values of 160 and 22 µM, with a high safety index (>22-fold), respectively (Willcox et al. 2004, Verdian-rizi 2009).

FUTURE PERSPECTIVES FOR RESEARCH

Lots of efforts are still ongoing for the discovery of anti-malarial compounds based on artemisinin and its mechanism of action. It is known that artemisinin requires hemoglobin digestion and the release of iron containing heme, which induces oxidative stress (Klonis et al. 2011). Klayman (1985) describes few other natural products with an endoperoxide, and this peroxide also offers an opportunity to make a new anti-malarial drugs (Charman et al. 2011).

There is great concern among the international health community regarding the development of artemisinin resistance in malarial parasite. The synthesized artemisinin has been used for less than 20 years and already the first cases of parasite resistance have been identified. However, *Artemisia annua* tea has been used in China for the last 1000 years without development of any resistance. It could be due to the fact that artificial artemisinin has sometimes been wrongly used as a monotherapy, while *A. annua*, contains other anti-malarial substances, such as artemetin, casticin, cirsilineol, chrysoplenetin, sesquiterpenes, and flavonoids. The synergistic effect of these compounds could be responsible for reducing the possibility of parasite developing resistance (Willcox 2004). Hence, there is a need to focus on the other bioactive substances in future in drug discovery programme.

Major funding organizations have to deal not only with the cost of the drug, but must also strengthen health-system infrastructure and improve provision for diagnosis and treatment in the field. Keeping the cost of the drug down means not only ensuring a sustainable supply of the raw material, but also putting pressure on manufacturers to keep the price of the final drug reasonable; both of which could be achieved by using subsidies to pre-order drugs years in advance. This, however, assumes that forecasting need becomes much better than is currently feasible.

CONCLUSION

Present compilation on *Artemisia annua* ascribed the status of artemisinin for the present scenario. The high demand and low availability is the major focus of the artemisinin research due to its incomparable potential as antimalarial drug. Now a days people are focusing on increasing the yield of artemisinin from natural resources, in which, synthetic biology by means of microbial production is the key factor for the researcher to satisfy the global demand. The other bioactive molecules from *A. annua* might be helpful for the treatment of malaria. Lots of efforts are required for better understanding the mechanisms and site of its action. This will explore and open up the success factor for the other alternative therapeutics in drug discovery programmes.

REFERENCES CITED

Abad MJ, Bedoya LM, Apaza L, Bermejo P. 2012. The *Artemisia* L. genus: A review of bioactive essential oils. *Molecules* 17: 2542-2566.

Allen PC, Lydon J, Danforth HD. 1997. Effects of components of *Artemisia annua* on coccidia infections in chickens. *Poultry Science* 76: 1156–1163.

Ashraf M, Hayat MQ, Jabeen S, Shaheen N, Khan MA, Yasmin G. 2010. *Artemisia* L. species recognized by the local community of northern areas of Pakistan as folk therapeutic plants. *Journal of Medicinal Plants Research* 4: 112–119.

Bhakuni RS, Jain DC, Sharma RP, Kumar S. 2001. Secondary metabolites of *Artemisia annua* and their biological activity. *Current Science* 80: 35-48.

Bhat JA, Kumar M, Bussmann RW. 2013. Ecological status and traditional knowledge of medicinal plants in Kedarnath Wildlife Sanctuary of Garhwal Himalaya, India. *Journal of Ethnobiology and Ethnomedicine* 9: 1-18.

Charles DJ, Simon JE, Wood KV, Heinstein P. 1990. Germplasm variation in artemisinin content of *Artemisia annua* using an alternative method of artemisinin analysis from crude plant extracts. *Journal of Natural Products* 53: 157-159.

Charman SA, Arbe-Barnes S, Bathurst IC, Brun R, Campbell M, Charman WN, Chiu FCK, Chollet J, Craft JC, Creek DJ, Dong Y, Matile H, Maurer M, Morizzi J, Nguyen T, Papastogiannidis P, Scheurer C, Shackleford DM, Sriraghavan K, Stingelin L, Tang Y, Urwyler H, Wang X, White KL, Wittlin S, Zhou L, Vennerstrom JL. 2011. Synthetic ozonide drug candidate OZ439 offers new hope for a single-dose cure of uncomplicated malaria. *Proceedings of the National Academy of Sciences of the United States of America* 108: 4400–4405.

Chopra RN, Nayar SL, Chopra IC. 1956. Glossary of Indian medicinal plant, New Delhi. Published by Council of Scientific Industrial Research, CSIR, New Delhi, India pp.1-330.

Das S. 2017. *Artemisia annua* (Qinghao): a pharmacological review. *International Journal of Pharmaceutical Sciences and Research* 3: 4573-4577.

Delabays N, Benakis A, Collet G. 1993. Selection and breeding for high artemisinin (Qinghaosu) yielding strains of *Artemisia annua*. *Acta Horticulturae (ISHS)*. 330: 203-208.

Devi U, Seth MK, Sharma P, Rana JC. 2013. Study on ethnomedicinal plants of Kibber Wildlife Sanctuary: A cold desert in Trans Himalaya, India. *Journal of Medicinal Plants Research* 7: 3400–3419.

Eckstein-Ludwig U, Webb RJ, Van Goethem ID, East JM, Lee AG, Kimura M, O'Neill PM, Bray PG, Ward SA, Krishnaet S. 2003. Artemisinins target SERCA of *Plasmodium falciparum*. *Nature* 424: 957-961.

Gairola S, Sharma Y, Bedi YS.2014. A cross-cultural analysis of Jammu, Kashmir and Ladakh (India) medicinal plant use. *Journal of Ethnopharmacology* 155: 925–986.

Hazra PK, Rao RR, Singh DK, Uniyal BP. 1995. Flora of India vol.12. Asteraceae (Anthemideae-Helmintheae), Botanical Survey of India. Calcutta.

Hommel M. 2008. The future of artemisinins: natural, synthetic or recombinant? *Journal of Biology* 7: 38.1-38.5.

Huang YP, Ling YR. 1996. Economic Compositae in China. In P.D.S. Caligari & D.J.N. Hind (eds). Compositae: Biology & Utilization. Proceedings of the International Compositae Conference, Kew, 1994 (D.J.N. Hind, Editor-in-Chief), vol. 2. pp. 431-451. Royal Botanic Gardens, Kew.

Jain DC, Mathur AK, Gupta MM, Singh AK, Verma RK, Gupta AP, Kumar S. 1996. Isolation of high artemisinin-yielding clones of *Artemisia annua*. *Phytochemistry* 5: 993-1001.

Jain N, Srivastava SK, Agarwal KK, Kumar S, Shyamsunder KV. 2002. Essential oil composition of *Artemisia annua* L. 'Asha' from the plains of Northern India. *Journal of Essential Oil Research* 14: 305-307.

Karthikeyan S, Sanjappa M, Moorthy S. 2009. Flora of India series 4. Flowering plants of India dicotyledons. vol.1. Botanical Survey of India, Howrah., India.

Kaul MK, Bakshi JK. 1984. Studies on the genus *Artemisia*. L. in North-West Himalayas with particular reference to Kashmir. *Folia Geobotanica et Phyto-taxonomica* 19: 299-316.

Klayman DL, Lin AJ, Acton N, Scovill JP, Hoch JM, Milhous WK, Theoharides AD, Dobek AS. 1984. Isolation of artemisinin (qinghaosu) from *Artemisia annua* growing in the United States. *Journal of Natural Products* 47: 715-717.

Klayman DL. 1985. Qinghaosu (artemisinin): an antimalarial drug from China. *Science* 228: 1049-1055.

Klonis N, Crespo-Ortiz MP, Bottova I, Abu-Bakar N, Kenny S, Rosenthal PJ, Tilley L. (2011). Artemisinin activity against *Plasmodium falciparum* requires hemoglobin uptake and digestion. *Proceedings of the National Academy of Sciences of the United States of America* 108: 11405–11410.

Koul B, Taak P, Kumar A, Khatri T, Sanyal I. 2018. The *Artemisia* genus: A review on traditional uses, phytochemical constituents, pharmacological properties and germplasm conservation. *Journal of Glycomics and Lipidomics* 7: 142

Krungkrai J, Burat D, Kudan S, Krungkrai S, Prapunwattana P. 1999. Mitochondrial oxygen consumption in asexual and sexual blood stages of the human malarial parasite *Plasmodium falciparum*. *The Southeast Asian Journal of Tropical Medicine and Public Health* 30: 636-42.

Kumar GP, Gupta S, Murugan MP, Singh SB. 2009. Ethnobotanical studies of Nubra Valley—A cold arid zone of Himalaya. *Ethnobotanical Leaflets* 13: 752–765.

Kumari P, Singh BK, Joshi GC, Tewari LM. 2009. Veterinary ethnomedicinal plants in Uttarakhand Himalayan Region, India. *Ethnobotanical Leaflets* 13: 1312–1327.

Li W, Mo W, Shen D, Sun L, Wang J, Lu S, Gitschier MJ, Zhou B. 2005. Yeast model uncovers dual roles of mitochondria in action of artemisinin. *PLoS Genetics* 1: 0329-0334.

Liersch R, Soicke H, Stehr C, Tullner HU. 1986. Formation of artemisinin in *Artemisia annua* during one vegetation period. *Planta Medica* 52: 387-390.

Mehra A, Bajpai O, Joshi H. 2014. Diversity, utilization and sacred values of ethno-medicinal plants of Kumaun Himalaya. *Tropical Plant Research* 1: 80–86.

Meier-zu-Biesen C. 2011. The rise to prominence of *Artemisia annua* L.–the transformation of a Chinese plant to a global pharmaceutical. *African Sociological Review* 14: 24–46.

Miller LH, Su X. 2011. Artemisinin: discovery from the Chinese herbal garden. *Cell* 146: 855-858.

Nobel Media AB. Youyou Tu – facts. Stockholm: Nobel Media AB; 2015. http://www.nobelprize.org/nobel_prizes/medicine/laureates/2015/tu-facts.html.

Pande PC, Tiwari L, Pande HC. 2007. Ethnoveterinary plants of Uttaranchal—A review. *Indian Journal of Traditioanal Knowledge* 6: 444–458.

Pandey AV, Tekwani BL, Singh RL, Chauhan VS. 1999. Artemisinin, an endoperoxide antimalarial, disrupts the hemoglobin catabolism and heme detoxification systems in malarial parasite. *Journal of Biological Chemistry* 274: 19383-19388.

Penissi AB, Giordanob OS, Guzmánc JA, Rudolphd MI, Piezzia RS. 2006. Chemical and pharmacological properties of dehydroleucodine, a lactone isolated from *Artemisia douglasiana*. *Molecular Medicinal Chemistry* 10: 1–11.

Rana CS, Sharma A, Kumar N, Dangwal LR, Tiwari JK. 2010. Ethnopharmacology of some important medicinal plants of Nanda Devi National Park (NDNP) Uttarakhand, India. *Nature and Science* 8(11): 9–14.

Rau MA. 1975. High altitude flowering plants of West Himalaya. Botanical Survey of India, Howrah, India pp. 1-234.

Sah S, Lohani H, Narayan O, Bartwal S, Chauhan NK. 2010. Volatile constituents of *Artemisia maritima* Linn. grown in Garhwal Himalaya *Journal of Essential Oil Bearing Plants.* 13: 603–606.

Semwal RB, Semwal DK, Mishra SP, Semwal R. 2015. Chemical composition and antibacterial potential of essential oils from *Artemisia cappillaris, Artemisia nilagirica, Citrus limon, Cymbopogon flexuosus, Hedychium spicatum* and *Ocimum tenuiflorum. The Natural Products Journal* 5: 199–205.

Simon JE, Charles E, Cebert L, Grant J, Janick J, Whipkey A. 1990. *Artemisia annua* L: a promising aromatic and medicinal. In Advances in New Crops: Proceeding of the First National Symposium New Crops: Research, Development, Economics Portland: Timber Press, Incorporated. 522-526.

Singh A, Vishwakarma RA, Husain A. 1988. Evaluation of *Artemisia annua* strains for higher artemisinin production. *Planta Medica* 64: 475-476.

Singh V, Gaur RD, Bohra B. 2008. A survey of fodder plants in mid-altitude Himalayan rangelands of Uttarakhand, India. *Journal of Mountain Science* 5: 265–278.

Stewart RR. 1972. Flora of West Pakistan, An Annonated catalogue of vascular plants of West Pakistan & Kashmir (Eds.) E. Nasir & S.I. Ali. Fakhri Printing Press, Karachi. pp.1-1028.

Towler MJ, Weathers PJ. 2015. Variations in key artemisinic and other metabolites throughout plant development in *Artemisia annua* L. for potential therapeutic use. *Industrial Crops and Products* 67: 185–191.

Trigg EI. 1989. Qinghaosu (artemisinin) as an antimalarial drug. *Economic and Medicinal Plant Research* 3: 19-55.

Tu Y. 2011. The discovery of artemisinin (qinghaosu) and gifts from Chinese medicine. *Nature Medicine* 17: 1217–1220.

Verdian-rizi MR. 2009. Chemical composition and antimicrobial activity of the essential oil of *Artemisia annua* from Iran. *Pharmacognosy Research* 1: 21-24.

Wang J, Huang L, Li J, Fan Q, Long Y, Li Y, Zhou B. 2010. Artemisinin directly targets malarial mitochondria through its speciûc mitochondrial activation. PLoS One 5: e9582.

White NJ. 2008. Qinghaosu (artemisinin): the price of success. *Science.* 320: 330-334.

Willcox M, Bodeker G, Bourdy G, Dhingra V, Falquet J, Ferreira JF. 2004. *Artemisia annua* as a traditional herbal antimalarial. *Traditional Medicinal Plants and Malaria* 43–59.

Woerdenbag HJ, Pras N, Chan NG, Bang BT, Bos R, van Uden W, Van YP, van Boi NV, Batterman S, Lugt CB. 1994. Artemisinin related sesquiterpenes and essential oil in *Artemisia annua* during a vegetation period in Vietnam. *Planta Medica* 60: 272-275.

World Health Organization Global Malaria Programme: Proceedings of the meeting on the production of artemisinin and artemisinin-based combination therapies: 6-7 June 2005; Arusha Global Malaria Programme; 2006.

World Health Organization. 2006. WHO monograph on good agricultural and collection practices (GACP) for *Artemisia annua* L. Geneva.

Zhu C, Cook SP. 2012. A concise synthesis of (+)-artemisinin. *Journal of American Chemical Society* 134: 13577-13579.

11

Threatened Subsistence of Economically Important *Spinifex littoreus* (Burm.f.) Merr. Family-Poaceae

Amal Kumar Mondal[†], Tamal Chakraborty and Sanjukta Mondal Parui*

†Plant Taxonomy, Biosystematics and Molecular Taxonomy, UGC-DRS-SAP-II & DBT-BOOST-WB Funded Department, Department of Botany & Forestry Vidyasagar University, Midnapur- 721102, West Bengal, INDIA
*Post Graduate Department of Zoology, Lady Brabourne College, P1/2 Suhrawardy Avenue, Kolkata -700 017, West Bengal, INDIA
Email: *amalcaebotvu@gmail.com, akmondal@mail.vidyasagar.ac.in tamal.bot@gmail.com, sanjuktaparui@gmail.com

ABSTRACT

The deeply rooted coastal grass *Spinifex littoreus* (Burm.f.) Merr. is a potent soil binder having capability to form a strong and stable dunes. However detailed features of this species has not been thoroughly studied till date. In this communication, salient morphological features along with different crucial micro-morphological peculiarities have been discussed. The comparative analysis of gradual degradation of population of *Spinifex littoreus* and native dune vegetation of the coastal belt in East Medinipur of West Bengal and adjacent Balasore district of Odisha are provided, along with different biodiversity indices. Plants were characterized as stout, dioecious, stoloniferous grass upto 90 cm tall, with rigid and sharp spiny leaves. Spikelets in racemes subtended by large bract-like spatheoles which are fascicled into large capitate spiny structures 30-37 cm in diameter. *S. littoreus* is fairly common along sandy shores and dunes. Plants are useful as a sand binder in unstable coastal dunes ecosystem.

Keywords: *Spinifex littoreus*, Micro-morphological Characters, Dune Vegetation, Biodiversity, Simpson's Diversity Index, Conservation

INTRODUCTION

The coastal zones constitute one of the most dynamic and sensitive parts of the earth surface. The coastal sand dune vegetations are the natural protector of the inland. Coastal sand dunes protect the coastal environment by absorbing energy from wind, tide and wave action (Jacqes 1991). It is continuously undergoing both gradual and sudden changes with many physical processes such as tidal flooding, sea level rise, land subsidence, volcanic activity and erosion-sedimentation (Maiti and Bhattacharya 2009). The scale and character of coastal dune system depends on the combination of physical factors such as wind and wave regime, sand supply from beach and offshore bars, and biotic controls such as plant succession and grazing pressure. Science has provided convincing evidence that human activity has had a growing and potentially unsustainable impact on different aspects of planet earth (Narpat 2012). Paul Crutzen (2002), the Nobel Prize-Winner in Atmospheric Chemist suggested that human have entered in new epoch- anthropocene because of global environmental effects of increased human population and economic development.

The Coastline of West Bengal is approximately 351 km long and a major part of the coastline lies within the delta region of river Ganges covering more than 50% area throughout the coastline. West Bengal coast adjacent to tide dominated Ganges is heavily populated and facing the serious problem due to unplanned tourism activities, exploitation of natural resources, construction along the sea shore combined with natural events such as storm damage, climate change, sea level rise and coastal erosion. From geographical point of view this 158 km coastline (Banerjee 1994) within the East Midnapur (Purba Medinipur) district of West Bengal and adjoining Balasore district of Odisha is covered with coastal dune vegetation under direct influence of sea waves, destructive tidal currents, flooding, beach erosion and siltation at jetties and storms which all together are making this region to vulnerable and sensitive type of ecosystem. The dunes along this coastal belt include the primary and the secondary dunes, the later being formed when stabilized dunes are activated. Small, phytogenic, semi-spherical or elongated embryo dunes are mostly found on these beaches. Incipient foredunes lies nearest the beach. A dune has a distinct windward slope (stoss slope or proximal slope), where sand is tightly packed by wind and a leeward slope (distal slope) that consists of loose strata (Helleman 1999). Behind these lie a intermediate and established dunes. The sand dune favour growth of different types of plants but all the plants possess stress tolerance and soil binding capacity (Chakraborty et al. 2012). The coastal plant species are playing a vital role in protecting the coast from erosion and flooding (Desai 2000). The coastal belt of West Bengal is rich in plant resources, which harbour many economically important and medicinal plant species as well as several natural soil binders. The stoloniferous deeply rooted coastal grass *Spinifex littoreus* is a potent soil binder as this species can create low angle, strong and stable coastal dunes.

The present paper deals with dwindling coastal sand dune vegetation (CSD) along a stretch of 76 km from Khejuri of East Midnapore district of West Bengal to Udaypur of Balasore district in Odisha over a period of six years (2008-2013) laying special emphasis on *Spinifex littoreus* (Burm.f.) Merr., a dominant member of the CSD (Figure 1a,b & c). The study has been restricted to the more affected fore-dunes and the embryo dunes. Some salient morphological features along with different crucial micromorphological peculiarities playing a vital role for self-protection against the herbivores in this species and the comparative analysis about the gradual degradation of the population of this potent soil binder grass particularly at the most affected beach of Mandarmoni of East Midnapore district of West Bengal has been discussed. The necessity of biodiversity conservation and judicious utilization of the coastal plant wealth is discussed as they have become threatened by over exploitation, clearing of forest for industrialization, rapid urbanization, pisciculture, human settlements, etc. This paper also evoked the present alarming scenario of the biodiversity of this particular area using biodiversity indices like Simpson's Biodiversity index.

Fig. 1a: Habitat of *Spinifex littoreus* with numerous male inflorescence in costal sand dune area

Fig. 1b: Inflorescence (Male) of *Spinifex littoreus* in coastal dune areas

Fig. 1c: Inflorescence (Female) of *Spinifex littoreus* with prominent feathery
stigma in coastal dune areas

Survey Sites

The study was carried out along the belt of Bay of Bengal from Khejuri of East
Midnapur (Purba Medinipore) district of West Bengal to the Talsari and Udaipur
region of adjacent Balasore district of Odisha, which lies between 21°362 502.2
"N and 19°482' 002.2" "N latitude and 85°52.2; 402.2" E and 87° 372 002.2 E
longitude covering 76 km of coast line. Beaches along this belt including
Mandarmoni, Sankarpur, Tajpur, Digha and New-Digha of West Bengal and
Talsari and Udaypur of Odisha (Table 1) were visited several times in different

growing seasons from 2008 to 2013 and the data were well collected including the records of native sand dune species composition.

Table 1: Data showing the soil parameters of the study areas.

State	Coastal site	Sand Quality	pH	Organic matter
West Bengal (East Midnapur district)	Mandarmoni	Medium to fine	7.5-7.6	0.07-0.23
	Shankarpur	Medium	7.3-7.4	0.07-0.21
	Tazpur	Coarse	7.1-7.2	0.01-0.14
	New Digha	Coarse	7.1-7.2	0.02-0.12
	Digha	Coarse	7.1-7.2	0.02-0.14
Odisha (Balasore district)	Talsari	Medium to fine	7.5-7.6	0.08-0.21
	Udaypur	Medium	6.8-7.0	0.07-0.15

Pattern of Survey

The sanddune zones were selected initially and fixed for sampling using quadrat method for study of species composition and relative cover as this method of sampling with quadrat plots of a standard size is a widely applicable technique. Only the native plant species were randomly counted using quadrat ($0.25m^2$) along transects perpendicular to the shore. The invasive plant species which are fast encroaching in some areas were excluded from the study. A transect line of 150 m was made and a quadrat was placed at every 6m interval. 150m line were divided by 25 to get the number of sections (i.e. 6m), and at every 6m a quadrat was placed. Quadrats were positioned along the transect line at intervals in the middle of the section. Each quadrat was divided into 16 squares (Fig. 2).

Fig. 2: Survey team at coastal area in West Bengal

The study was restricted to the embryo dunes and foredunes as these lie in the low-lying, mesotidal tropical coast of Bay of Bengal and are most vulnerable to erosion due to natural calamites such as cyclonic storms, waves, tides, shore drift and human activities including rampant use of dune fields for construction of hotels and restuarant.

Calculation and Use of Biodiversity Indices

A diversity index is a quantitative measure that reflects different types of species present in datasets, and simultaneously takes into an account of how evenly the basic entities (such as individuals) are distributed among those counts. The value of a diversity index increases both when the number of types increases and when evenness increases. For a given number of types, the value of a diversity index is maximized when all types are equally abundant.

Table 2: Mean distribution of the different species along the coastal stretch of East Midnapur, West Bengal and adjoining Odisha during the period of 2008 to 2013

S.No.	Species	Mean number of individuals per quadrat					
		2008	2009	2010	2011	2012	2013
1.	*Canavalia rosea*	04	04	02	02	02	01
2.	*Clerodendrum inerme*	09	09	09	09	08	08
3.	*Ipomoea pes-caprae*	10	10	09	08	06	06
4.	*Launaeas armentosa*	02	02	02	01	00	00
5.	*Opuntia stricta*	03	03	04	03	03	03
6.	*Pandanus fascicularis*	09	09	07	07	06	06
7.	*Paspalum vaginatum*	12	12	11	10	08	08
8.	*Sesuvium portulacastrum*	01	01	01	01	01	01
9.	*Spinifex littoreus*	06	05	04	04	03	03

Alpha Diversity (α-diversity) is the biodiversity within a particular area, community or ecosystem, and is usually expressed as the species richness of the area. It can be measured by counting the number of taxa (distinct groups of organisms) within the ecosystem (e.g. family, genera etc.). However, such estimates of species richness are strongly influenced by sample size, so a number of statistical techniques can be used to correct sample size to get comparable values. The area was sampled for calculaton of various indices are provided recorded. The number of individuals of each species present in the samples were noted. The diversity of the ground flora as well as some trees which were in a small number at the dune vegetation sites especially at the fore dunes were noted down while sampling random quadrats at the six beaches i.e. Mandarmoni, Sankarpur, Tajpur, Digha, New-Digha, Talsari and Udaypur.

The number of plant species within each quadrat, as well as the number of individuals of each species were recorded. A year wise mean data of the quadrat sampling of the six beaches during the period of the study has been presented in this paper (Table 2).

Simpson's Diversity Index

It takes into account the number of species present, as well as the abundance of each species (Simpson 1949).

$$D = \frac{\sum_i n_i(n_i - 1)}{N(N-1)}$$

Where D =Simpson's Diversity Index

N is the total percentage cover or total number of organisms,

n_i is the percentage cover of a species or number of organisms of species i.

As index increases, diversity decreases. The value of **D** ranges between 0 and 1. With this index, 0 represents infinite diversity and 1, no diversity. That is, the bigger the value of D, the lower the diversity. This is neither intuitive nor logical, so to get over this problem, D is often subtracted from 1 to give a logical value.

Simpson's Index of Diversity or Dominance Index

$$1 - \left(\frac{\sum_i n_i(n_i - 1)}{N(N-1)}\right)$$

Where N is the total percentage cover or total number of organisms,

n_i is the percentage cover of a species or number of organisms of species i.

The value of this index also ranges between 0 and 1, but now, the greater the value, the greater the sample diversity. In this case, the index represents the probability that two individuals randomly selected from a sample will belong to different species.

Another way of overcoming the problem of the counter-intuitive nature of Simpson's Index is to take the reciprocal of the Index.

Reciprocal Simpson Index = 1/D

$$\frac{1}{\left(\dfrac{\sum_i n_i^2}{N^2}\right)}$$

As index increases, diversity increases. The value of this index starts with 1 as the lowest possible figure. This figure would represent a community containing only one species. The higher the value, the greater the diversity. The maximum value is the number of species in the sample.

A special study was done at Mandarmoni in East Midnapore district of West Bengal starting from 2006 to 2012. Geomorphologically, this area (21° 38.964'N latitude and 87° 38.845'-87° 45.254'E longitude) has relatively low waves than nearer beach of Digha. However, still this beach is characterized by deposition with formation of neo dunes in several areas especially around Dadanpatrabar. This beach was specially selected for the study as this was a virgin beach unaffected by tourism or human interference during the initial period of study and was rich in biodiversity with greater number of species than the other areas studied. *Spinifex littoreus* was particularly dominant at this beach. But unfortunately over the years this beach has been the most affected site by tourism. It is a relatively new resort destination and much of the facilities are still unplanned and unauthorized. Mandarmani boasts of a 14 km long motorable beach, probably the longest motorable beach road in India which has greatly affected the vegetation on this beach. In the present paper data of two years i.e. 2006-07 and 2011-12 has been presented to show the drastic deterioration in the overall vegetation in this particular area (Table 3 & 4).

Table 3: Distribution of the different species along the coastal stretch of Mandarmoni, East Midnapur during the years 2006-07 and 2011-12

S.No.	Species	Mean number of individuals per quadrat	
		2006-07	2011-12
1.	*Acanthus illicifolius*	7	5
2.	*Canavalia rosea*	7	3
3.	*Clerodendrum inerme*	15	13
4.	*Gisekia pharnaceoides*	6	5
5.	*Ipomoea pes-caprae*	61	28
6.	*Launaea sarmentosa*	18	11
7.	*Opuntia stricta*	29	22
8.	*Pandanus fascicularis*	18	14
9.	*Phyla nodilora*	30	11
10.	*Sesuvium portulacastrum*	31	17
11.	*Spinifex littoreus*	49	20

Table 4: Biodiversity indices used for Alpha biodiversity of CSD at Mandarmoni of East Midnapur, West Bengal during the period of 2006-07 and 2011-12

Biodiversity index	2006-2007	2011-2012
Total Number of plants:	149	271
Average population size:	13.55	24.64
Total Number of Regions:	1	1
Total Number of Species:	11	11
Decimal Accuracy:	4	4
Total Number of Region Sets:	1	1
Simpson Index $\frac{\sum_i n_i(n_i-1)}{N(N-1)}$	0.22	0.26
Dominance Index $1 - \left(\frac{\sum_i n_i(n_i-1)}{N(N-1)} \right)$	0.77	0.73
Reciprocal Simpson Index $\left(\frac{1-\frac{1}{\sum_i n_i^1}}{N^2} \right)$	4.45	3.82
Simpson Index Approximation $\left(\frac{\sum_i n_i^2}{N^2} \right)$	0.11	0.13

Identification and Preservation of Vouchers

Several field surveys and tours were undertaken for collection of information on various uses of the coastal sand dune species in different parts of the coast of West Bengal and adjacent Odisha based on conversation and discussion with ethnic groups adjacent to the dune vegetation, native traditional healers, Ayurvedic practitioners and botanists dealing with medicinal wild plants in selective coastal villages. Each plant materials were assigned a field note book number and documented according to binomials with family, local name, part used and therapeutic uses. Plant parts that were identified as useful in ethnobotany were collected and compressed, and voucher specimens were collected and identified by referring to standard Flora (Hooker 1884). The voucher specimens have been maintained in the herbarium at Department of Botany and Forestry (VUH), Vidyasagar University, Midnapur, West Bengal, India.

Leaves of *Spinifex littoreus* were collected from different coastal regions and studied under the microscope for stomatal characters. The leaves of fresh plant specimens were fixed in FAA solution (acetic acid: alcohol: formalin: water in the ratio 2:5:1:12) for 24 hours and washed in 70% ethanol. Pieces of samples were then cut from the leaf surface and the samples boiled in 5% aqueous solution of KOH for 5-10 minutes. Epidermal peals were then scraped off and stained with saffranin: aqueous ethanol (1% in 50%) and mounted in glycerine. The slides were then observed under Light Microscope (Olympus model – CH 20i BIFM No. 8G13840) and Phase Contrast Microscope (Leica DM-LED-1000).

SALIENT FEATURES OF *SPINIFEX LITTOREUS*

Taxonomy

This species is an efficient sand binder, forming large colonies and stabilizing dunes as it is deeply rooted into the sand as well as rooting and copiously branching at nodes, having stoloniferous stem, more or less hard, stout and multi-node structure, flowering shoots ascending to 28–110 cm and internodes farinose.

Fig. 3a: Inflorescence (female) of *Spinifex littoreus* in the coastal dune

Fig. 3b: Inflorescence (male) of *Spinifex littoreus* in the coastal dune

Fig. 3c: Close- view of inflorescence (male) of *Spinifex littoreus* in the coastal dune

Leaf sheaths are broad, rounded on back and imbricate; leaf blades are distichous, involute-subulate, curved, 7–20 × 0.2–0.3 cm, margins scabrous, apex spiny and ligules densely ciliate.

Plants are dioecious; female inflorescence is large globose spiny head, staminate inflorescence of 2–5 clustered heads 5–10 × 6–8 cm; racemes 3–6 cm, bearing 5–10 loosely imbricate spikelets (Figure 3a,b and c). Staminate spikelets lanceolate, 7–12 mm; glumes oblong-lanceolate, 7–9-veined, lower glumes 1/2 spikelet length, upper glumes 2/3 spikelet length; lower lemma 5-veined, 7–10 mm, palaea having wing like ciliated keels. Female inflorescence globose, 18–36 cm in diameter; racemes unispiculate with needle-like rachis of 10-16 cm in length. Female spikelets lanceolate-oblong, 10–20 mm, acuminate; glumes oblong-lanceolate, lower glume having multi veined structure, upper glumes 7–9-veined; lower lemma ovate-lanceolate, 5-veined, palea absent; upper lemma lanceolate, yellowish.

Foliar Epidermal Characters

During the microscopic examination of leaves two regions were distinctly seen i.e. coastal region (midrib) and intercostals region. Stomata are found on the abaxial lower surface, which is generally rolled inward, measuring 14-17 μm in breadth and 28-31 μm in length (Table 5) and subsidiary cells are domed but the guard cells are overlapped by the inter-stomatal cells (Figure 6). Silica bodies are very sharp. To examine the variation of stomata, the samples were collected from three different regions namely Mandermoni, Tazpur and Udaypur and the measurement of stomatas were recorded, but no drastic variation occurs (Figure 4).

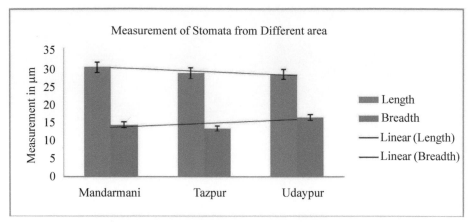

Fig. 4: Graph showing measurement of stomata of *Spinifex littoreus*
collected from different areas

However all the leaf samples of *Spinifex littoreus* collected from the areas adjacent to human habitation like hotels and temporary residential shanties and shops at all three places possesses comparatively larger stomata than the samples collected from the beach front area where human influences are comparatively lesser. Mostly large hotels constructed on the dunes and a large number of makeshift hotels and shops in the low-lying areas do not have any sewage system.

The sewage water runs over the beach sand and ultimately seeps beneath the embankments. So these pollutants or excretory substances drained from the hotels may have direct effect on the modification of micro-morphological features such as stomatal size of the studied plant as revealed from the present study. Rate of photosynthesis, stomatal conductance and chlorophyll content has been found to decrease in plants grown in sewage sludge-amended soil as compared to those grown in unamended soil (Singh and Agrawal 2007). However further studies are required to come to more contructive conclusions on this. The roughness or spine like appearance felt when fingers are drawn from apex to base is due to presence of prickle hairs having inflated base bearing a short barb directed towards the apex. Prickle-hairs (Metcalfe 1960) may be hook-shaped or straight in the Panicoides.

In *Spinifex littoreus,* it is hook shaped and 15 -18 µn in length while 7-9 µn at the inflated base (Figure 5). Hooked trichomes are abundant on the margins at the lower surface of leaf blade, empty glumes, and lemma. These trichomes are thick-walled and may have the silica deposit in lumen of their cell as they glittered under microscope.

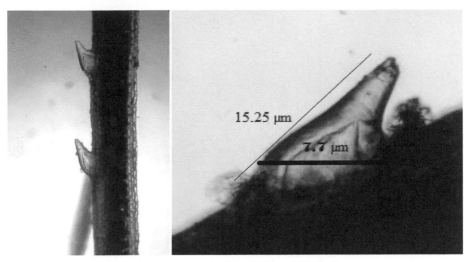

Fig. 5: Arrangement of prickle hairs in *Spinifex littoreus*

Flowering and Fruiting Season

Mainly from April to October

Seed Dispersal

The seeds within are dispersed at the spiky head along the coast by wind and sea.

Current Status of Biodiversity at the Vulnerable Coastal Sites

The Mandarmoni coastal site was found to be richer in species diversity as compared to the other coastal sites studied, with 11 native species. *Ipomoea pes-caprae* and *Spinifex littoreus* were the two most dominating species. *Acanthus ilicifolius, Gisekia pharnaceoides* and *Phyla nodiflora* were found to grow only at Mandarmoni. *Phyla nodiflora* was also among one of the dominating species on this beach but the other two species were found in rare numbers. A comparative study of the dominance index (Table 6) between 2006-07 and 2011-12 also showed a decrease in the number of individuals under each species and distribution, although the number of species remained the same. *Launaea sarmentosa* however totally disappeared from this beach after 2012. In 2006, the survey data indicates the existence of several patches of *Spinifex littoreus* at this beach covering an area of about 6720 square feet. But by 2011, the rapid development of tourism in this part of the coastal belt of West Bengal has led to the construction of hotels, markets etc. and this has caused shrinkage and destruction of the *Spinifex littoreus* vegetation which are now been restricted to an area of just 1220 square feet at this particular beach. So, unfortunately the vegetation is disappearing gradually along with the other members of the coastal vegetation which is a growing concern in the recent years.

Table 5: Measurement of stomata from the samples collected from different areas

Place	Length of Stomata (μm)						Breadth of Stomata (μm)				
	Length	Avg.	SE(yEr±)	SD(yEr±)	Var.	Breadth	Avg.	SE(yEr±)	SD(yEr±)	Var.	
Mandarmoni	28	29.66	0.88	1.52	2.33	14	14.66	0.66	1.15	1.33	
	31					14					
	30					16					
Tazpur	27	28	0.57	1	1	13	13.66	0.33	0.57	0.33	
	29					14					
	28					14					
Udaypur	27	27.66	0.33	0.57	0.33	16	16	0.57	1	1	
	28					15					
	28					17					

Destruction of Native Vegetation

It has been observed that the loss of habitat possess greater impact than the loss of individual species, because of the ecological interactions between species leading to a chain of impacts. Many invasive species such as *Ipomoea fistulosa*, *Lantana camara* and *Calotropis gigantea* are found disturbing the native flora and fast encroaching over the native vegetation.

Anthropogenic activities

Several anthropogenic activities are responsible for habitat destruction. These include the construction of coastal protection, land reclamation, recreation and developments including hotels, fishing-harbours, industries and sand transportation (figures 9a,b&c). The huge tourist pressure possesses a significant threat to the coastal habitats at West Bengal. Artificial embankment with boulders and concrete seawall had been undertaken by the State Government to protect this coast during the last one-decade. But this measure did not prove to be yielding any effective benefit against erosion of the coast but has led to the loss of natural vegetation. Construction of seawall has also had a negative impact on the adjoining coastal areas with the dunes fast retreating landward. The State Government has made some attempts to conserve the neodune fields by wire fencing and plantation of certain exotic species like *Casuarina equisetifolia*, *Ipomoea fistulosa*, *Lantana camara*, *Calotropis gigantea*, bamboo, etc., but these invasive species have had a negative impact on the native vegetation and fast encroaching over them, leading to loss of some very rare species. The beach is under severe erosional threat from natural forcings of cyclonic storms, waves, tides and longshore drift.

CONCLUSION

The deeply rooted coastal grass *Spinifex littoreus*, a potent soil binder having capability to form strong, stable dunes is thus fast disappearing from the coastal belt of West Bengal and adjacent Odisha due to the anthropogenic activities mainly the boom in tourism in recent years. The invasive species have also had a negative impact on the native vegetation by fast encroaching over them, leading to loss of some very rare species.

Abbreviations

CSD: Coastal Sand dune

ACKNOWLEDGEMENTS

Authors are indebted to the UGC-DRS-SAP Laboratory for instrument facilities Department of Botany and Forestry, Vidyasagar University, Midnapore-721102

West Bengal, India, for providing the other necessary facilities for carrying out the work.

REFERENCES CITED

Banerjee LK. 1994. Conservation of coastal plant communities in India. *Bulletin of Botanical Survey of India* 6(1-4): 160-165.

Chakraborty T, Mondal AK, Mondal S. 2012. Studies on the prospects and problems of sand dune vegetation. *African Journal of Plant Sciences* 6 (2): 48-56.

Crutzen PJ. 2002. Geology of mankind. *Nature,* 415: 23.

Desai KN. 2000. Dune vegetation: need for a reappraisal, Coastin: A Coastal Policy. *Research Newsletter* 3-7.

Hellemaa P. 1999. The Development of Coastal Dunes and their vegetation in Finland. Helsinginyliopistonverkkojulkaisut.

Hooker JD. 1884. *The Flora of British India.* L. Reeve and Co. Kent, London.

Jacqes CJ. 1991. The sand dunes and their vegetation along the Mediterranean coast of France.Their likely response to climate change. *Landscape Ecology.* 6 (1&2): 65-75.

Maiti S, Bhattacharya A. 2009. Shoreline change analysis and its application to prediction: a remote sensing and statistics based approach. *Marine Geology* 257: 11-23.

Metcalfe CR. 1960. *Anatomy of the monocotyledons. I. Gramineae.* At the Clarendon Press. Oxford, England, p. 731.

Narpat S, Shekhawat et al. 2012. Bioresearches of Fragile Ecosystem/Desert. *Proceeding of the National Academy of Sciences, India. Sectrion B. Biological Sciences* 82(S2): 319–334.

Simpson EH. 1949. Measurement of diversity. *Nature* 163: 688.

Singh RP, Agrawal M. 2007. Effects of sewage sludge amendment on heavy metal accumulation and consequent responses of Beta vulgaris plants. *Chemosphere* 67(11): 2229-2240.

12

Plant Growth Promoting Bacteria as a Potent Tool in Amelioration of Salinity Stress: A Review

Prachi Sharma and Ratul Baishya*

Department of Botany, University of Delhi, Delhi-110007, INDIA
*Email: rbaishyadu@gmail.com

ABSTRACT

Soil salinity is one of the most serious environmental issues that limit agricultural productivity. Most of the food crops belong to glycophytes and hence, they are not able to withstand salinity stress. Salt stress significantly affects the various morphological, biochemical and physiological processes of the plant which leads to poor plant growth. Therefore, there is an emergent need to find out mitigation strategies to cope with the deleterious impacts of salt stress. Plant breeding techniques, efficient resource management and genetic engineering have been employed to cope with salt stress but these were of limited success as salt tolerant trait is complex both genetically and physiologically. Therefore, using plant growth promoting bacteria seems to be a cost effective and sustainable approach for salinity stress management. Hence plant growth promoting bacteria could play a significant role in alleviating salt stress. This chapter endeavor to provide an overview of the various important mechanisms such as facilitation of nutrient uptake, production of phytohormones, ACC deaminase production and siderophore production employed by the plant growth promoting bacteria to enhance the plant growth under salinity stress.

Keywords: Plant Growth Promoting Bacteria, Salinity Stress, Phytohormones, ACC deaminase, Siderophore

INTRODUCTION

Soil salinity is a serious environmental problem which is reducing the agricultural productivity worldwide. Scarcity of water resources, environmental pollution and increased salinization of soil are the major environmental issues prevailing in the world (Shrivastava & Kumar 2015). Increasing human population and reduction in the land available for cultivation are the two main threats for sustainable agriculture (Shahbaz & Ashraf 2013). According to Barot and Patel (2011), approximately 7 million hectares of land is covered by saline soil in India and the area under salinity is increasing day by day due to availability of irrigation in new areas. The world population is going to increase by 2.3 billion people till 2050 and production of 70% more food crop will be a major challenge (FAO 2009). Salinity stress is the major problem which limits agricultural production for the growing population. Plants are divided into two types on the basis of whether they can tolerate salinity stress or not. Halophytes can withstand salinity and glycophytes are not able to withstand salinity stress. Majority of the food crops fall in the second category that is the glycophytes and therefore, salinity stress limits their productivity (Gupta & Huang 2014).

Soil salinity is a condition in which the soluble salts are present in very high concentrations. Soils are defined as saline when the Electrical Conductivity (EC) value is 4ds/m or more, which is approximately equal to 40Mm NaCl. Most of the crop plants show reduction in their growth at 4ds/m EC and some crops show reduction even at lower EC value than 4ds/m (Munns 2005). Approximately 20% of the total cultivated and 33% of the irrigated land is affected by high salinity (Jamil et al. 2011). Due to low precipitation, high surface evaporation, weathering of native rocks, poor irrigation practices the soil salinity is increasing at a rate of 10% annually in new areas (Shrivastava & Kumar 2015). Soil salinity involves changes in various physiological, biochemical and metabolic processes of the soil (Rozema & Flowers 2008). Salt stress causes interruption in the osmotic balance of the plant, as in the initial stages of the stress, the water absorption capacity of the root system decreases abruptly and water loss from the leaves is accelerated and thus, salinity is considered as a hyperosmotic stress. Na^+ and Cl^- ions accumulate in the plant tissue which is exposed to soil having high NaCl concentration and this situation becomes highly detrimental to the plant (Munns 2005). Reactive oxygen species (ROS) is produced in the plant cells in response to salinity stress and this ROS can lead to oxidative damage to the proteins, lipids and DNA found in various cellular compartments, and thereby, interrupting various important functions of the plant cell (Gupta & Huang 2014).

There are two types of causes for salinity stress, one is primary and the other is secondary. Primary causes includes natural geological, hydrological and pedological processes, while the secondary causes include deforestation, industrial emissions, salinization due to contamination with chemicals and problem of overgrazing in arid and semi- arid regions (Yadav et al. 2011). There are two ways in which the salt stress affects the plant growth. Firstly it becomes difficult for the plant roots to extract water from salt affected soils, and therefore, high salt concentration is toxic to the plant in many ways. Secondly, the growth and the metabolism of root cells which are in direct contact with toxic levels of salt are highly affected (Munns & Tester 2008). Salinity stress results in secondary stresses such as osmotic and oxidative stress as well as ionic stress responsible for the plant poor growth (Gill & Tuteja 2010). The major deleterious effects of salinity on plants are: osmotic imbalance, lack of nutrients, reduces the photosynthetic ability and interrupts with other major physiological processes in the soil (Parida & Das 2005).

Many economic, scientific and technological measures have been designed to support farmers and increase food production in developed countries. However, it is difficult to find evidence of successful implementation of these approaches in developing countries. Implementation is usually a problem due to challenges related to financing, technical issues, policy management, climatic and environment conditions (Raimi et al. 2017). The salt tolerance trait is complex both genetically and physiologically and hence, several approaches such as conventional breeding programs, interspecific hybridization and use of transgenic plants to improve the salt tolerance of the plants were of limited success (Flowers 2004). Moreover, these strategies are long drawn and cost intensive and thus, there is a need to develop simple and low cost methods for the management of various abiotic stress like salinity stress and can be used on short term basis (Grover et al. 2010). Various strategies have been developed in order to decrease the detrimental effects of salt stress on plant growth and development which mainly includes plant genetic engineering and recently includes the use of plant growth promoting bacteria (PGPB) (Dimkpa et al. 2009). The use of PGPB for salt stress alleviation is found to be cost- effective and sustainable approach to facilitate plant growth in saline conditions. PGPB are free living soil bacteria that uses various mechanisms like fixed nitrogen, phytohormones production and phosphate solubilization and for increasing resistance of plants towards various environmental stresses (Mayak et al. 2004). These PGPB also acts as a unique model for understanding the salt tolerance, adaptation and response mechanism and thus, this information can be utilized in genetic engineering of new crop plants that can withstand various abiotic stresses present in the environment induced by climate change.

PLANT GROWTH PROMOTING BACTERIA (PGPB)

Plant growth promoting bacteria (PGPB) includes bacteria that inhabits the soil rhizosphere and facilitates plant growth through various direct and indirect mechanisms (Ahmed & Kibret 2014). Direct mechanisms include enhanced availability of nutrients like nitrogen, phosphorus, production of phytohormones while the indirect mechanisms involve suppression of pests and pathogens with the help of antibiotics and lytic enzymes, and induced systemic resistance (ISR) (Glick 2014). PGPB induces various physiological and biochemical changes in the plant that induces salt tolerance through elicitation of induced systemic tolerance (Pieterse et al. 2009). Table 1 shows different mechanisms employed by PGPB to alleviate salt stress in plants. PGPB are very potent tools for providing insoluble phosphorus from the organic and inorganic forms via solubilization and mineralization processes and thus, aids in the proper uptake of nutrients and minerals from the soil (Chen et al. 2006). This induced systemic tolerance involves changes in the phytohormone levels, antioxidant defense and production of bacterial exopolysaccharides (Cohen et al. 2015).

Table 1: Mechanisms used by bacterial species to alleviate salt stress

S.No.	Bacterial Species	Mechanism	References
1.	*Microbacterium oleivorans*	Improved overall plant growth and increased chlorophyll content.	Halm et al. (2017)
2.	*Brevibacterium iodinum*		
3.	*Rhizobium massiliae*	Siderophore production and ACC deaminase activity	
4.	*Pseudomonas aeruginosa*	Improved plant growth and photosynthetic efficiency	Tank and Saraf (2010)
5.	*Pseudomonas stutzeri*		
6.	*Brevibacterium iodinum*	Produces ACC deaminase	Siddikee et al. (2011)
7.	*Bacillus licheniformis*		
8.	*Zhihengliuela alba*		
9.	*Bacillus firmus*	Produces ACC deaminase and has phosphate solubilizing activity	Gururani et al. (2012)
10.	*Chryseobacterium sp.*	Siderophore Production	
11.	*Enterobacter sp.*	ACC deaminase activity	Habib et al. (2016)
12.	*Bacillus megaterium*		
13.	*Bacillus pumilus*	ACC deaminase activity	Bharti et al. (2012)
14.	*Exiguobacterium oxidotoluans*	Produces Exopolysaccharide	
15.	*Pseudomonas fluorescens*	ACC deaminase activity	Tank and Saraf (2010); Grover et al. (2010)
16.	*Serratia proteamaculans*	Improves plant growth and fix nitrogen in legume plants.	Han and Lee (2005)
17.	*Rhizobium leguminosarum*		
18.	*Bacillus amylolequifacians*	Restricted Na^+ influx	Grover et al. (2010)
19.	*Scytonema* sp.	Produces gibberellic acid and extra cellular products.	

Contd.

20.	*Azospirillum brasilense*	Enhanced the mineral uptake of the plant.	Spark (2010)
21.	*Pseudomonas mendocina*	ACC deaminase activity and induces high level of antioxidant enzymes.	Kohler et al. (2009)
22.	*Achromobacter piechaundii*	ACC deaminase activity	Mayak et al. (2004)
23.	*Bacillus subtilis*	Solubilize Potassium and Phosphorus by the production of organic acids, produces IAA and Exopolysaccharide.	Anjanadevi et al. (2015)
24.	*Bacillus megaterium*		
25.	*Bacillus pumilis*	Solubilize Potassium and thus helps in plant growth promotion.	Jha (2017)
26.	*Pseudomonas pseudoalcaligens*		
27.	*Pseudomonas putida*	ACC deaminase activity and increases the chlorophyll content.	Zahir et al. (2009)
28.	*Pseudomonas aeruginosa*		
29.	*Serratiaproteamaculans*		
30.	*Bacillus polymyxa*	Promoted plant growth and nutrient uptake.	Dimkpa et al. (2009)
31.	*Mycobacterium phlei*		
32.	*Pantoea agglomerans*	Solubilizes Potassium	Yaghoubi Khanghahi et al. (2017)

Salinity stress is root borne stress and thus, causes ionic imbalance and water deficit conditions which mainly affect the photosynthetic process. PGPB causes modifications in the root system architecture under salt stress conditions which is a stress defensive mechanism (Postma & Lynch 2011). Modification in root system architecture includes increase in the root tip number and root surface area which would facilitates water and proper nutrient conductance (Kaushal & Wani 2016).These bacteria not only causes changes in the physiology and metabolism of the plant but also has the ability to induce changes in the production of defensive proteins which helps the plant during various types of environmental stresses (Glick 1999). Different types of mechanisms are employed by these bacteria to facilitate plant growth promotion such as (1) fulfilling nutrient requirement of the plant by facilitating in the acquisition of different important nutrients like nitrogen, phosphorus and iron, (2) by serving as an effective biocontrol agents by synthesizing antibiotics and different types of lytic enzymes (3) or by PGPB phytohormones like auxin, cytokinin and reducing the levels of ethylene which can induce senescence in the plant tissue during the stress conditions and this is done with the help of the enzyme 1- aminocyclopropanc-1-carboxylate (ACC) deaminase (Glick 2007). During the salinity stress the levels of reactive oxygen species increases and thus, the level of anti-oxidant enzyme increases.

(Han & Lee 2005) reported decreased activity of anti-oxidant enzymes in the PGPB inoculated lettuce plant as compared to the control plant and thus show that inoculation with PGPB helps in salt stress alleviation. PGPB produces exopolysaccharides (EPS) constituting homo and hetero polysaccharides. Polysaccharide composition varies from species to species, in which common monomers include glucose, galactose and mannose (Kaushal & Wani 2016).

EPS production by PGPB strains have an important role in supporting plant growth in salt stress conditions, as it forms hydrophilic biofilms colonizing roots of the plant and thus protect it from desiccation (Rossi et al. 2012). EPS producing bacteria induces development of extensive root system and increases shoot growth in plants during salt stress condition (Kaushal & Wani 2016). Microorganisms can impart tolerance to the plants towards various abiotic stresses such as salinity, heavy metal, and drought. According to the recent research, bacteria belonging to genera *Rhizobium, Bacillus, Pseudomonas, Pantoea, Paenibacillus, Burkholderia, Achromobacter, Azospirillum, Microbacterium, Ethylobacterium, Variovorax* and *Enterobacter* have been reported to facilitate plant growth promotion during abiotic stress condition (Grover et al. 2010). Some of the PGPB produces IAA, gibberellins and results in increased plant growth by root length, root surface area leading to enhanced water and nutrient uptake under saline conditions (Shrivastava & Kumar 2015).

MECHANISMS OF PGPB THAT AIDS IN ALLEVIATING SALT STRESS

Nitrogen Fixing Bacteria

Most of the agricultural soils lack sufficient amount of nutrients like fixed nitrogen, phosphorus and iron. These important elements are provided by the plant growth promoting bacteria and by facilitating nutrient uptake these bacteria help in the plant growth promotion. Most of the farmers are dependent on chemical fertilizers for providing nutrients to the agricultural soils but these are costly and very hazardous to the environment in many ways (Glick 2012). Certain microorganisms are able to symbiotically fix atmospheric nitrogen by converting nitrogen gas from the atmosphere into biologically useful forms. Nitrogenase is the major enzyme which is utilized by bacteria to reduce nitrogen gas to ammonia (Dixon et al. 1986).

A variety of organisms have the ability to fix the nitrogen, but only a small number of species are able to do so. Approximately 87 species in 2 genera of archaea, 38 genera of bacteria and 20 genera of cyanobacteria are able to fix nitrogen (Zahran 1999). Plant growth promoting bacteria are very useful in agriculture as they include nitrogen fixing bacteria, mycorrhiza, plant disease suppressive bacteria and stress tolerant microbes. (Dhanasekar & Dhandapani 2012) reported in sunflower that efficient strains of *Azotobacter, Azospirillum, Phosphobacter* and *Rhizobacter* provide sufficient amount of nitrogen to the plant and also leads to increase in various growth parameters like plant height, number of leaves and seed weight. *Azotobacter* is an important nitrogen fixing bacteria which plays major role in the nitrogen cycle (Sahoo et al. 2013). Besides the fact that it provides nitrogen to the plants, it also produces phytohormones like IAA, gibberellins, cytokinins and vitamins such as thiamine and riboflavin (Revillas et al. 2000).

Azotobacter chroococcum inhibits pathogenic microorganisms around the root system and thus, improves the plant growth by increasing the seed germination. *Azospirillum* is another nitrogen fixing bacteria which is free living, motile and has the ability to survive in flooding conditions (Sahoo et al. 2014). Inoculation of plant by *Azospirillum* enhances the plant growth as it can modulate the root morphology by producing siderophore (Bhattacharyya & Jha 2011). It can also increase the number of roots and promotes root hair formation in order to facilitate nutrient and water uptake from the soil. Another very important nitrogen fixing bacteria is *Rhizobium.* This bacteria enters the root hairs and form nodules by multiplying as it is resistant to different temperature ranges (Nehra et al. 2007*).* *Rhizobium* inoculants were found to be effective in increasing the grain yields of groundnut, soybean, pea and Bengal gram (Patil & Medhane 1974). The strains of *Rhizobium* isolated from rice plant, provide nitrogen and enhance its growth and development (Peng et al. 2008). Photosynthetic ability of the plant is enhanced by one of the species of *Rhizobium* by increasing the endogenous level of the plant hormone (Bhardwaj et al. 2014). Two strains of *Azotobacter chroococcum* were tested for their plant growth promotion activities like nitrogen fixing ability, production of siderophores, Exopolysaccharide production and IAA production and it was found that the two strains were highly tolerant to salt stress and were able to alleviate the negative effects of salt stress on tomato plant. Beneficial rhizospheric bacteria like *Azospirillum* play a major role in plant growth promotion under stress conditions by several mechanisms. Among those mechanisms, essential role is played by increasing the mineral uptake in plants when inoculated with these bacteria (Bashan & Levanony 1991). These properties enhance tolerance towards various abiotic stress (Viscardi et al. 2016).

Phosphate Solubilizing Bacteria

Soil microorganisms across different genera which are able to convert insoluble phosphorus into soluble form which is accessible to the plants are known as phosphate solubilizing microorganisms and these have been found as best ecofriendly option for providing inexpensive phosphorus to the plants (Sharma et al. 2013). Phosphate solubilizing bacteria convert the insoluble form of phosphorus like tricalcium phosphate and iron phosphate to soluble form of phosphorus (Gupta et al. 2004). Among all mechanisms adopted by the microorganisms, the most common and well recognized method is the secretion of low molecular weight organic acids for phosphorus solubilization (Maliha et al. 2004, Khan et al. 2010). The quality of the organic acid is more important for the process of phosphate solubilization as the efficiency of phosphate solubilization depends on the type of organic acid produced along with its concentration (Khan 2016).

There are several benefits to plants from phosphate solubilizing bacteria which include increase in seed germination rate, root growth, yield, leaf area, chlorophyll content, nutrient uptake, protein content, hydraulic activity, tolerance to abiotic stress, shoot and root weights, biocontrol, and delayed senescence (Adesemoye & Kloepper 2009). Other beneficial effects of phosphate solubilizing bacteria include enhancing phosphorus availability, sequestering iron for plants by production of siderophore,producing plant hormones such as gibberellins, cytokinins, and auxins; and synthesizing the enzyme 1-amino cyclopropane-1-carboxylate (ACC) deaminase, which lowers plant levels of ethylene, thereby reducing environmental stress on plants (Glick et al. 2007, Adesemoye & Kloepper 2009). Phosphate solubilizing bacteria constitute 1-50% of the total population present in the soil (Chen et al. 2006). Whitelaw (1999) reported that bacterial strains from genera *Pseudomonas, Bacillus, Rhizobium* and *Enterobacter* are the effective phosphate solubilizers present in the soil. Phosphate solubilization involves lowering pH of the surroundings by secretion of organic acids and insoluble form of phosphate is converted to soluble form via chelation of H^+ ions in the environment (Shahid et al. 2012).

Potassium Solubilizing Bacteria

Potassium is one of the most essential macronutrient required for the growth and development of crop plants. The availability of potassium in the soil for plant uptake is made through a number of factors which includes exchangeable and non- exchangeable potassium and rate of weathering of potassium containing minerals like Feldspar and mica (Sparks 1987). In many areas of India, soil is deficient in potassium both in terms of available and non- available forms. But since, potassium is one of the most important nutrient required by the plants, it largely effects the yield and productivity of the food crop. In order to provide the required amount of potassium to the plants, chemical fertilizers are used. The chemical potassic fertilizers are costly and also not eco-friendly in nature, thus, there is an emergent need to find out alternative approach that is cost effective and environmental friendly (Sheng 2005). Potassium solubilizers play a significant role in providing sufficient amount of potassium from the fixed forms of potassium such as clay minerals. Among the various bacterial strains *Bacillus mucilaginosus* and *Bacillus edaphicus* are described as potent potassium solubilizers (Anjanadevi et al. 2015).

Meena et al. (2016) reported that *B. mucilaginosus, B. circulanscan* and *B. edaphicus* are potent potassium solubilizers, which can solubilize the silicate rocks. There are certain factors like pH, oxygen, and the type of potassium mineral that influences the microbial solubilization of potassium in the soil (Etesami et al. 2017). Potassium solubilizing bacteria protects the plant from the deleterious effects of salt stress by enhancing the rate of stomatal conductance, maintaining osmotic balance and lipid peroxidation during salt stress conditions (Bhattacharyya & Jha 2011). The plants that were inoculated with potassium solubilizing bacteria showed greater tolerance towards salinity stress. The plants that were inoculated with *Pseudomonas pseudoalcaligenes* and *Bacillus pumilus* showed enhanced rate of photosynthesis in salt stress conditions and also inhibits leaf chlorosis (Jha 2017). Figure 1 shows the impact of salinity stress on plant and various mechanisms employed by PGPB to combat it.

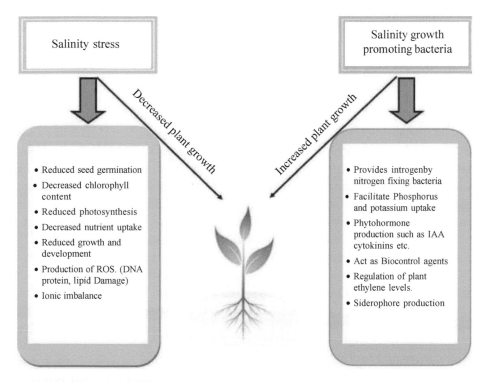

Fig. 1: Impact of salinity stress on plant growth and various mechanisms of plant growth promoting bacteria which facilitates in plant growth promotion.

Phytohormone Production by PGPB

Phytohormone includes auxins, gibberellins, cytokinins, ethylene and abscisic acid. Plant growth promoting bacteria are capable of producing these phytohormones and thus, helps in the plant growth promotion during stress conditions (Patten & Glick 1996). These phytohormones lead to enhancement in the plant cell size and cell division. Indole -3- acetic acid (IAA) is an auxin produced by many PGPB and it is well known for increasing the plant growth (Glick 2014). Since, IAA is known for inducing rapid cell elongation, cell division and differentiation, therefore, the plants treated with IAA showed highly developed roots and helps in proper uptake of water and minerals (Goswami et al. 2016). IAA known to stimulate root growth, helps in ameliorating salt stress, interaction between legume and rhizobia, plant-pathogen interaction. They are employed in stimulating induced systemic resistance against various pathogens but it is basically involved in increasing the surface area and length of lateral roots and thus, help the plant in uptake of water and soil minerals (Egamberdieva 2009). Bacterial IAA increases tolerance of the plant towards different stress by stimulating different physiological changes under different abiotic stress conditions (Ahmed 2014). Increase in the rate of seed germination was observed in canola and lettuce when inoculated with IAA producing strain of *Rhizobium leguminosarum* (Khan 2016). Egamberdieva (2009) reported significant increase in the root and shoot growth of the wheat seedling, when inoculated with bacterial strains producing IAA under saline conditions.

PGPB as a Potent Biocontrol

The major indirect mechanism of plant growth promotion in rhizobacteria is through acting as biocontrol agents. PGPB uses various mechanisms such as inducing systemic resistance, production of antifungal metabolites like producing fungal cell wall degrading enzymes and also (Ahemad & Kibret 2014). Moreover, these bacteria mediate biological control indirectly by eliciting induced systemic resistance against a number of plant diseases. Application of some phosphate solubilizing bacterial strains to seeds or seedlings has also been found to lead to a state of induced systemic resistance in the treated plant (Behera et al. 2014). Some bacteria produce a wide range of low-molecular-weight metabolites with antifungal potential. The best known is hydrogen cyanide (HCN), to which the producing bacterium, usually a pseudomonad, is resistant. HCN produced by bacteria can inhibit the black root rot pathogens of tobacco (Hillel et al. 2005).

Regulation of Plant Ethylene Levels

Plant releases ethylene as a physiological response under different types of stress and thus, ethylene is also known as stress hormone (Grichko & Glick 2001). Ethylene has role in various physiological processes in the plant cell which facilitates growth and development in plants. At high concentration it causes reduction in plant growth and development by inducing senescence and chlorosis in the plant tissue thus, having inhibitory effect on the plant growth (Khan 2016). The rate of ethylene biosynthesis increases during salinity stress as the stress conditions leads to an increase in the levels of 1-aminocyclopropane-1-carboxylic acid (ACC) which is an immediate precursor of ethylene in plants (Hontzeas et al. 2004). Therefore, it becomes important to control the levels of ACC for better plant growth under stress conditions. Plant growth promoting bacteria are known to produce ACC deaminase enzyme, which has the ability to hydrolyze the endogenous levels of ACC into α-ketobutyrate and ammonia and thus, reducing the levels of plant ethylene (Tank & Saraf 2010). Report shows that lettuce and maize plant inoculated with ACC producing PGPB leads to increase in the level of chlorophyll content (Han & Lee 2005). As a result of producing ACC deaminase enzyme, plant growth promoting bacteria are able to reduce ethylene levels in the plant under a wide range of environmental stresses and thus, plants inoculated with these bacteria have longer roots and shoots and are more resistant to the inhibitory effects of ethylene at higher concentration (Huang et al. 2005). ACC deaminase producing bacteria induces resistant in the plant towards various bacterial and fungal infections such as inhibition of growth in cucumber plant by *Pythium ultimum* a root pathogen is prevented by these bacteria (Wang et al. 2000). A lot of attention has been given to develop salinity resistant varieties of plant through genetic engineering but proved to be of a little success (Flowers 2004).

Plants inoculated with PGPB producing ACC deaminase seems to be an effective alternative approach to combat problems associated with the inhibitory effects of salinity stress in plants. Bacterium *Achromobacter piechaudii* effectively reduced the levels of ethylene in the tomato plants grown in high salt concentration and thus, prevents the plant from the inhibitory effects of ethylene (Mayak et al. 2004). ACC deaminase enzyme is produced in limited amount in *Rhizobia* strains as compared to free living bacteria. Therefore, there are two types of ACC deaminase producing bacteria. One is a free living bacteria which non-specifically binds to plant tissue and produces dramatically high concentration of ACC deaminase enzyme. On the other hand, *Rhizobia* strains specifically binds to the roots of certain plants, producing low level of enzyme which promotes nodulation by locally reducing the ethylene levels (Glick 2007). It is well documented that ACC deaminase producing PGPB aids in promoting plant growth under different types of stress conditions. Such as plants inoculated with *Pseudomonas* and *Serratia* strains producing ACC deaminase are effective in improving growth of wheat plant under salt stress conditions (Zahir et al. 2009).

Siderophore production

Iron is a vital nutrient required for almost all the important physiological processes such as respiration, photosynthesis and nitrogen fixation as in most of the enzymatic reactions it acts as a cofactor (Payne 1994). It is one of the most abundant element present in the soil but still it is not available to the plants and microorganisms. Iron is basically present in the form of Fe^{3+}, this oxidized form of iron reacts with insoluble oxides and hydroxides and thus, becomes inaccessible to plants (Goswami et al. 2006). So, the bacteria releases low molecular mass iron chelators which are known as siderophores and these molecules can readily form complex with iron as they have high association constant for forming complex with iron. These siderophores are mostly water soluble and are of two types that is extracellular siderophores and intracellular siderophores (Khan et al. 2009). Rhizobacteria are able to utilize siderophores of the same genus while other bacteria can utilize siderophores produced by different genera. Iron in Fe^{3+} siderophore complex is reduced to Fe^{2+} on bacterial membrane in both the Gram positive and Gram negative bacteria and it is released in the cell from it via the gating mechanism between the inner and the outer membranes. Therefore, siderophore aids in providing iron under iron deficient conditions (Ahmed & Kibret 2014).

There are three types of iron ligating groups which can be classified as (i) hydroxamate, (ii) catecholates (iii) hydroxyl acids/carboxylates. The type of ligand used by the siderophore is the basis for the classification of the siderophores and some of these may have more than one of the three chelating groups. Siderophore produced by rhizobia facilitates nodule formation by acquisition of iron from the soil and hence, aids growth promotion of nodulated legumes in the field condition. *Pseudomonas* strain produced siderophores which enhanced the symbiotic nitrogen fixation in clover (Khan 2016). Siderophore producing isolates of *Streptomyces* enhance the growth and development of wheat plant under saline stress conditions as reported by Sadeghi et al. (2012).

CONCLUSION

Various abiotic stresses such as salinity, drought and temperature are the major limitations in crop plant productivity. Among these stress, salinity stress is the most serious environmental issue as the land under salinity is increasing day by day and thus, limits the area under cultivation. There is an emergent need to develop salt tolerant plant species as to feed the ever increasing population. Several adaptations and mitigation strategies are employed to cope up with this issue. Among those are plant breeding and genetic engineering techniques. But these strategies are long drawn and cost intensive. Therefore, there is a need for a simple and cost effective approach. Thus, plant growth promoting bacteria play important role in this respect. PGPB show various mechanisms such as nitrogen fixation, phosphate and potassium solubilization by which it enhance plant growth under stress condition. Therefore, the use of plant growth promoting bacteria is considered to be a sustainable approach in providing tolerance to plants against abiotic stress. Further research should be focused on finding out various new bacterial strains that are tolerant to salt stress, and thus, can be used as inoculum for plants that are salt sensitive.

Abbreviations

ACC   : 1-Amino Cyclopropane 1- Carboxylate

EC   : Electrical Conductivity

EPS   : Exopolysaccharide

HCN   : Hydrogen Cyanide

IAA   : Indole – 3 – Acetic Acid

ISR   : Induced Systematic Resistance

KSB   : Potassium Solubilizing Bacteria

NaCl   : Sodium Chloride

PGPB   : Plant growth promoting bacteria

PSB   : Phosphate Solubilizing Bacteria

ACKNOWLEDGMENTS

The authors are deeply thankful to Department of Botany, University of Delhi, Delhi for providing necessary research facilities. The first author is also thankful to University Grants Commission, Delhi for providing necessary financial support in the form of UGC-Non NET fellowship and R&D research grants from University of Delhi.

REFERENCES

Adesemoye A, Kloepper J. 2009. Plant–microbes interactions in enhanced fertilizer-use efficiency. *Applied Microbiology and Biotechnology* 85: 1-12.

Ahemad M, Kibret M. 2014. Mechanisms and applications of plant growth promoting rhizobacteria: Current perspective. *Journal of King Saud University - Science* 26: 1-20.

Ahemad M. 2014. Phosphate-solubilizing bacteria-assisted phytoremediation of metalliferous soils: a review. *3 Biotech* 5: 111-121.

Anjanadevi I, John N, John K, Jeeva M, Misra R. 2015. Rock inhabiting potassium solubilizing bacteria from Kerala, India: characterization and possibility in chemical K fertilizer substitution. *Journal of Basic Microbiology* 56: 67-77.

Barot C, Patel V. 2011. Comparative Study of Seasonal Variation in Physicochemical Properties of Selected Wetlands of Mehsana Districts, North Gujarat. *Indian Journal of Applied Research* 4: 44-47.

Bashan Y, Levanony H. 1991. Alterations in membrane potential and in proton efflux in plant roots induced by *Azospirillum brasilense*. *Plant and Soil* 137: 99-103.

Behera B, Singdevsachan S, Mishra R, Dutta S, Thatoi H. 2014. Diversity, mechanism and biotechnology of phosphate solubilising microorganism in mangrove—A review. *Biocatalysis and Agricultural Biotechnology* 3: 97-110.

Bhardwaj D, Ansari M, Sahoo R, Tuteja N. 2014. Biofertilizers function as key player in sustainable agriculture by improving soil fertility, plant tolerance and crop productivity. *Microbial Cell Factories* 13: 66.

Bharti N, Yadav D, Barnawal D, Maji D, Kalra A. 2012. *Exiguobacteriumoxidotolerans*, a halotolerant plant growth promoting rhizobacteria, improves yield and content of secondary metabolites in *Bacopamonnieri* (L.) Pennell under primary and secondary salt stress. *World Journal of Microbiology and Biotechnology* 29: 379-387.

Bhattacharyya P, Jha D. 2011. Plant growth-promoting rhizobacteria (PGPR): emergence in agriculture. *World Journal of Microbiology and Biotechnology* 28: 1327-1350.

Chen YP, Rekha PD, Arun AB, Shen FT, Lai WA, Young CC. 2006. Phosphate solubilizing bacteria from subtropical soil and their tricalcium phosphate solubilizing abilities. *Applied Soil Ecology* 34: 33–41.

Cohen AC, Bottini R, Pontin M, Berli FJ, Moreno D, Boccanlandro H, Travaglia CN, Piccoli PN. 2015. *Azospirillum brasilense* ameliorates the response of *Arabidopsis thaliana* to drought mainly via enhancement of ABA levels. *PhysiologiaPlantarum* 153: 79–90.

Dhanasekar R, Dhandapani R. 2012. Effect of biofertilizers on the growth of *Helianthus annuus*. *International Journal of plant, Animal and Environmental Sciences* 2: 143–147.

Dimkpa C, Weinand T, Asch F. 2009. Plant-rhizobacteria interactions alleviate abiotic stress conditions. *Plant, Cell & Environment* 32: 1682-1694.

Dixon ROD, Wheeler CT. 1986. *Nitrogen fixation in plants*. Blackie, Glasgow, United Kingdom.

Egamberdieva D. 2009. Alleviation of salt stress by plant growth regulators and IAA producing bacteria in wheat. *ActaPhysiologiae Plantarum* 31: 861–864.

Etesami H, Emami S, Alikhani H. 2017. Potassium solubilizing bacteria (KSB): Mechanisms, promotion of plant growth, and future prospects a review. *Journal of Soil Science and Plant Nutrition* 17: 897-911.

FAO 2009. High Level Expert Forum-How to Feed the World in 2050, Economic and Social Development, Food and Agricultural Organization of the United Nations, Rome, Italy.

Flowers TJ. 2004. Improving crop salt tolerance. *Journal of Experimental Botany* 55: 307–319.

Gill S, Tuteja N. 2010. Reactive oxygen species and antioxidant machinery in abiotic stress tolerance in crop plants. *Plant Physiology and Biochemistry* 48: 909-930.

Glick B, Cheng Z, Czarny J, Duan J. 2007. Promotion of plant growth by ACC deaminase-producing soil bacteria. *European Journal of Plant Pathology* 119: 329-339.

Glick B. 2012. Plant Growth-Promoting Bacteria: Mechanisms and application. *Scientifica* 2012: 1-15.

Glick B. 2014. Bacteria with ACC deaminase can promote plant growth and help to feed the world. *Microbiological Research* 169: 30-39.

Glick BR, Patten CL, Holguin G, Penrose DM. 1999. *Biochemical and genetic mechanisms used by plant growth promoting bacteria*. London: Imperial College Press.

Goswami D, Thakker J, Dhandhukia P. 2016. Portraying mechanics of plant growth promoting rhizobacteria (PGPR): A review. *Cogent Food & Agriculture* 2: 1127500.

Grichko VP, Glick BR. 2001. Amelioration of flooding stress by ACC deaminase containing plant growth-promoting bacteria. *Plant Physiology and Biochemistry* 39: 11–17.

Grover M, Ali S, Sandhya V, Rasul A, Venkateswarlu B. 2010. Role of microorganisms in adaptation of agriculture crops to abiotic stresses. *World Journal of Microbiology and Biotechnology* 27: 1231-1240.

Gupta B, Huang B. 2014. Mechanism of Salinity Tolerance in Plants: Physiological, Biochemical, and Molecular Characterization. *International Journal of Genomics* 2014: 1-18.

Gupta D, Rai U, Sinha S, Tripathi R, Nautiyal B, Rai P, Inouhe M. 2004. Role of *Rhizobium* (CA-1) Inoculation in Increasing Growth and Metal Accumulation in *Cicerarietinum* L. Growing under Fly-Ash Stress Condition. *Bulletin of Environmental Contamination and Toxicology* 73: 424-431.

Gururani M, Upadhyaya C, Baskar V, Venkatesh J, Nookaraju A, Park S. 2012. Plant Growth-Promoting Rhizobacteria Enhance Abiotic Stress Tolerance in *Solanum tuberosum* Through Inducing Changes in the Expression of ROS-Scavenging Enzymes and Improved Photosynthetic Performance. *Journal of Plant Growth Regulation* 32: 245-258.

Habib S, Kausar H, Saud H. 2016. Plant Growth-Promoting Rhizobacteria Enhance Salinity Stress Tolerance in Okra through ROS-Scavenging Enzymes. *BioMed Research International* 2016: 1-10.

Hahm M, Son J, Hwang Y, Kwon D, Ghim S. 2017. Alleviation of Salt Stress in Pepper (*Capsicum annum* L.) Plants by Plant Growth-Promoting Rhizobacteria. *Journal of Microbiology and Biotechnology* 27: 1790-1797.

Han HS, Lee KD. 2005. Physiological responses of soybean-inoculation of *Bradyrhizobium japonicum* with PGPR in saline soil conditions. *Research journal of agriculture and biological sciences* 1: 216–221.

Hillel D, Hatfield J, Powlson D, Rosenzweig C, Scow K. 2005. *Encyclopedia of soils in the environment. Volume 2. Fe-M.* Oxford: Elsevier

Hontzeas N, Saleh SS, Glick BR. 2004. Changes in gene expression of canola roots induced by ACC-deaminase containing plant growth promoting bacteria. *Molecular Plant Microbe Interaction* 17(8): 865-871.

Huang X.-D, El-Alawai Y, Gurska J, Glick B. R, Greenberg B. M. 2005. A multi-process phytoremediation system for decontamination of persistent total petroleum hydrocarbons (TPHs) from soils. *Microchemical Journal* 81: 139–147.

Jamil A, Riaz S, Ashraf M, Foolad M.R. 2011. Gene expression profiling of plants under salt stress. *Critical Reviews in Plant Sciences* 30 (5): 435–458.

Jha Y. 2017. Potassium mobilizing bacteria: enhance potassium intake in paddy to regulate membrane permeability and accumulate carbohydrates under salinity stress. *Brazilian Journal of Biological Sciences* 4: 333-344.

Kaushal M, Wani S. 2016. Rhizobacterial-plant interactions: Strategies ensuring plant growth promotion under drought and salinity stress. *Agriculture, Ecosystems & Environment* 231: 68-78.

Khan M. 2016. *Phosphate solubilizing microorganisms.* Springer International Pu.

Khan MS, Zaidi A, Ahemad M, Oves M, Wani PA. 2010. Plant growth promotion by phosphate solubilizing fungi-current perspective. *Archives of Agronomy and Soil Science* 56: 73–98.

Khan MS, Zaidi A, Wani PA, Oves M. 2009. Role of plant growth promoting rhizobacteria in the remediation of metal contaminated soils. *Environmental Chemistry Letters* 7: 1–19.

Kohler J, Hernández J, Caravaca F, Roldán A. 2009. Induction of antioxidant enzymes is involved in the greater effectiveness of a PGPR versus AM fungi with respect to increasing the tolerance of lettuce to severe salt stress. *Environmental and Experimental Botany* 65: 245-252.

Maliha R, Samina K, Najma A, Sadia A, Farooq L. 2004. Organic acid production and phosphate solubilization by phosphate solubilizing microorganisms under in vitro conditions. *Pakistan Journal of Biological Sciences* 7: 187-196.

Mayak S, Tirosh T, Glick B. 2004. Plant growth-promoting bacteria confer resistance in tomato plants to salt stress. *Plant Physiology and Biochemistry* 42: 565-572.

Meena VS, Maurya BR, Verma JP, Meena RS. 2016. *Potassium solubilizing microorganisms for sustainable agriculture*. Springer.

Munns R, Tester M. 2008. Mechanisms of Salinity Tolerance. *Annual Review of Plant Biology* 59: 651-681.

Munns R. 2005. Genes and salt tolerance: bringing them together. *New Phytologist* 167: 645-663.

Nehra K, Yadav SA, Sehrawat AR, Vashishat RK. 2007. Characterization of heat resistant mutant strains of *Rhizobium* sp. [Cajanus] for growth, survival and symbiotic properties. *Indian Journal of Microbiology* 47: 329–335.

Parida AK, Das AB. 2005. Salt tolerance and salinity effects on plants. *Ecotoxicology and Environmental Safety* 60: 324–349.

Patil P, Medhane N. 1974. Seed inoculation studies in gram (*Cicerarietinum* L.) with different strains of *Rhizobium* sp. *Plant and Soil* 40: 221-223.

Patten C, Glick B. 1996. Bacterial biosynthesis of indole-3-acetic acid. *Canadian Journal of Microbiology* 42: 207-220.

Payne SM. 1994. Detection, isolation, and characterization of siderophores. *Methods in Enzymology* 235: 329–344.

Peng G, Yuan Q, Li H, Zhang W, Tan Z. 2008. *Rhizobium oryzae* sp. nov., isolated from the wild rice *Oryzaalta*. *International Journal of Systematic and Evolutionary Microbiology* 58: 2158-2163.

Pieterse C, Leon-Reyes A, Van der Ent S, Van Wees S. 2009. Networking by small-molecule hormones in plant immunity. *Nature Chemical Biology* 5: 308-316.

Postma JA, Lynch JP. 2011. Root cortical aerenchyma enhances growth of *Zea mays* L. on soils with suboptimal availability of nitrogen, phosphorus and potassium. *Plant Physiology* 156: 1190–1201.

Raimi A, Adeleke R, Roopnarain A. 2017. Soil fertility challenges and Biofertiliser as a viable alternative for increasing smallholder farmer crop productivity in sub-Saharan Africa. *Cogent Food & Agriculture* 3: 1400933.

Revillas JJ, Rodelas B, Pozo C, Martinez-Toledo MV, Gonzalez LJ. 2000. Production of B-Group vitamins by two Azotobacterstrainswith phenolic compounds as sole carbon source under diazotrophicandadiazotrophic conditions. *Journal of Applied Microbiology* 89: 486–493.

Rossi F, Potrafka RM, Pichel FG, De Philippis R. 2012. The role of the exopolysaccharides in enhancing hydraulic conductivity of biological soil crusts. *Soil Biology and Biochemistry* 46: 33–40.

Rozema J, Flowers T. 2008. Ecology: Crops for a Salinized World. *Science* 322: 1478-1480.

Sadeghi A, Karimi E, Dahaji PA, Javid MG, Dalvand Y, Askari H. 2012. Plant growth promoting activity of an auxin and siderophore producing isolate of *Streptomyces* under saline soil conditions. *World Journal of Microbiology and Biotechnology* 28: 1503–1509.

Sahoo RK, Ansari MW, Dangar TK, Mohanty S, Tuteja N. 2013. Phenotypic and molecular characterization of efficient nitrogen fixing *Azotobacter* strains of the rice fields. *Protoplasma* 251: 511-523.

Sahoo RK, Ansari MW, Pradhan M, Dangar TK, Mohanty S, Tuteja N. 2014. Phenotypic and molecular characterization of efficient native *Azospirillum* strains from rice fields for crop improvement. *Protoplasma* 251: 943-953.

Shahbaz M, Ashraf M. 2013. Improving Salinity Tolerance in Cereals. *Critical Reviews in Plant Sciences* 32: 237-249.

Shahid M, Hameed S, Imran A, Ali S, Van Elsas J. 2012. Root colonization and growth promotion of sunflower (*Helianthus annuus* L.) by phosphate solubilizing *Enterobacter* sp. Fs-11. *World Journal of Microbiology and Biotechnology* 28: 2749-2758.

Sharma SB, Sayyed RZ, Trivedi MH, Gobi TA. 2013. Phosphate solubilizing microbes: sustainable approach for managing phosphorus deficiency in agricultural soils. *Springer Plus* 2: 587.

Sheng XF. 2005. Growth promotion and increased potassium uptake of cotton and rape by a potassium releasing strain of *Bacillus edaphicus*. *Soil. Biology and Biochemistry* 37: 1918–1922.

Shrivastava P, Kumar R. 2015. Soil salinity: A serious environmental issue and plant growth promoting bacteria as one of the tools for its alleviation. *Saudi Journal of Biological Sciences* 22: 123-131.

Siddikee M, Glick B, Chauhan P, Yim W, Sa T. 2011. Enhancement of growth and salt tolerance of red pepper seedlings (*Capsicum annuum* L.) by regulating stress ethylene synthesis with halotolerant bacteria containing 1-aminocyclopropane-1-carboxylic acid deaminase activity. *Plant Physiology and Biochemistry* 49: 427-434.

Sparks D. 2010. *Advances in Agronomy*. Burlington: Academic Press.

Sparks DL. 1987. Potassium dynamics in soils. *Advances in Soil Science* 6: 1–63.

Tank N, Saraf M. 2010. Salinity-resistant plant growth promoting rhizobacteria ameliorates sodium chloride stress on tomato plants. *Journal of Plant Interactions* 5: 51-58.

Viscardi S, Ventorino V, Duran P, Maggio A, De Pascale S, Mora M, Pepe O. 2016. Assessment of plant growth promoting activities and abiotic stress tolerance of Azotobacter chroococcum strains for a potential use in sustainable agriculture. *Journal of Soil Science and Plant Nutrition* 16: 848-863.

Wang C, Knil E, Glick BR, De´fago G. 2000. Effect of transferring 1- aminocyclopropane-1-carboxylic acid (ACC) deaminase genes into *Pseudomonas fluorescens* strain CHA0 and its gacA derivative CHA96 on their growth promoting and disease-suppressive capacities. *Canadian Journal of Microbiology* 46: 898–907.

Whitelaw M. 1999. Growth promotion of plants inoculated with phosphate-solubilizing fungi. *Advances in Agronomy* 69: 99–151

Yadav S, Irfan M, Ahmad A, Hayat S. 2011. Causes of salinity and plant manifestations to salt stress: A review. *Journal of Environmental Biology* 32: 667-685.

Yaghoubi Khanghahi M, Pirdashti H, Rahimian H, Nematzadeh G, Ghajar Sepanlou M. 2017. Potassium solubilising bacteria (KSB) isolated from rice paddy soil: from isolation, identification to K use efficiency. *Symbiosis* 2017: 1-11.

Zahir Z, Ghani U, Naveed M, Nadeem S, Asghar H. 2009. Comparative effectiveness of *Pseudomonas* and *Serratia* sp. containing ACC-deaminase for improving growth and yield of wheat (*Triticumaestivum* L.) under salt-stressed conditions. *Archives of Microbiology* 191: 415-424.

Zahran HH. 1999. Microbiology and molecular biology reviews. *American Society for Microbiology* 63: 968-989.

13

Systematic Studies on Starch Grain of Selected Indian Pulses: Potential Health Benefit and Commercial Values

Shilpa Dinda[1] and Amal Kumar Mondal[*]

Department of Biological Sciences, Midnapore City College, Kuturia, Bhadutata, Midnapore, Paschim Medinipure-721129, West Bengal, INDIA
[2]Plant Taxonomy, Biosystematics and Molecular Taxonomy Laboratory UGC-DRS-SAP & DBT-BOOST-WB Funded Department, Department of Botany and Forestry, Vidyasagar University, Midnapore-721102, West Bengal, INDIA
[*]Email: amalcaebotvu@gmail.com, shilpadindabotvu@gmail.com

ABSTRACT

The people meet their need by taking pulses as their daily meal and therefore, the pulse play a significant role as food supplements. Pulses are called 'Nutrient Powerhouse' and are edible food of plants in the legume family. It contains high protein, fibers, various vitamins and amino acids. Starch is the main form in which plants store carbon. It occurs as semi crystalline granules composed of two polymers of glucose called amylose and amylopectin, depending on the plant organ. Starch granules are characterized by internal growth rings. Starch grain is a very useful taxonomic tool in plant systematic. In this investigation we atudied the morphological variations of starch grains in seeds. These starch grain contain the composition of carbon to form different structural diversity of seeds like simple i.e. two types: concentric and eccentric, semi-compound and compound. Polarized microscopy (PM) was used for better micro-morphological (Maltase cross) study. These micro-characters are carried out by UPGMA and construction dendrogram for visual appreciations of taxonomic relationship.

Keywords: Starch Grain, Numerical Analysis, Commercial Uses, Phylogeny Dendogram

INTRODUCTION

Leguminaceae is one of the largest families of flowering plants whose members widely distributed throughout the world. The characteristics feature of this family is its fruit in the form of pod. Based on variation of floral and inflorescence characters, family is divided into three sub-families, *viz.*, Papilionaceae, Mimosaceae and Caesalpiniaceae. The importance of micro-morphological feature for the taxonomic consideration of Angiosperms in established (Ramagya 1972, Tomlison 1979). Micro-morphological parameters of different plants have been used as aids in the taxonomic characters from a biosystematics point of view and taxonomic studies of a number of families were conducted across the world.

Pulses are edible seeds of plants in the legume family. This is the most important legume for nearly 300 million people living in developing countries. This crop is known as "meat of the Poor" (Romero-Arenas et al. 2013). The dry common beans (*Phaseolus vulgaries* L.) are widely consumed throughout the world (Juhi et al. 2010). They are rich protein, fibre, various vitamins and amino acids. As nutrient dense foods, pulses offer a wide range of health benefits. They are: rich in dietary fiber, low fat high protein sources, packed with essential micronutrients. Starch grains are small granules found in the leaves, roots, fruits, and seeds of plants. These grains serves as energy reserves for plants. People consumes starch grain for energy. The structure of starch is made up of 2 different chains known as linear helical amylase and branched amylopectin. Amylase is soluble while Amylopectin is insoluble. Starch is located in most plant tissues, particularly in storage organ such as rhizomes, tubers and grain. According to Whisttler and Paschall (1965), starches can be incorporated into a scheme of classification that marks it possible to determine their origin from particular species of plants.

Schimper believed that the young starch grains are composed of homogeneous dense substance .The grains increases in size and a weakly refractive region develop in middle called hilum. The formation of the hilum causes a reduction in the strain on the surrounding starch with the result that a loose layer forms between two denser ones. Starch molecules arranged themselves in the plant in semi-crystalline granules. Each plant species has a unique starch granular size. Starch becomes soluble in water when heated. The granules swell and burst. The semi crystalline structure is lost and the smaller amylase molecules start leaching out of the granules, forming a network that holds water and increasing the misture's viscosity.

Composition and Structure of Starch Grain

Starch or amyllum is a carbohydrate consisting of a large number of glucose units joined by glycosidic bonds. This polysaccharide is produced by most green plants as an energy store. Starch is basically a polymer of glucose, chemically at least two types of polymers are distinguishable: amylase and amylopectin. The molecular weight of amylase is around 250.000 (1500 glucose molecules) but varies widely. Amylose is mostly linear α-(1-4)- linked glucose polymer with a degree of polymerization (DP) of 1000-5000 glucose unit. The structure of this polymer was assumed to be mainly linear, but this appears to be the only 20-25 glucose molecules long. Amylopection has a molecular weight of 108. The ratio of amylose to amylopectin is relatively constant, which is 23. Amylopection is a much large glucose polymer (DP 105-106) in which α-(1-4) linked glucose polymers are connected by 5-6% α-(1-6)-linkage. Starch molecules arrange themselves in the plant in the semi-crystalline granules. Each plant species has a unique starch granular size.

Morphology and Location of Starch Grain

There are several morphological forms of starch grain in seed coat of seed in Leguminaceae and the number of starch grains per cell may also vary. The size and shape of starch granules are species specific. Microscopic techniques are widely used to characterize the size and the shape of starch granules. Classification of starch grain is invariably based on their shapes; for example, simple, semi-compound and compound starch grain and it also can be classified based on presence of hilum i.e. ecentric and concentric. The most common type of starch grain in legume family's seed is simple eccentric or concentric. Commonly round, oval shaped starch grain is found in leguminaceae. Sometime star shaped starch grain also recorded.

TYPES OF STARCH GRAIN

There are basically five types of starch grain

Simple starch grain

Starch grain has smooth wall and ovoid, round or renifrom shaped. That is two types-

Concentric simple starch grain

Striation of starch grain has been start from central position.

Eccentric simple starch grain

Striation of starch grain has been start from a side.

Semi-compound starch grain

Irregular shaped starch grain is semi- compound starch grain.

Compound starch grain

Polyhedral or aggregated starch grain is compound starch grain.

SAMPLE COLLECTION FOR INVESTIGATION

The selected plant material, seeds, are mostly collected from the study area of Purba and Paschim Medinipur, West Bengal, India at different day and time. The species identification of the selected material were determined according to standard literatures and it was done in the month from January to May (Table 1).

Table 1: Selected plant material (seeds), place and date of collection

Sl. No.	Scientific name of the plant species	Place of collection	Date of collection
1.	*Cajanus cajan* DC.	Dantan	17.02.2016
2.	*Cajanus scarabaeoides* (L.) Thouars	V.U. Campus	15.02.16
3.	*Cicer arietinum* L.	Tamluk	27.02.2016
4.	*Lathyrus sativus* L.	Tamluk	16.01.2016
5.	*Lens culinaris* L.	Tamluk.	15.01.2016
6.	*Phaseolus mungo* L.	Dantan	17.02.2016
7.	*Phaseolus rediatus* L.	Dantan	17.02.2016
8.	*Phaseolus vulgaris* L.	Tamluk.	17.02.2016
9.	*Paisum sativum* L.	Contai	18.01.2016
10.	*Vigna unguiculata* (L.) Walp.	Tamluk.	27.02.2016

Isolation of Starch Grain

The seed of leguminous plant isolated from fresh plant specimen. Seeds were crushed and these were added in 80% ethanol followed by shaking and centrifugation. The supernatants were discarded. Extraction procedure with ethanol was repeated. A small drop of water placed on one side of a standard microscope slide. Use a narrow pointed spatula or dissecting needle to transfer about 5 mg of sample on to the water. Mix thoroughly to disperse the starch. Place a cover slip over the suspension taking care to avoid entrapment of air bubbles; bleed or wick off excess water with a small piece of tissue paper held at the edge of the cover slip to obtain a thin film. Prepare a second mount on the other half of the same slide using glycerol or mineral oil. Compare appearances of the specimen under bright and polarized light in both media. Numerical analysis with the help of computer software (Origin-6.1) and dendrogram constructing with the help of software (NTss, Ver-2.2) (Table 2).

Table 2: Commercial importance of leguminaceae plant species for human health

Scientific Name	Uses of pulses
Cajanus cajan	i) Boost energy ii) Aids immunity iii) Digestive health iv) Healthy heart v) Prevent anemia
Cajanus scarabaeoides	i) Diabetes ii) Oral ulcer and inflammations. iii) Staunch blood, as an analgesic and to kill parasite
Cicer arietinum	i) Good source of protein, carbohydrate, minerals and vitamins b ii) Stomach ache
Lathyrus sativus	i) Pea seeds are used as food in many dishes ii) ODPA lis the main component iii) ODPA is responsible for making the crop tolerant to drought and flooding.
Lens culinaris	i) The unripe and green pods are used as vegetables. ii) The dried plants are used as fooder. iii) Some timesthe pulse is used as an excellent substitute of meat iv) Easy to digest and good for weight loss
Phaseolus mungo	i) The ripe seeds are used as dal, either whole or split in two. ii) It is easily digestable and is given during stomach disorder and during convalescence. iii) The plant is useful for crop rotation and is restoring the fertility of soil.
Phaseolus rediatus	i) Vitamin K and Copper rich ii) Anti-oxident
Phaseolus vulgaris	i) Anti-platelet ii) Anti-cancer iii) Anti-leukemia iv) Anti-diabetic
Paisum sativum	i) Good sources of carbohydrate and few amount of saccharose ii) Rich in minerals and Vitamins C, K, B1, B2 and B6
Vigna unguiculata	i) Anti-agening and anti-oxident ii) Vitamine K and C sources iii) Cell protection and boost immunity iv) Cardiovascular health

Analysis of Starch Grain: From seeds and seed coats of selected plant species

The present study is done over investigation on 7 genera and 10 species of sub-family Papilionoideae. In the sub family the micro-morphological character, starch grain show more divergence. Most of the species have mature and immature, granular, small, broad in size, round shaped, simple, types of starch grain. And some species have semi-compound, compound and polygonal compound types of starch grain in both seed and seed coat. Compound starch grains are aggregated with 2 or multi compound starch grain and semi-compound are irregular shaped. In the sub-family, immature starch grains are very abundantly rather than mature starch grain in both seed and seed coat. The striation over starch grain is absent in most of species of this sub-family but it found in fewer with two basic types striation viz. concentric and ecentric. Various types of hilum are found like cleft hilum, punctate hilum, linear hilum, stellate hilum, and lombdoidal hilum. In the sub-family, ovoid and reniform shaped starch grain are also present in seed and seed coat. Starch grains are abundently distributed in seed more than seed coat of the family. Mature starch grains are very much present more than immature starch grain in seed (Plate-I to V).

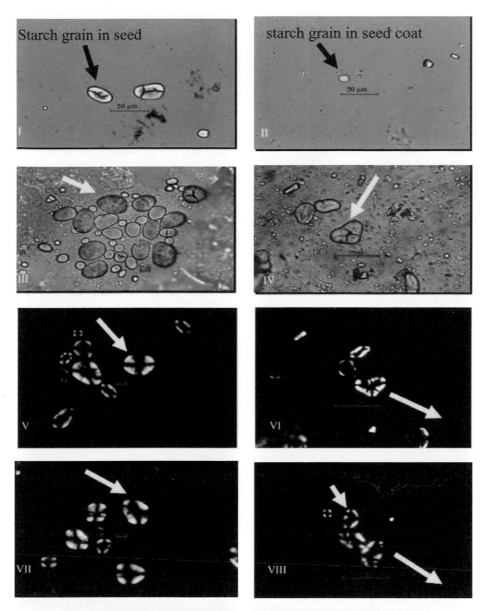

Plate- 1: Analysis of starch grain of *Cajanus cajan* under Light Microscope
(I, II, III, IV,) and Polarized microscope (V, VI, VII, VIII)

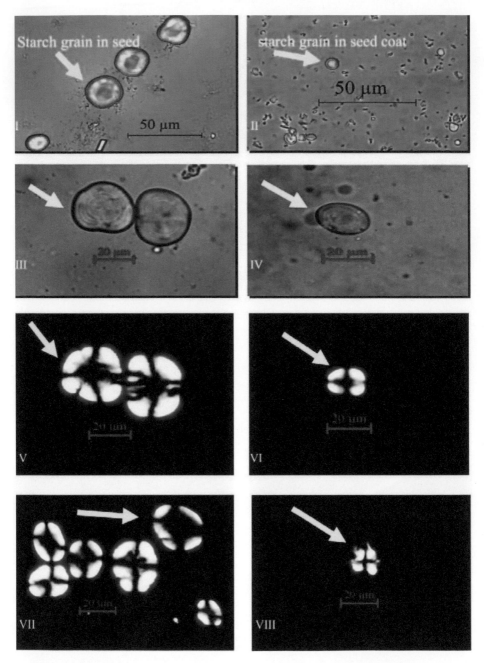

Plate - 2: Analysis of starch grain of *Cajanus scorobocoides* under Light Microscope
(I, II, 111, IV) and Polarized microscope (V, VI, VII, Vlll)

Plate - II

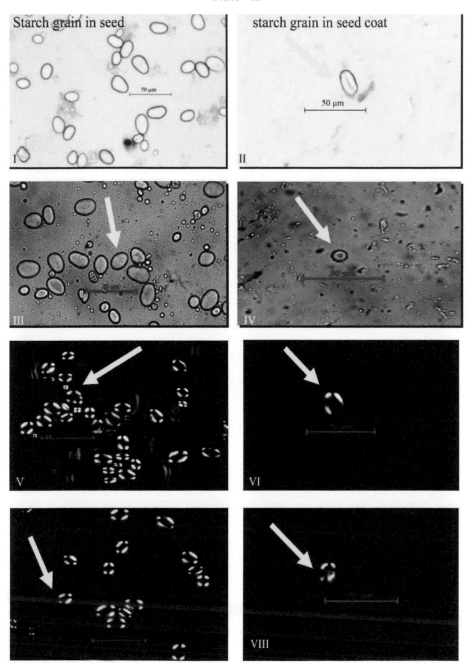

Plate 3: Analysis of starch grain of Cicer *arieninum* under Light Microscope (I,II,III,IV) and Polarized Microscope (V,VI,VII,VIII)

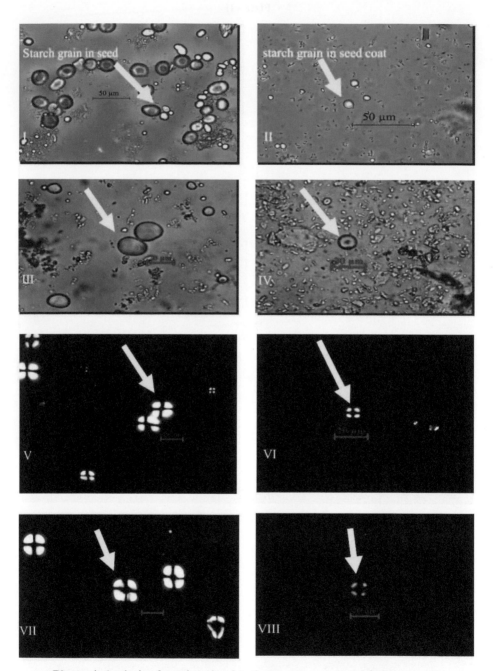

Plate - 4: Analysis of starch grain of *Lathyrus sativus* under Light Microscope
(I, II, III, IV) and Polarized Microscope (V, VI, VII, VIII)

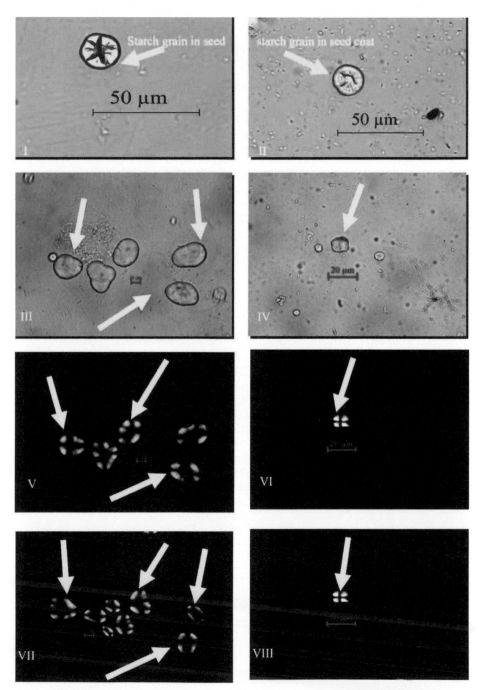

Plate-5: Analysis of starch grain of *Pisum sativum* under Light Microscope (I, II, III, IV) and Polarized Microscope (V, VI, VII, VIII)

Qualitative and Quantitative Micro-morphological Characters of Starch Grains

The qualitative and quantitative micro-morphological features of starch grain in seed of the selected take are summarized in tables which included starch grain such as measurements of including length and breadth (Table 3-6).

Table 3: Statistical analysis to length of seed

Scientific names	Mean ± Se	SD(yEr±)	Min.(Y)	Max.(Y)
Cajanus cajan	38.07 ± 0.24	0.59	31.49	39.01
Cajanus scarabaeoides	21.84 ± 1.62	3.62	16.75	25.80
Cicer arietinum	20.43 ± 1.40	2.96	17.44	24.53
Lathyrus sativus	24.93 ±1.86	4.16	21.16	32.08
Lens culinaris	22.98 ± 1.94	4.77	18.62	29.38
Phaseolus mungo	26.14 ± 1.20	2.70	23.58	29.41
Phaseolus rediatus	25.62 ± 1.46	3.88	21.18	31.81
Phaseolus valgaris	26.50 ± 1.33	3.27	22.8	31.5
Pisum sativum	39.91 ± 2.46	6.02	33.43	46.63
Vigna unguiculata	20.54 ± 1.12	2.75	15.97	24.12

Table 4: Statistical analysis to breadth of seed

Scientific names	Mean(Y)	SD (yEr±)	Min.(Y)	Max.(Y)
Cajanus cajan	27.28 ± 1.24	3.04	24.39	32.87
Cajanus scarabaeoides	19.56 ± 2.14	4.79	13.3	25.43
Cicer arietinum	13.72 ± 0.51	1.02	12.88	15.21
Lathyrus sativus	20.71 ± 0.73	1.64	8.7	22.78
Lens culinaris	17.00 ± 1.12	2.76	12.69	20.43
Phaseolus mungo	17.96 ± 0.54	1.22	16.38	19.58
Phaseolus rediatus	18.13 ± 0.68	1.80	15.91	20.68
Phaseolus valgaris	17.99 ± 1.25	3.06	15.84	23.68
Pisum sativum	21.08 ± 1.07	2.62	18.63	25.01
Vigna unguiculata	12.91 ± 0.59	1.44	11.05	15.19

Table 5: Statistical analysis to length of seed coat

Scientific names	Mean (Y)	SD (yEr±)	Min.(Y)	Max.(Y)
Cajanus cajan	27.79 ± 7.49	14.98	14.39	42.02
Cajanus scarabaeodies	7.27 ± 0.42	1.03	5.85	8.36
Cicer arietinum	20.68 ± 2.68	4.64	17.02	25.91
Lathyrus sativus	5.17 ± 0.52	1.04	4.02	6.25
Lens culinaris	7.76 ± 0.47	1.05	6.5	9.17
Phaseolus mungo	22.72 ± 1.42	3.19	18.42	26.42
Phaseolus rediatus	24.43 ± 1.53	3.43	20.5	28.45
Phaseolus valgaris	7.15 ± 0.50	1.05	6.00	8.25
Pisum sativum	15.48 ± 1.11	2.49	12.93	19.00
Vigna unguiculata	20.89 ± 2.47	3.49	18.42	23.36

Table 6: Statistical analysis of breadth of seed coat

Scientific names	Mean(Y)	SD(yEr±)	Min.(Y)	Max.(Y)
Cajanus cajan	18.12 ± 4.30	8.61	10.25	26.88
Cajanus scarabaeoides	6.53 ± 0.41	1.01	4.82	7.72
Cicer arietinum	15.41 ± 1.31	2.24	12.95	17.43
Lathyrus sativus	6.12 ± 0.46	0.92	5.25	6.99
Lens culinaris	6.17 ± 0.54	1.21	5.24	8.19
Phaseolus mungo	14.94 ± 0.53	1.19	13.46	16.34
Phaseolus rediatus	17.70 ±1.11	2.49	14.97	21.1
Phaseolus valgaris	4.5 ± 0.13	0.27	4.23	4.88
Pisum sativum	14.84 ± 1.24	2.77	12.02	19.34
Vigna unguiculata	17.05 ± 0.85	1.20	16.2	17.91

Grapli-l: Graphical representation of length and breadth of starch grain in the seeds under the sub-family Papilionoideae

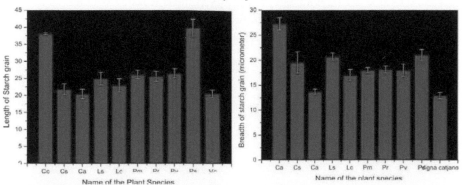

Graph-2: Graphical representation of length and breadth of starch grain in the seed coat nder the sub-family Papilionoideae

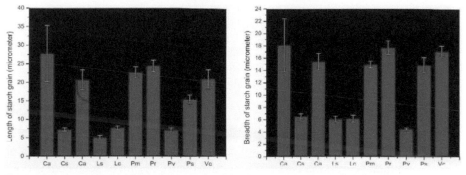

Ca-*cajanus cajan* Name of the olant soecies CS- *Cajanus scarabaeoides*, Ca- *Cicer arietinum*, Ls- *Lathyrus salivas*, Lc -*Lens culiiuiris*, Pm-*phaseolus mungo*, Pr-*phaseolus rediatus*, Pv-*Phaseolus vulgaris*, Ps-*Pisuin sativum*, Vc-*Vigna unguiculata*

The present study revealed that micro-morphological features of starch grain have a considerable taxonomic value. Starch grain may be present in the seed and seed coat leaves, stem etc. seed and seed coat's starch grain are simple, semi compound, compound concentric and ecentric type and mature and immature types of starch grain. *Pisum sativum* found highest length (39.91 ± 2.46 μm) and *Cajanus cajan* found highest breadth (27.28 ±1.24 μm) of starch grain in seeds. *Cicer arietinum* found lowest length (13.72 ± 0.51 μm) and *Vigna unguiculata* found lowest breadth (20.89 ± 2.47 μm) of starch grain in seeds.

In Figure 1, the linear regression curve of length of starch grain of seeds is done where R value is 0.1151, t value is 12.31176 and probability value is 6.18688 which is significant in 0.05 levels but not significant in 0.001 levels. The linear regression curve of breadth of starch grain of seeds is done where R value is 0.47337, t value is 14.59985 and probability value is 1.42639 which is significant in 0.05 levels but not significant in 0.001 levels. The highest length (27.79 ± 7.49 μm) and breadth (18.12 ± 4.30 μm) are found in *Cajanus cajan* in the starch grain of seed coat.

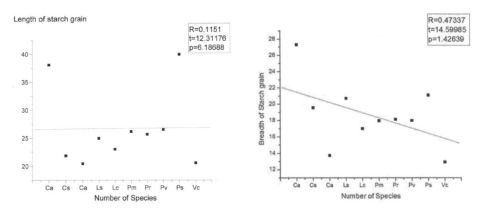

Fig. 1: Graphical representation of length and breadth of starch grain with Regrassion, t value and probability value (linear fit) in the seeds

In Figure 2, the linear regression curve of length of starch grain of seed coat is done where R value is 0.001, t value is 5.97063 and probability value is 2.09938 which is significant in 0.05 levels but not significant in 0.001 levels. In Graph No. **4** the linear regression curve of breadth of starch grain of seed coat is done where R value is 0.12374, t value is 6.90224 and probability value is 7.04934 which is significant in 0.05 levels but not significant in 0.001 levels.

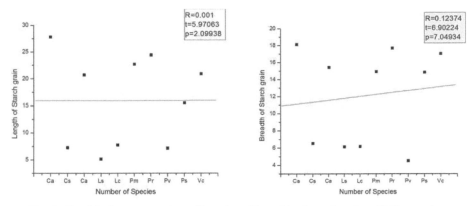

Fig. 2: Graphical representation of length and breadth of starch grain with Regrassion, t value and probability value (linear fit) in the seed coat

Dendrogram based on 54 micro-morphological characters of starch grain

The dendrogram constracted through similarity matrix using UPGMA (Unweighted Pair Groups Method with Arithmetic average) any the investigated species and showing the parrining affinity value. The mutualistic characters reveilles the four cluster. Cluster- 1 consist four species *Cajanus cajan, Lens culinaris, Phaseolus mungo* and *Phaseolus vulgaris*. In this cluster *Lens culinaris* and *Phaseolus vulgaris* were closely related because both the species the type of starch grain were similar and co-efficient value is 86%. Then closely related species were *Cajanus scarabaeoides* and *Lathyrus sativus* and co-efficient value is 81%. *Cicer arietinum* and *Phaseolus vulgaris* were closely related and their co-efficient value is 63%. Cluster number four *Paisum sativum* and *Vigna unguiculata* were closely related and co-efficient value is 60% (Figure 3).

CONCLUSION

Pulses are one of the most important crops in India and it is not expensive for the poor people. Pulses are considered as second source of protein and they are especially important for the nutrition of women and children. That's why they are called 'Nutrient Powerhouse'. Adequate amounts of protein are required every day for regulated body function and proper growth. They are great economic sources for small farmers. Pulses are easily digestible human food and starch grain also helps to digest food. The present study shows that the starch grains are important micro-morphological characters that helps regulate the body function. We are showing in this paper the pairing affinity value and constructed dendrogram on the basis of micro-morphological characters of starch grain of pulses.

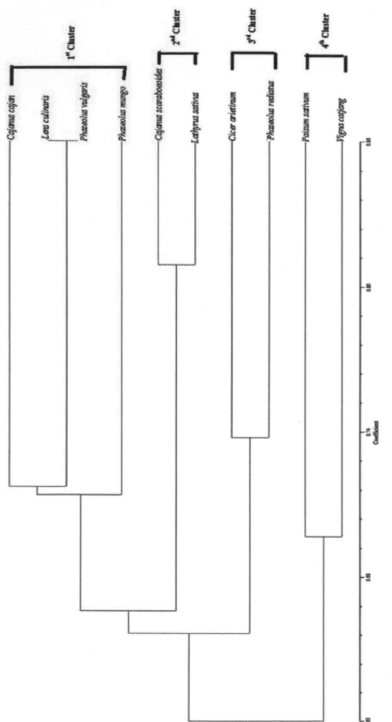

Fig. 3: Dendrogram showing co-relation among the 10 species of pulses based on micro-morphological character of starch grain.

Abbreviations

LM : Light Microscope

PM : Polarized Microscopy

SD : Standard deviation

SE : Standard error

yEr : Error in Y axis

UPGMA : Unweighted Pair Groups Method With Arithmetic Average.

Conflict of Interest

No conflict of interest

ACKNOWLEDGEMENTS

I am highly greatful to UGC-DRS-SAP, New Delhi for their financial support in the form of research project and also thanks to all of the research Scholar of the Plant Taxonomy, Biosystematic and Molecular Taxonomy Laboratory, UGC-DRS-SAP and DBT-BOOST-WB Funded Department, Department of Botany & Forestry, Vidyasagar University, Midnapore-721102, West Bengal, India. I am also grateful to the Technical staffs of University Science Instrumentation Centre (USIC), Vidyasagar University as well as the University of Burdwan and Bose Institute, Kolkata.

REFERENCES CITED

Juhi M, Rattan DS, Vikas SJ, Manju GA. 2010. Assesment of phenolic components and antioxidative activitites of *Phaseolus vulgaries* L. *International Journals of Integrative Biology* 9(1): 26-30.

Albert S, Sharma B. Comparative foliar micromorphological studies of some Bauhinia (Leguminosae) species. *Turkish Journal of Botany* 37:276-281.

Romero-Arenas O, Damian Huato MA, Rivera Tapia JA, Baez Simon A, Huerta Lara M, Cabrera Huerta E. 2013. The Nutritional Value of Beans (Phaseolus vulgaris L.) and its importance for Feeding of Rural communities In Puebla-Mexico. *International Journal of Biological Sciences* 2(8): 59-65.

Whister LR, Paschall EF. 1965. Starch: Chemistry and Technology. USA.

14

Improvement of Commercially Valuable Plants Through Genetic Manipulations

Ujjal Jyoti Phukan[1] and Vigyasa Singh[2]

[1]School of Life Sciences, Jawaharlal Nehru University, New Delhi-110067, INDIA
[2]Department of Molecular Bioprospection, CSIR-CIMAP, Lucknow-226015, INDIA
Email: ujjwal.phukan@gmail.com, vigyasa105@gmail.com

ABSTRACT

Commercially valuable plants are one of the main sources of economy. Growth, development and yield of these plants are severely affected by various environmental factors, which depend on dynamic fluctuations in different parameters. To counteract these responses, plants try to adapt evolutionarily but one form of adaptation is never going to suffice the simultaneous attack of multiple stresses. Crossing and plant breeding approaches came to practice, which is quite beneficial but requires lot of screening and is manpower consuming. In recent era new approaches have been developed combining recombinant DNA technology with tissue culture techniques to obtain improved varieties with specific desired traits. This method had an edge over other technologies but it still carries lot of complications. It is evident that genes do not act alone but act cooperatively and remain in continuous influence of other factors. So to target a particular gene for improved variety development requires a lot of detailed micro-details. In this communication, an attempt have been made to discuss mainly how transcription factors (TFs) have been targeted for this purpose. TFs in many occasions act as master regulators or central players of a particular pathway. This is attained by their ability to interact with multiple *cis*-elements present in the promoters of downstream stress or metabolism responsive genes. This behavior sometime leads to occurrence of some unwanted traits, which should be kept in mind to develop transgenics. Here, we focused mainly on important TF families such as ERF, WRKY, NAC and bHLH. We have been an attempt to analyzed how these TFs influence growth, yield and stress response of commercially valuable plants such as rice, wheat, tomato and cotton. This section will help to understand the genetic manipulation of plants can be targeted for improved variety development.

Keywords: Commercial Plants, Genetic Manipulation, ERF, WRKY, NAC.

INTRODUCTION

Plants and plant-derived products are the major source of our livelihood (Beyene et al. 2016). Whether they are used for foods such as rice or used for commercial purposes such as cotton, they proved to be an irreplaceable aspect of our lives. Most importantly they provide us oxygen to breathe and consume harmful CO_2. One important characteristic of plants is that they are sessile in nature because of which they are constantly exposed to harmful environmental factors. Growth, development and yield of commercially important plants are severely affected by either biotic (pathogens, insects and herbivores) or abiotic (drought, water-logging, salinity, cold and irradiation) stresses (Phukan et al. 2018).

During the course of time plants have evolved to counteract these adversities by various adaptations parameters. These adaptations include morphological alterations and physiological modifications. The point that is to be noted here is these adaptations vary from genus to genus (Phukan et al. 2014). Even at species level plants display significant difference in their ability to cope up with a particular stress (Ahsan et al. 2007). This happens because of their genetic architecture and how such is regulated. Keeping in mind the nature of these genetic variations, researchers have focused on the exploration of these factors for improvement of economically important plants. These factors include transcription factors (TFs) and stress/metabolism responsive genes. These genes positively or negatively regulate a particular plant response. An over-expression or knock-down of that gene facilitates plants with traits required to survive or surpass the respective stress. Regulation of a particular response by these is a complicated network and involves various integrated pathways. A thorough understanding of the complete grid is required to proceed with any genetic alteration of plants. Therefore it is the utmost need of the time that we should focus on these aspects of plant improvement.

In this communication, an attempt have been made to discuss how plants can be genetically altered for its better yield and stress tolerance. We will mainly target TFs because they are reported to regulate various developmental as well as stress responses and acts as central hub in many responses. We have also focused on those systems where the gene has been over-expressed or silenced in the native plant species. This is important because regulation of a particular gene differs even in closely related species and functional attributes tend to change because of different genetic architecture of the host plant. Till now working on *Arabidopsis* model is preferred because of short life span, knowledge of entire genome and availability of functional mutants. Also transformation procedures are very easy in *Arabidopsis* in comparison to others plants. Because of this for characterization of a gene we make transgenic in *Arabidopsis*, which is not suitable. Therefore in this chapter we will focus on only those economically important plants where transformation and regeneration protocols have been established.

IMPROVEMENT OF EDIBLE PLANT SPECIES

Plants are greatest source of food and widely cultivated all around the world. To meet the ever-increasing demand of rapid food supply to the growing market, it is necessary to increase the yield and the quality of products. To address these issues genetic manipulation of plants can be targeted. We will mainly discuss important food crops such as rice, wheat and tomato.

RICE CROPS

Transcription Factor Mediated Modulation

ERFs

Rice cultivation is greatly affected by various environmental factors. It possesses many genetic factors that can be explored for generation of improved breeds with better yield and quality. ERFs (Ethylene response factors) belong to AP2/ ERF family of TFs that are involved in various developmental and stress responses through interacting with the *cis*-elements present in the promoters of down-stream genes (Phukan et al. 2017). Therefore these TFs are key regulators of these complex signaling pathways. These interactions do not necessarily yield a positive or beneficial response each time. In many cases in order to attain a desired response they impart negative regulation on another response such as constitutive expression of drought tolerant ERF impart dwarf phenotype under normal conditions (Agarwal et al. 2006). Therefore for normal metabolism and plant survival auto-/cross regulation of these TFs at transcriptional, translational and domain level should be tightly checked. These are activated by different or interconnected signaling pathways and recognise the various *cis*-elements such as DRE (Dehydration Response Element), GCC box and JARE (Jasmonic Acid Response Element). As these TFs show specificity or plasticity to single or multiple *cis*-elements validate them as useful assets for genetic manipulation of commercially important crops. Also because of this unusual behaviour of interaction they can simultaneously regulate various downstream genes involved in different responses. Nevertheless the mechanism of this unique regulation is not universal and it may vary from genus to genus or even species to species (Phukan et al. 2014). One of the main concerns that come with these TFs is the unwanted associated traits, which could only be averted by further study and exploration of ERFs. There are 139 ERFs in rice (Nakano et al. 2006).

Based on DNA binding domain (DBD), AP2/ERFs are classified into AP2, RAV (related to Abscisic acid insensitive3/Viviparous1), DREB (subgroup A1-A6), ERF (subgroup B1-B6) and others (Sakuma et al. 2002). Some of the studies where ERFs have been targeted for improved plant responses are described. Constitutive expression of OsEREBP1 activates JA as well as ABA signaling cascades and provides better survival of rice under biotic and abiotic stresses (Jisha et al. 2015). Overexpression of an ERFOsAP37 in rice provided enhanced tolerance to drought and better grain yield under drought stress (Oh SJ et al. 2009). Recent reports demonstrate that OsERF71 provides enhanced tolerance to drought through modulation of ABA signaling and proline biosynthesis (Ahn et al. 2017, Li et al. 2018). During drought OsERF71 modifies root structure including larger aerenchyma and radial root growth to sustain drought response (Lee et al. 2017). OsERF71 is also reported to activate OsXIP (XIP-type rice xylanase inhibitor) and promote defence responses *via* a JA-mediated signaling pathway (Zhan et al. 2017). On the other hand over-expression of OsERF109 in rice negatively regulates biosynthesis of ethylene and drought tolerance (Yu et al. 2017). RD1 (Rice Drought-responsive) and RD2 also provide drought tolerance in rice (Samota et al. 2017). OsDRAP1 (DREB family protein) provides drought tolerance in rice (Huang et al. 2018). During submergence Sub1A (Submergence1A) ensures survival of rice (FR13A variety) under water by conserving its carbohydrate reserve, which could be reutilised once the stress is passed. Sub1A inhibit elongation by inducing expression of DELLA family GA signaling repressors SLR1 (Slender Rice 1) as well as SLRL1 (SLR1 Like 1) and restricting GA functioning (Fukao and Bailey-Serres 2008). Some rice varieties such as C9285 possesses ERFs SK1 (Snorkel1) and SK2 induce gibberellin-mediated internode elongation (by post-mitotic expansion of differentiating cells) to rapidly escape submergence tolerance (Hattori et al. 2009). Therefore these ERFs can be targeted for development of varieties that can withstand flooding stress by adapting different strategies. OsERF922 is reported to negatively regulate resistance to blast disease in rice (Wang et al. 2016).

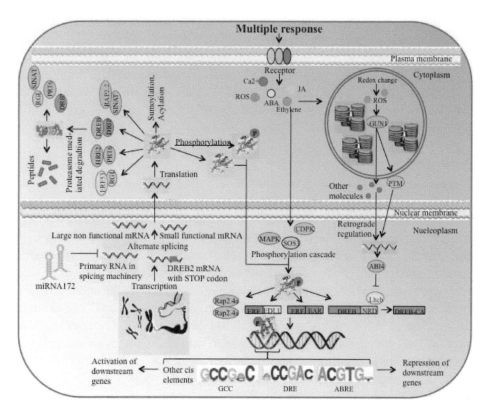

Fig. 1: Regulation of ERFs to modulate downstream stress and developmental responses. Various stimuli activate cell surface receptors, which activates MAPK phosphorylation cascade. This activates ERFs, which then interact with the cis-elements such as DRE, GCC, JARE present in the promoters of downstream genes to provide stress tolerance and promote development.

Another ERF OsEATB inhibits internode elongation process by repressing GA responsiveness through down-regulation of OsCPS2, a GA biosynthetic gene (Qi et al. 2011). Various ERFs are involved in vegetative growth development such as MFS1 (Multi-Floret Spikelet1) determines spikelet meristem and flower organ identity. MFS1 regulates these responses by up-regulating expression of various genes such as LSL (Long Sterile Lemma), IDS1 (Indeterminate Spikelet1), SNB (Supernumerary Bract) and OsIDS1 (Osindeterminate Spikelet 1, Ren et al. 2013). An auxin responsive ERF TF CRL5 (Crown Rootless 5) leads to increased expression of OsRR1 (a type-A response regulator of cytokinin signaling) and positively regulates emergence of crown roots in rice (Kitomi et al. 2011). HL6 (Hairy Leaf 6) physically interacts with OsWOX3B (a homeodomain-containing protein) and enhances expression of some auxin-related genes such as OsYUCCA5. This regulatory network promotes trichome formation in rice (Sun et al. 2017).

The Interaction between OsERF3 and WOX11 (WUSCHEL-related homeobox gene) Promotes Crown Root Development by Regulating Gene Expression Involved in Auxin as well as Cytokinin Signaling such as RR2 (Zhao et al. 2015). OsERF115 directly interacts with GCC box present in the promoter of OsNF-YB1 (NUCLEAR FACTOR Y B1 TF) and specifically expressed in the aleurone layer of developing endosperm. OsNF-YB1 regulates grain filling and endosperm development (Xu et al. 2016).

WRKYs

Plants naturally try to cope up with the required necessity through various evolutionary adaptations but these developmental, physiological and biochemical changes does not provide solution to simultaneous responses. On contrary certain TFs such as WRKYs are capable of regulating various responses simultaneously though complicated network of genes. So orchestration of WRKYs at the genetic level might open up the window of most anticipated solution for simultaneous multiple responses (Phukan et al. 2016). Regulation of WRKYs and regulation by WRKYs are tightly regulated at transcriptional, translational and domain level. These WRKYs are also regulated at epigenetic, retrograde and proteasomal level. As WRKYs are able to attain such a dynamic cellular homeostatic reprogramming, they proved themselves as the most assuring prospect of crop improvement. They mainly target W-box and W-box like sequences present in the promoters of downstream responsive genes. The paradox that exists with these kinds of TFs is that they regulate various factors, which might positively or negatively respond to a particular condition. So these unwanted traits and factors associated with these traits should be explored and dealt carefully for successful transgenic development. WRKYs are an important group of plant TFs with 109 members in rice (Rushton et al. 2010). Their DBD is made of four-stranded α-sheet containing approximately 60 amino acids. WRKYs with 2 DBDs are categorized into group I, those with single DBD as well as different C2H2 zinc finger are categorized into group II and those with single DBD as well as C2H2 zinc finger are categorized into group III. Group II is not monophyletic and based on the primary amino acid sequence this group is further divided into IIa, IIb, IIc, IId and IIe (Rushton et al. 2010). WRKYs have been reported to be activated by various stimuli. Then these TFs can further process the signal through multiple, interconnected and complex network or pathway to obtain the desired character.

Although it is not possible to fully control such an interconnected network but as many times these TFs can act as central or master regulator of various responses, we can minimize the unwanted traits. Till now full proof improved variety development genetic manipulation is still in its early phase. But certain inspiring steps have been undertaken to ensure success in long-term projects. These TFs are potential targets to enhance the tolerance of rice towards different stresses but for successful development of transgenic lines upstream and downstream components need to be studied carefully. OsWRKY11 expressed under HSP101 promoter provided drought tolerance in rice by maintaining more chlorophyll content and low leaf wilting (Wu et al. 2009).

Drought tolerance is provided by OsWRKY30 and OsWRKY47 when over-expressed in rice (Shen et al. 2012, Raineri et al. 2015). In a study OsWRKY24/45 negatively and OsWRKY72/77 positively regulates an ABA-inducible promoter which can be engineered to promote abiotic stress response in rice (Xie et al. 2005). Constitutive expression of OsWRKY76 in rice positively regulated cold tolerance while negatively regulated resistance to pathogens such as blast fungus *Magnaporthe oryzae* and the leaf blight bacterium *Xanthomonas oryzae* (Yokotani et al. 2013). OsWRKY62 also negatively regulates disease response (Liu et al. 2016). On the other hand OsWRKY89 transgenic rice lines led to UV-B tolerance and disease resistance by enhanced wax accumulation (Wang et al. 2007). It is reported that resistance to necrotrophic sheath blight fungus *Rhizoctonia solani* in rice is mediated by OsWRKY4 (Wang et al. 2015). OsWRKY80 is reported to regulate expression of OsWRKY4 by interacting with W-box (TTGAC[C/T]) or W-box like (TGAC[C/T]) *cis*-elements present in its promoter (Peng et al. 2016). OsWRKY6 also positively regulates disease tolerance in over-expressed rice (Choi et al. 2015). OsWRKY67 provide resistance to *M. oryzae* and *X. oryzae* through regulation of defense responsive genes such as PR1a, PR1b, PR4, PR10a and PR10b (Vo et al. 2018).

Tolerance to phosphate starvation in rice is also regulated by WRKY known as OsWRKY74 (Dai et al. 2016). OsWRKY6 provides increased disease resistance to pathogens in rice by specifically interacting with the promoter of *OsPR10a* and *OsICS1* (Choi et al. 2015). Two transcriptional repressors OsWRKY13 and OsWRKY42 and activator WRKY45-2 forms a transcriptional regulatory cascade that provides resistance to fungal pathogen *M. oryzae* (Cheng et al. 2015). Not only pathogen response but also WRKYs are also involved in herbivore-induced defense responses in rice such as OsWRKY53 (Hu et al. 2016). It is reported that OsWRKY45 negatively regulates resistance of rice to the brown planthopper *Nilaparvata lugens* and sheath blight caused by *R. solani* (Huangfu et al. 2016). Various WRKYs have been identified in a transcriptome study that might regulate responses to *X. oryzae* infection, which can be targeted for improved variety development (Jiang et al. 2017).

Recent findings show that OsWRKY11 specifically interact with the promoter of Chitinase2 (a defense responsive gene) and positively regulates pathogen response in rice. OsWRKY11 also interacts with promoter of RAB21 (a drought-responsive gene) to provide drought tolerance in rice (Lee et al. 2018). OsMADS57 involved in rice tiller growth directly interacts and activates OsWRKY94 during cold stress while suppresses its activity under normal conditions. On the other hand OsTB1 interacts and suppresses OsWRKY94 under both normal and cold stress. OsMADS57 and OsTB1 also targets organogenesis gene D14. Together they play an important role of cold tolerance in rice through modulating organogenesis (Chen et al. 2018).

NACs

There are various other TFs such as NAC (NAM for no apical meristem, ATAF1-2 and CUC2 for cup-shaped cotyledon) that regulate different pathways and processes in plants that significantly influence growth, development and stress responses. There are 151 NAC TFs discovered in rice (Nuruzzaman et al. 2013). Shim et al. 2018 reported that OsNAC14 provides drought tolerance in rice by interacting with the promoter of OsRAD51A1 (involved in homologous recombination in DNA repair system). OsNAC6 also provides drought tolerance by regulating various genes involved in root structural adaptations and nicotianamine biosynthesis (Lee et al. 2017). ONAC095 positively regulates cold tolerance while negatively regulates drought tolerance (Huang et al. 2016). Constitutive expression of ONAC022 in rice improved tolerance towards drought and salt stress (Hong et al. 2016).

Another NAC TF SNAC3 is reported to regulate heat and drought tolerance in rice by regulating ROS accumulation during stress (Fang et al. 2015). Rice ONAC011 accelerates senescence of leaf and promotes heading time (when panicle is fully visible) in rice (Mannai et al. 2017). While ONAC106 inhibits leaf senescence and promotes tolerance to salt stress (Sakuraba et al. 2015). Another NAC TF OsNAC2 promotes premature leaf senescence via ABA biosynthesis. OsNAC2 induces expression of chlorophyll degradation genes (OsSGR and OsNYC3) and ABA biosynthetic genes (OsNCED3 and OsZEP1) while represses expression of ABA catabolic gene such as OsABA8ox1 (Mao et al. 2017, Shen et al. 2017). OsNAC2 also positively regulates salt-induced cell death probably by targeting promoters of OsCOX11 (ROS scavenger) and OsAP37 (caspase-like protease, Mao et al. 2018. ONAC020 and ONAC023 are reported to interact with ONAC026 to regulate seed size in rice (Mathew et al. 2016). NAC29/31 activates MYB61 that further activates CESA (Cellulose synthase gene) involved in rice internode development. This pathway is blocked by SLR1 (Slender rice1), a DELLA repressor of GA signaling and NACs (Huang et al. 2015).

bHLHs

Another major TF group that mediates various responses in plant is bHLH (Basic Helix Loop Helix) having 167 members in rice. OsMYC2 a bHLH TF promotes expression of various defense responsive genes by Jasmonic acid (JA) signaling (Ogawa et al. 2017). It is shown that bHLH142 regulates pollen fertility and development through coordination with TDR1 (Tapetum Degeration Retardation1). bHLH142 and TDR1 interact with each other to form a heterodimer to bind to the EAT1 promoter (Ji et al. 2013, Fu et al. 2014, Ko et al. 2014, Ko et al. 2017, Ranjan et al. 2017). EAT1 (Eternal Tapetum1) then enhances expression of AP37 and AP25 that induce stapetal programmed cell death (PCD). Diterpenoid phytoalexin factor (DPF) a bHLH TF regulates biosynthesis of phytocassanes through Copalyl Diphosphate Synthase2 (CPS2) and momilactones through Cytochrome P450 Monooxygenase 99A2 (CYP99A2). DPF is induced by various stress and mainly expressed in panicles as well as roots (Yamamura et al. 2015). Positive Regulator of Grain Length 1 (PGL1), PGL2 and Antagonist of PGL1 (APG) regulate size and weight of rice grains that can be targeted for increased yield (Heang and Sassa 2012).

Iron homeostasis is important for plant growth. OsIRO2 and OsIRO3 bHLH TFs are involved in regulation of the Fe homeostasis response in rice (Ogo et al. 2007, Zheng et al. 2010). Distribution of iron between roots and shoots of rice is mediated by OsbHLH133 (Wang et al. 2013). OsMAPK3 and OsMAPK6 play a major role in plant immunity. The bHLHRac Immunity1 (RAI1) is a direct target of these kinases which then activates various defense responsive genes to regulate disease responses (Kim et al. 2012).

MYBs

Myeloblastosis (MYB) TFs also play central roles in stress, developmental processes and in nutrient deficiency responses. A genome-wide analysis identified 155 members in rice (Katiyar et al. 2012). Over-expression of OsMYB5P in rice increases tolerance to phosphate starvation and increased biomass. OsMYB5P directly interacts with the MBS (MYB binding site) motifs of OsPT5 (plasma-membrane-localized Pi transporters) promoter to regulate increased Pi accumulation in shoots and roots (Yang et al. 2018).

A R2R3 MYB TF OsARM1 (Arsenite-Responsive MYB1) regulates arsenic-associated transporters genes such as OsLsi1, OsLsi2 as well as OsLsi6 and provides tolerance to arsenic stress in rice (Wang et al. 2017). OsMYBR1 over-expression imparted drought tolerance in transgenic rice and leads to decreased ABA sensitivity. Transgenic OsMYBR1 plants showed higher free proline and soluble sugar level (Yin et al. 2017). MYB1 from rice is involved in phosphate starvation signaling and GA biosynthesis. This R2R3 MYB TF also regulates root development in rice (Gu et al. 2017). OsMYB30 negatively regulates tolerance to cold stress. OsMYB30 mediates this response by physically interacting with OsJAZ9 to repress β-amylase (Lv et al. 2017). CEF1/ OsMYB103L is a R2R3 MYB TF regulates secondary walls formation and cellulose biosynthesis mainly by interacting with the promoters of CESA4 (Cellulose synthase A catalytic subunit), CESA7, CESA9 and BC1 (Brittle Culm1). OsMYB103L physically interacts with a DELLA repressor of GA signaling SLR1 (Slender Rice1) and regulates GA-mediated regulation of cellulose synthesis pathway (Ye et al. 2015). OsMYB91 over-expression regulates salt stress tolerance and plant growth (Zhu et al. 2015). All together these TFs can be targeted for improved rice variety development.

Non-TF Mediated Modulation

As we have discussed above TFs are an important regulator of various plant responses. Nevertheless non-TF mediated regulation of plant responses has been identified. A few of them have been described in this section. OsEXPA10 an expansin gene positively regulates plant height and grain size while negatively regulates tolerance to brown planthopper as well as *Magnaporthe grisea* (Tan et al. 2018). SPX1 (Syg1/Pho81/XPR1) and SPX2 are induced in response to phosphate (Pi) starvation that inhibits lamina joint cell elongation, restricts lamina joint size and induce leaf erectness in rice. SPX negatively regulates leaf inclination by physically interacting with Regulator of Leaf Inclination 1 (RLI1). This interaction inhibits RLI1 to further activate downstream genes Brassinosteroid Upregulated1 (BU1) and BU1-like 1 Complex1 (BC1) involved in enhancing leaf inclination (Ruan et al. 2018). Fruit-weight 2.2-like" (OsFWL) mediates the translocation of cadmium from roots to shoots (Xiong et al. 2018). Constitutive expression of OsLEA5 (Late Embryogenesis Abundant Protein) positively regulated drought and salt stress. OsLEA5 enhanced ABA accumulation by up-regulating expression of ABA biosynthetic genes (Huang et al. 2018b). OsLEA5 also assist ZFP36 (zinc finger transcription factor) to regulate seed germination in rice through OsAPX1 (ascorbate peroxidase, Huang et al. 2018a).

Chloroplast is an integral part of plant photosynthesis. Genes such as OsFLN1 (Fructokinase-like Proteins1) and OsFLN2 leads to chloroplast development and plant growth (He et al. 2018). OsJAZ1 (Jasmonate ZIM-domain) protein negatively regulates drought tolerance by modulating JA and ABA signaling in rice (Fu et al. 2017). RING (Really Interesting New Gene)-type E3 ligase protein DHS (Drought Hypersensitive) leads to low accumulation of cuticular wax and reduced drought tolerance. DHS mediates ubiquitin/26S proteasome-mediated degradation of ROC4 (homeodomain-leucine zipper IV protein) involved in biosynthesis of cuticular wax and drought tolerance (Wang et al. 2018). Members of Ribosomal protein gene family such as RPL23A are involved in drought and salt tolerance in rice. Therefore these genes can also be targeted for development of improved breeds of rice with better tolerance and yield.

WHEAT CROPS

Transcription Factor Mediated Modulation

Along with rice wheat is also an essential crop cultivated all around the globe. Various ERFs with potential applications have been identified which can be targeted for improved variety development. Constitutive expression of TaPIE1 leads to increased tolerance to freezing stress and enhanced resistance to fungal pathogen *Rhizoctonia cerealis*. TaPIE1 over-expressed lines displayed higher proline content as well as soluble sugar, while reduced H_2O_2 content and electrolyte leakage (Zhu et al. 2014).

Tolerance to drought and salt stress is provided byTaERF3, whose over-expression induced higher proline accumulation, while inhibited degradation of chlorophyll, formation of H_2O_2 and stomatal conductance. TaERF3 specifically interacted with the GCC-box present in the promoter of various stress-responsive genes such as *Chit1* (Chitinase 1), *TIP2* (Delta tonoplast intrinsic protein), *BG3* (β-glucans 3), *POX2* (Peroxidase 2), *RAB18* (ras-related protein 18), *LEA3* (Late embryogenesis abundant 3) and *GST6* (Glutathione S-Transferase 6, Rong et al. 2014). TaPIEP1 when constitutively expressed in wheat regulated expression of various defense-responsive genes of ET/JA pathway and provided enhanced tolerance to fungal agent *Bipolaris sorokiniana* (Dong et al. 2010). Durum wheat ERF TdERF1 provides drought and salt tolerance in transgenic lines (Makhloufi et al. 2014, Makhloufi et al. 2015).

TaWRKY49 negatively regulates resistance to HTSP (high-temperature seedling-plant resistance to Pst that is a fungal pathogen *Puccinia striiformis* f. *tritici*) while TaWRKY62 and TaWRKY70 positively regulates resistance to HTSP (Wang et al. 2017).

A NAC TF TaNAC30 also affects resistance of wheat to stripe rust (Wang et al. 2017). TaNAc69 over-expressed lines showed more biomass and more tolerance to salinity and dehydration (Xue et al. 2011). Constitutive expression of TaNAC69-1 produced longer primary seminal root as well as more above-ground biomass and grains. TaNAC69-1 suppresses expression of TaSHY2 and TaIAA7 (auxin/IAA transcriptional repressor gene family proteins involved in negative regulation of root growth) promoting root growth during drought (Chen et al. 2016). When TaNAC-S was over-expressed in wheat it led to stay-green phenotype (delayed leaf senescence) as well as higher nitrogen content in grains (Zhao et al. 2015). Over-expression of TaNAC2-5A increased root growth and nitrate influx rate of roots. Transgenic lines also had higher grain mass with higher nitrogen content. TaNAC2-5A could specifically interact with the promoters of nitrate transporter and glutamine synthetase genes (He et al. 2015). When TaRNAC1 was predominantly expressed in wheat roots it leads to increased root length and biomass as well enhanced drought tolerance (Chen et al. 2018).

Two bHLH TFs TaPpm1 (purple pericarp-MYB 1) and TaPpb1 (purple pericarp-bHLH 1) is reported to regulate dietary anthocyanins in purple pericarp of bread wheat (Jiang et al. 2018). TaMyc1 also enhances expression of anthocyanin biosynthesis structural genes in purple wheat grains (Zong et al. 2017, Shoeva 2018). TabHLH1 when over-expressed in wheat improved tolerance towards phosphate and nitrogen starvation as well as more biomass accumulation (Yang et al. 2016). There these transcription factors can be taken up for further studies to successfully develop improved varieties with better qualities.

TOMATO CROPS

Transcription Factor Mediated Modulation

ERFs

Tomatoes are daily component of dietary system and one significant disadvantage of tomato is its early ripening tendency. At first genetically modified FlavrSavr was approved by FDA (USA) for human consumption, which was made by silencing polygalacturonase gene and delaying ripening (Krieger et al. 2008). FlavrSavr production was discontinued later because of mounting costs. Nevertheless a lot of research has been done in this area particularly in the field of TFs such as ERFs. Potential role of ERFs in regulation of fruit ripening in tomato has been studied by Liu et al. 2016. JREs (Jasmonate-Responsive ERF TF) are reported to regulate accumulation of various defense related Steroidal glycoalkaloids. JRE4 acts as a master transcriptional regulator, specifically interacts with the GCC box-like P box and the GCC box present in the promoter of biosynthetic genes like Sterol Reductase as well as Glycoalkaloid Metabolism 5 and regulate their accumulation (Thagun et al. 2016, Abdelkareem et al. 2017, Nakayasu et al. 2018). SlERF52 regulates meristematic activities in pedicel abscission zones by modulating the expression of TFs such as Goblet, Wuschel homologue and Lateral suppressor (Nakano et al. 2014). Pti5 (Pto-interacting protein 5) positively regulates resistance potato aphids in tomato (Wu et al. 2015). Constitutive expression of SlERF5 led to enhanced salt as well as drought tolerance and increased resistance to bacterial pathogen

Ralstonia solanacearum (causes wilt). Relative water content and expression of PR5 (Pathogenesis related 5) was also more in SlERF5 over-expressed lines (Pan et al. 2012). Certain members of ERF group B3 such as SlERF.A1, SlERF.A3, SlERF.B4 and SlERF.C3 positively regulates resistance to *B. cinerea* (Ouyang et al. 2016). Over-expression of another ERF SlRAV2 in tomato provided increased resistance to *R. solanacearum* through enhanced expression PR5 and SlERF5, which suggest that SlERF5 probably acts downstream of SlRAV2 (Li et al. 2011). Over-expression of TERF2/LeERF2 regulates ethylene production as well as cold stress tolerance in tomato (Zhang and Huang 2010).

WRKYs

SolyWRKY41 and SolyWRKY54 positively regulate defense signaling pathway against TYLCV in tomato (Huang et al. 2016). Tomato WRKY TF SlWRKY3 when over-expressed in tomato showed enhanced tolerance to salt stress and reduced ROS as well as proline levels. SlWRKY3 transgenic lines showed relatively higher plant biomass and photosynthesis (Hichri et al. 2017). It is reported that SlWRKY16, SlWRKY17, SlWRKY53 and SlWRKY54 are involved in color change during tomato fruit ripening (Wang et al. 2017). SlWRKY45 negatively regulates resistance to root knot nematode *M. javanica* infection (Chinnapandi et al. 2017). WRKY TF SlDRW1 positively regulates resistance towards *B. cinerea* and tolerance towards oxidative stress (Liu et al. 2014). Silencing SlWRKY70 in tomato suppressed attenuated Mi-1(R gene) -mediated resistance against both *Macrosiphum euphorbiae* (potato aphid) and root-knot nematode *Meloidogyne javanica* (Atamian et al. 2012).

NACs

Tomato NAC TF NAC1 positively regulates plant defense responses. Ubiquitin ligase SINA3 (Seven In Absentia3) negatively regulates this response by processing NAC1 degradation through polyubiquitination (Miao et al. 2016). Another NAC Jungbrunnen1 positively regulates tolerance to drought in tomato (Thirumalaikumar et al. 2018). SlNAC61 shows resistance to TYLCV infection in tomato (Huang et al. 2017). A NAC TF Goblet affects leaf boundary and elaboration in tomato while SlNAM2 is involved in flower-boundary morphogenesis (Berger et al. 2009, Hendelman et al. 2013). SlNAC1 provided chilling tolerance in transgenic tomato lines by maintaining the higher maximal photochemical efficiency of photosystem II and oxygen-evolving activities (Ma et al. 2013). SlNAC1 also regulates tomato fruit ripening through ethylene and ABA dependent pathways (Ma et al. 2014). SlNAC48 and SlNAC19 also positively regulate fruit ripening in tomato (Kou et al. 2016). SlNAC4 positively regulates carotenoid accumulation and fruit ripening in tomato. SlNAC4 could physically interact with both RIN (ripening inhibitor) and NOR (non-ripening) proteins (Zhu et al. 2014a). SlNAC4 is also reported to positively regulate drought and salt stress responses in tomato (Zhu et al. 2014b). NAC TFs JA2 (Jasmonic acid2) and JA2L (JA2-like) differentially regulate bacterial pathogen *Pseudomonas syringae*-induced stomatal closure and reopening through distinct mechanisms (Du et al. 2014). SlSRN1 is reported to positively regulate disease resistance against *p. syringae* and *B. cinerea* while negatively regulate tolerance of tomato towards drought and oxidative stress (Liu et al. 2014).

bHLHs

The bHLH TF LeMYC2 negatively regulates the blue light mediated photomorphogenesis in tomato by specifically interacting with G-box of LeRBCS-3A promoter. Over-expressed LeMYC2 lines displayed more root length and lateral roots while reduces internode length with more branches (Gupta et al. 2014). bHLH TFs SlbHLH068 and FER interact with each other to form a heterodimer and activate transcription of LeFRO1 (ferric-chelate reductase 1) and LeIRT1 (iron-regulated transporter 1) in tomato and regulates accumulation of iron in roots and leaves (Du et al. 2015). The Hoffman's anthocyaninless gene is a cold inducible bHLH TF involved in anthocyanin biosynthesis and accumulation during young seedling stages (Qiu et al. 2016). SlPRE2 is a PRE (paclobutrazol-resistant)-like atypical bHLH TF that affects plant morphology. Over-expressed plants showed increased leaf angle as well as stem internode elongation, and rolling leaves with decreased chlorophyll content. SlPRE2 is a negative regulator of accumulation of fruit pigment (Zhu et al. 2017).

IMPROVEMENT OF NON-EDIBLE SPECIES

COTTON CROPS

Transcription Factor Mediated Modulation

Cotton is a soft and fluffy staple fiber made of cellulose that grows in a boll. The fiber is most often spun into yarn or thread and used to make a soft, breathable textile. ERF TF GhERF-IIb3 positively regulates JA pathway and provides increased resistance to bacterial blight in cotton (Cacas et al. 2017). GhWRKY6-like a WRKY TF from cotton was suppressed by VIGS it leads to reduced tolerance to salt and drought stress in cotton (Ullah et al. 2018). It is reported that GhMAP3K15 phosphorylates GhMKK4 that phosphorylates GhMPK6. GhMPK6 then phosphorylates and activates GhWRKY59 that interacts with the W-box present in GhDREB2 promoter and activate its transcription to promote drought responses (Li et al. 2017). Over-expression of a NAC TF GhNAC79 in cotton imparted enhanced drought tolerance (Guo et al. 2017).

Another NAC TF GhATAF1 when over-expressed in cotton led to increased salt tolerance but decreased resistance to fungal pathogens *Verticillium dahlia* and *B. cinerea*. GhATAF1 activated SA-mediated signaling while suppressed JA-mediated signaling in transgenic lines (He et al. 2016). Constitutive expression of GhNAC2 in cotton improved root growth and tolerance to drought stress (Gunapati et al. 2016). SNAC1 when over-expressed in cotton displayed increased tolerance drought and salt stress (Liu et al. 2014). Constitutive expression of GhbHLH171 in cotton activates JA synthesis as well as signaling pathway and enhances resistance to the fungus *V. dahlia*. GhJAZ2 acts as negative regulator of this response as it interact and inhibit GhbHLH171 (He et al. 2018). GhJAZ2 is reported to interact with another bHLH TF GhMYC2 and probably regulate cotton fibre initiation (Hu et al. 2016).

GhMYB108 interacts with a calmodulin-like protein GhCML11 and provides increased resistance to *V. dahlia* infection in cotton (Cheng et al. 2016). GbMYB5 provides enhanced drought tolerance in cotton. Silencing GbMYB5 led to lower proline level, reduced antioxidant enzyme activities and increased malondialdehyde (MDA) level in cotton under drought stress (Chen et al. 2015). It is reported that cotton fiber elongation is negatively regulated by R3-MYB TF GhCPC. GhCPC physically interacts with GhMYC1 that interacts with E-box *cis* element present in the promoter of a homeodomain-leucine zipper TF GhHOX3 involved in cotton fibre elongation (Liu et al. 2015). Therefore these TFs can be targeted for improved variety development of cotton with more and good quality fibre.

CONCLUSION AND FUTURE PERSPECTIVES

Environmental anomalies harboring biotic and abiotic stresses severely affects commercially valuable plants. It leads to severe loss of manpower, time and money if the problems are not taken care of. This effect is multiplied when several stresses act simultaneously. To address these issues without comprising yield or growth new approach towards development of improved varieties is needed. Classical breeding strategies take longer time to screen the desired trait and demand more manpower. Therefore improvement of varieties by recombinant DNA technology combined with efficient transformation and faster regeneration techniques have gained popularity recently. Transgenic approach though in its early infancy could be targeted for development of varieties with tolerance to multiple stresses.

In the present communication, we have summed up the current understanding and knowledge of how transgenic approach has been used to address the concerns associated with plant development in hazardous environment. TFs are the main focus in context of generating better breeds, as they are potential candidates for exploring interconnected network involved in stress and developmental response. TFs not only modulate developmental and stress responses they are also the key players of specialized metabolic pathways. As TFs can interact with multiple *cis*-elements simultaneously, they have the ability to modulate wide set of signaling pathways. The point is that with so many positive effects, they also negatively influence certain pathways necessary for normal physiology and metabolism. Sounder normal conditions to maintain normal cellular homeostasis modulation of these TFs and modulation of downstream signaling by these TFs is kept under tight regulation by different mechanisms to avoid unwanted response. We strongly believes that these should be analyzed in detail and particular attention should be given to the future research focusing on the integrated regulatory network of different TFs.

Following points should be undertaken to bring forward the idea of improvement of commercially valuable plants through genetic manipulation:

- Recent approaches of genetic manipulation by microRNA and small RNA should be targeted. As for example MiR156 and MiR172 regulate intricate floral organogenesis. MiR172 is also reported to be involved in nodulation and rhizobia nitrogen-fixation symbiosis in legumes. It is obvious that these do not play alone and other activators /suppressors need to be analyzed for understanding of the entire network. It is also reported that for processing, targeting, interaction and activity various other factors are required which needs to be studied. It is highly necessary that large-scale genome wide identification of these small RNAs is carried out.

- Advanced genetic tools such as CRISPR (Clustered Regularly Interspaced Short Palindromic Repeats) and CRISPR-associated 9 (Cas9) gene could be targeted to suppress stress responsive genes in various non-model plants where functional mutants are difficult to generate. CRISPR/cas9 genome editing tool would have an advantage over conventional tedious silencing approach (through pART vectors) or VIGS (through TRV vectors) in the coming era as traditional silencing or VIGS often produces some leaky expression of target genes. Therefore this technique could be employed for functional characterization of TFs before exploiting them for transgenic development. Also negative regulatory genes could be modulated for better stress tolerance response or more yield.

- These TFs would impart growth anomalies and other mulfunctions under normal condition if otherwise not regulated properly at post-transcriptional (capping, splicing and histone modifications) and post-translational level (phosphorylation, ubiquitination, SUMOylation). Therefore these factors should be studied at a large scale so that specific targets in the complicated pathway could be identified, that could be targeted for improved variety development.

- In genetically modified plants these TFs should be under proper regulation of promoters. To address this issue, constitutive promoters should not be used as they impart their influence under normal conditions also. In fact inducible promoters or tissue specific promoters should be used.

- These TFs undergo protein-protein and protein-DNA interactions to activate various responses. Protein DNA interaction includes specific or plastic interaction of TFs with *cis*-elements present in promoters of downstream genes. This is not only regulated by minor differences in DBD but also by the motifs flanking it. Therefore these motifs along with protein-protein interacting motifs should be studied in detail. Along with it signaling moieties, phosphorylating enzymes and interacting partners should be studied.

- TFs are also involved in retrograde signaling. Signaling from chloroplast or mitochondria to nucleus and its feedback should be properly analyzed. Along with orthologs and paralogs of these TFs should be studied to identify probable activator and target of these ERFs in a bigger level. Also TF mediated interconnected network in which other TFs interplay should be studied.

- Extensive and repeated long-term field-testing of improved breeds should be carried out for accurate observation of modified traits. Issues related to horizontal gene transfer should be kept in mind during field-testing. Commercialization of improved breeds should be endorsed with utmost precautions and care.

- Advanced high throughput transcriptomic, proteomic, metabolomic profiling along with microarray analysis should be undertaken to identify new development and stress responsive TFs from plant species which have naturally adapted to combat and thrive in stressful environmental conditions. These aspects would give a deeper understanding of the plant stress responses and allow plants of agricultural and commercial value to propagate under environmental anomalies.

Abbreviations

ABA : Abscisic acid

AP2 : Apetala 2

APG : Antagonist of PGL1

APX1   :   Ascorbate peroxidase 1

BC1   :   BU1-like 1 Complex1

BG3   :   β-glucans 3

bHLH   :   Basic Helix Loop Helix

BU1   :   Brassinosteroid Upregulated1

Cas9   :   CRISPR-associated 9

CEF1   :   Culm easily fragile1

CESA   :   Cellulose synthase

Chit1   :   Chitinase 1

CML   :   Calmodulin-like protein

CPC   :   Caprice

CPS2   :   Copalyl Diphosphate Synthase2

CRISPR :   Clustered Regularly Interspaced Short Palindromic Repeats

CRL5   :   Crown Rootless 5

DBD   :   DNA binding domain

DHS   :   Drought Hypersensitive

DPF   :   Diterpenoid phytoalexin factor

DRE   :   Dehydration response element

DREB   :   Dehydration response element binding protein

DRW1   :   Defense-related WRKY1

ERF   :   Ethylene Response factor

ET   :   Ethylene

EXPA   :   Expansin

FDA   :   Food and Drug Administration

FLN1   :   Fructokinase-like Proteins1

FRO1   :   Ferric-chelate reductase 1

FWL   :   Fruit-weight 2.2-like

GA   :   Gibberellic acid

GST6	:	Glutathione S-Transferase 6
HL6	:	Hairy Leaf 6
HOX	:	Homeodomain-leucine zipper
HTSP	:	High-temperature seedling-plant
IDS1	:	Indeterminate Spikelet 1
IDS1	:	Indeterminate Spikelet1
JA	:	Jasmonic acid
JARE	:	Jasmonic acid response element
JAZ	:	Jasmonate Zim-Domain
JRE	:	Jasmonate-Responsive ERF
LEA5	:	Late Embryogenesis Abundant Protein
LeIRT1	:	Iron-regulated transporter 1)
LSL	:	Long Sterile Lemma
MAPK	:	Mitogen Activated Protein Kinase
MFS1	:	Multi-Floret Spikelet1
NAC	:	NAM for no apical meristem, ATAF1-2 and CUC2 for cup-shaped cotyledon
NAM2	:	Nuclear Accommodation of Mitochondria
NF-YB1	:	NUCLEAR FACTOR Y B1
NOR	:	Non-ripening
NYC3	:	Non-yellow coloring1
PCD	:	Programmed cell death
PGL1	:	Positive Regulator of Grain Length 1
PIE1	:	Pathogen-Induced ERF1
POX2	:	Peroxidase 2
Ppb1	:	purple pericarp-bHLH 1
Ppm1	:	purple pericarp-MYB 1
PR	:	Pathogenesis responsive
PRE	:	Paclobutrazol-resistant

Pti5   :   Pto-interacting protein 5

RAB18   :   ras-related protein 18

RAI1   :   bHLHRac Immunity1

RAV   :   Related to Abscisic acid insensitive3/Viviparous1

RD1   :   Rice Drought-responsive

RIN   :   Ripening inhibitor

RING   :   Really Interesting New Gene

RLI1   :   Regulator of Leaf Inclination 1

SGR   :   Stay green

SK1   :   Snorkel1

SLR1   :   Slender rice1

SNB   :   Supernumerary Bract

SPX1   :   Syg1/Pho81/XPR1

SRN1   :   Stress related NAC1

Sub1A   :   Submergence1A

SUMO   :   Small Ubiquitin-like Modifier

TB1   :   Teosinte branched 1

TDR1   :   Tapetum Degeration Retardation1

TF   :   Transcription factors

TIP2   :   Delta tonoplast intrinsic protein

TYLCV :   Tomato yellow leaf curl virus

VIGS   :   Virus induced gene silencing

ZFP   :   Zinc finger protein

Competing Interest

Authors declare that they have no competing interest.

ACKNOWLEDGEMENTS

Authors would like to thanks, Prof. Ashis Kumar Nandi, JNU, New Delhi and Dr. M. P. Darokar, CSIR-CIMAP, Lucknow for their help and support.

REFERENCES CITED

Abdelkareem A, Thagun C, Nakayasu M, Mizutani M, Hashimoto T, Shoji T. 2017. Jasmonate-induced biosynthesis of steroidal glycoalkaloids depends on COI1 proteins in tomato. *Biochemical and Biophysical Research Communications* 89: 206-210.

Agarwal PK, Agarwal P, Reddy MK, Sopory SK. 2006. Role of DREB transcription factors in abiotic and biotic stress tolerance in plants. *Plant Cell Reports* 25: 1263-74.

Ahn H, Jung I, Shin SJ, Park J, Rhee S, Kim JK, Jung W, Kwon HB, Kim S. 2017. Transcriptional Network Analysis Reveals Drought Resistance Mechanisms of AP2/ERF Transgenic Rice. *Frontiers in Plant Science* 15: 8:1044.

Ahsan N, Lee DG, Lee SH, Kang KY, Bahk JD, Choi MS, Lee IJ, Renaut J, Lee BH. 2007. A comparative proteomic analysis of tomato leaves in response to waterlogging stress. *Physiologia Plantarum* 131: 555-570.

Atamian HS, Eulgem T, Kaloshian I. 2012. SlWRKY70 is required for Mi-1-mediated resistance to aphids and nematodes in tomato. *Planta* 235: 299-309.

Berger Y, Harpaz-Saad S, Brand A, Melnik H, Sirding N, Alvarez JP, Zinder M, Samach A, Eshed Y, Ori N. 2009. The NAC-domain transcription factor GOBLET specifies leaflet boundaries in compound tomato leaves. *Development* 136: 823-832.

Beyene B, Beyene B, Deribe H. 2016. Review on Application and Management of Medicinal Plants for the Livelihood of the Local Community. *Journal of Resources Development and Management* 22: 33-39.

Cacas JL, Pré M, Pizot M, Cissoko M, Diedhiou I, Jalloul A, Doumas P, Nicole M, Champion A. 2017. GhERF-IIb3 regulates the accumulation of jasmonate and leads to enhanced cotton resistance to blight disease. *Molecular Plant Pathology* 18: 825-836.

Chen D, Chai S, McIntyre CL, Xue GP. 2018. Overexpression of a predominantly root-expressed NAC transcription factor in wheat roots enhances root length, biomass and drought tolerance. *Plant Cell Reports* 37: 225-237.

Chen D, Richardson T, Chai S, Lynne McIntyre C, Rae AL, Xue GP. 2016. Drought-Up-Regulated TaNAC69-1 is a Transcriptional Repressor of TaSHY2 and TaIAA7, and Enhances Root Length and Biomass in Wheat. *Plant and Cell Physiology* 57: 2076-2090.

Chen L, Zhao Y, Xu S, Zhang Z, Xu Y, Zhang J, Chong K. 2018. OsMADS57 together with OsTB1 coordinates transcription of its target OsWRKY94 and D14 to switch its organogenesis to defense for cold adaptation in rice. *New Phytologist* 218: 219-231.

Chen T, Li W, Hu X, Guo J, Liu A, Zhang B. 2015. A Cotton MYB Transcription Factor, GbMYB5, is Positively Involved in Plant Adaptive Response to Drought Stress. *Plant and Cell Physiology* 56: 917-929.

Cheng H, Liu H, Deng Y, Xiao J, Li X, Wang S. 2015. The WRKY45-2 WRKY13 WRKY42 transcriptional regulatory cascade is required for rice resistance to fungal pathogen. *Plant Physiology* 167: 1087-1099.

Cheng HQ, Han LB, Yang CL, Wu XM, Zhong NQ, Wu JH, Wang FX, Wang HY, Xia GX. 2016. The cotton MYB108 forms a positive feedback regulation loop with CML11 and participates in the defense response against Verticilliumdahliae infection. *Journal of Experimental Botany* 67: 1935-1950.

Chinnapandi B, Bucki P, Braun Miyara S. 2017. SlWRKY45, nematode-responsive tomato WRKY gene, enhances susceptibility to the root knot nematode; M. javanica infection. *Plant Signaling & Behavior* 12: e1356530.

Choi C, Hwang SH, Fang IR, Kwon SI, Park SR, Ahn I, Kim JB, Hwang DJ. 2015. Molecular characterization of Oryza sativa WRKY6, which binds to W-box-like element 1 of the Oryza sativa pathogenesis-related (PR) 10a promoter and confers reduced susceptibility to pathogens. *New Phytologist* 208: 846-859.

Dai X, Wang Y, Zhang WH. 2016. OsWRKY74, a WRKY transcription factor, modulates tolerance to phosphate starvation in rice. *Journal of Experimental Botany* 67: 947-960.

Dong N, Liu X, Lu Y, Du L, Xu H, Liu H, Xin Z, Zhang Z. 2010. Overexpression of TaPIEP1, a pathogen-induced ERF gene of wheat, confers host-enhanced resistance to fungal pathogen Bipolarissorokiniana. *Functional & Integrative Genomics* 10: 215-226.

Du J, Huang Z, Wang B, Sun H, Chen C, Ling HQ, Wu H. 2015. SlbHLH068 interacts with FER to regulate the iron-deficiency response in tomato. *Annals of Botany* 116: 23-34.

Du M, Zhai Q, Deng L, Li S, Li H, Yan L, Huang Z, Wang B, Jiang H, Huang T, Li CB, Wei J, Kang L, Li J, Li C. 2014. Closely related NAC transcription factors of tomato differentially regulate stomatal closure and reopening during pathogen attack. *Plant Cell* 26: 3167-3184.

El Mannai Y, Akabane K, Hiratsu K, Satoh-Nagasawa N, Wabiko H. 2017. The NAC Transcription Factor Gene OsY37 (ONAC011) Promotes Leaf Senescence and Accelerates Heading Time in Rice. *International Journal of Molecular Sciences* 18: E2165.

Fang Y, Liao K, Du H, Xu Y, Song H, Li X, Xiong L. 2015. A stress-responsive NAC transcription factor SNAC3 confers heat and drought tolerance through modulation of reactive oxygen species in rice. *Journal of Experimental Botany* 66: 6803-6817.

Fu J, Wu H, Ma S, Xiang D, Liu R, Xiong L. 2017. OsJAZ1 Attenuates Drought Resistance by Regulating JA and ABA Signaling in Rice. *Frontiers in Plant Science* 8: 2108.

Fu Z, Yu J, Cheng X, Zong X, Xu J, Chen M, Li Z, Zhang D, Liang W. 2014. The Rice Basic Helix-Loop-Helix Transcription Factor TDR INTERACTING PROTEIN2 Is a Central Switch in Early Anther Development. *Plant Cell* 26:1512-1524.

Fukao T, Bailey-Serres J. 2008. Submergence tolerance conferred by Sub1A is mediated by SLR1 and SLRL1 restriction of gibberellin responses in rice. *Proceedings of the National Academy of Sciences of the United States of America* 105: 16814–16819.

Gu M, Zhang J, Li H, Meng D, Li R, Dai X, Wang S, Liu W, Qu H, Xu G. 2017. Maintenance of phosphate homeostasis and root development are coordinately regulated by MYB1, an R2R3-type MYB transcription factor in rice. *Journal of Experimental Botany* 68: 3603-3615.

Gunapati S, Naresh R, Ranjan S, Nigam D, Hans A, Verma PC, Gadre R, Pathre UV, Sane AP, Sane VA. 2016. Expression of GhNAC2 from G. herbaceum, improves root growth and imparts tolerance to drought in transgenic cotton and Arabidopsis. *Scientific Reports* 6: 24978.

Guo Y, Pang C, Jia X, Ma Q, Dou L, Zhao F, Gu L, Wei H, Wang H, Fan S, Su J, Yu S1. 2017. An NAM Domain Gene, GhNAC79, Improves Resistance to Drought Stress in Upland Cotton. *Frontiers in Plant Science* 8: 1657.

Gupta N, Prasad VB, Chattopadhyay S. 2014. LeMYC2 acts as a negative regulator of blue light mediated photomorphogenic growth, and promotes the growth of adult tomato plants. *BMC Plant Biology* 14: 38.

Hattori Y, Nagai K, Furukawa S, Song XJ, Kawano R, et al. 2009. The ethylene response factors SNORKEL1 and SNORKEL2 allow rice to adapt to deep water. *Nature* 460: 1026–1030.

He L, Zhang S, Qiu Z, Zhao J, Nie W, Lin H, Zhu Z, Zeng D, Qian Q, Zhu L. 2018. FRUCTOKINASE-LIKE PROTEIN 1 interacts with TRXz to regulate chloroplast development in rice. *Journal of Integrative Plant Biology* 60: 94-111.

He X, Qu B, Li W, Zhao X, Teng W, Ma W, Ren Y, Li B, Li Z, Tong Y. 2015. The Nitrate-Inducible NAC Transcription Factor TaNAC2-5A Controls Nitrate Response and Increases Wheat Yield. *Plant Physiology* 169: 1991-2005.

He X, Zhu L, Wassan GM, Wang Y, Miao Y, Shaban M, Hu H, Sun H, Zhang X. 2018. GhJAZ2 attenuates cotton resistance to biotic stresses via the inhibition of the transcriptional activity of GhbHLH171. *Molecular Plant Pathology* 19: 896-908.

He X, Zhu L, Xu L, Guo W, Zhang X. 2016. GhATAF1, a NAC transcription factor, confers abiotic and biotic stress responses by regulating phytohormonal signaling networks. *Plant Cell Reports* 35: 2167-2179.

Heang D, Sassa H. 2012. An atypical bHLH protein encoded by Positive Regulator of Grain Length 2 is involved in controlling grain length and weight of rice through interaction with a typical bHLH protein APG. *Breeding Science* 62: 133-141.

Hendelman A, Stav R, Zemach H, Arazi T. 2013. The tomato NAC transcription factor SlNAM2 is involved in flower-boundary morphogenesis. *Journal of Experimental Botany* 64: 5497-5507.

Hichri I, Muhovski Y, •i•ková E, Dobrev PI, Gharbi E, Franco-Zorrilla JM, Lopez-Vidriero I, Solano R, Clippe A, Errachid A, Motyka V, Lutts S. 2017. The Solanuml ycopersicum WRKY3 Transcription Factor SlWRKY3 Is Involved in Salt Stress Tolerance in Tomato. *Frontiers in Plant Science* 8: 1343.

Hong Y, Zhang H, Huang L, Li D, Song F. 2016. Overexpression of a Stress-Responsive NAC Transcription Factor Gene ONAC022 Improves Drought and Salt Tolerance in Rice. *Frontiers in Plant Science* 7: 4.

Hu H, He X, Tu L, Zhu L, Zhu S, Ge Z, Zhang X. 2016. GhJAZ2 negatively regulates cotton fiber initiation by interacting with the R2R3-MYB transcription factor GhMYB25-like. *Plant Journal* 88: 921-935.

Hu L, Ye M, Li R, Lou Y. 2016. OsWRKY53, a versatile switch in regulating herbivore-induced defense responses in rice. *Plant Signaling & Behavior* 11: e1169357.

Huang D, Wang S, Zhang B, Shang-Guan K, Shi Y, Zhang D, Liu X, Wu K, Xu Z, Fu X, Zhou Y. 2015. A Gibberellin-Mediated DELLA-NAC Signaling Cascade Regulates Cellulose Synthesis in Rice. *Plant Cell* 27: 1681-1696.

Huang L, Hong Y, Zhang H, Li D, Song F. 2016. Rice NAC transcription factor ONAC095 plays opposite roles in drought and cold stress tolerance. *BMC Plant Biology* 16: 203.

Huang L, Jia J, Zhao X, Zhang M, Huang X, E Ji, Ni L, Jiang M. 2018. The ascorbate peroxidase APX1 is a direct target of a zinc finger transcription factor ZFP36 and a late embryogenesis abundant protein OsLEA5 interacts with ZFP36 to co-regulate OsAPX1 in seed germination in rice. *Biochemical and Biophysical Research Communications* 495: 339-345.

Huang L, Wang Y, Wang W, Zhao X, Qin Q, Sun F, Hu F, Zhao Y, Li Z, Fu B, Li Z. 2018. Characterization of Transcription Factor Gene OsDRAP1 Conferring Drought Tolerance in Rice. *Frontiers in Plant Science* 9: 94.

Huang L, Zhang M, Jia J, Zhao X, Huang X, Ji E, Ni L, Jiang M. 2018. An Atypical Late Embryogenesis Abundant Protein OsLEA5 Plays A Positive Role In ABA-Induced Antioxidant Defense In Oryza Sativa L. *Plant and Cell Physiology* doi: 10.1093/pcp/pcy035. [Epub ahead of print]

Huang Y, Li MY, Wu P, Xu ZS, Que F, Wang F, Xiong AS. 2016. Members of WRKY Group III transcription factors are important in TYLCV defense signaling pathway in tomato (Solanumlycopersicum). *BMC Genomics* 17: 788.

Huang Y, Li T, Xu ZS, Wang F, Xiong AS. 2017. Six NAC transcription factors involved in response to TYLCV infection in resistant and susceptible tomato cultivars. *Plant Physiology and Biochemistry.* 120: 61-74.

Huangfu J, Li J, Li R, Ye M, Kuai P, Zhang T, Lou Y. 2016. The Transcription Factor OsWRKY45 Negatively Modulates the Resistance of Rice to the Brown PlanthopperNilaparvatalugens. *International Journal of Molecular Sciences* 17: E697.

Ji C, Li H, Chen L, Xie M, Wang F, Chen Y, Liu YG. 2013. A novel rice bHLH transcription factor, DTD, acts coordinately with TDR in controlling tapetum function and pollen development. *Molecular Plant* 6: 1715-1718.

Jiang C, Shen QJ, Wang B, He B, Xiao S, Chen L, Yu T, Ke X, Zhong Q, Fu J, Chen Y, Wang L, Yin F, Zhang D, Ghidan W, Huang X, Cheng Z. 2017. Transcriptome analysis of WRKY gene family in Oryza officinalis Wall ex Watt and WRKY genes involved in responses to Xanthomonas oryzae pv. oryzae stress. *PLoS One* 12: e0188742.

Jiang W, Liu T, Nan W, ChamilaJeewani D, Niu Y, Li C, Wang Y, Shi X, Wang C, Wang J, Li Y, Gao X, Wang Z. 2018. Two Transcription Factors TaPpm1 and TaPpb1 Co-Regulate the Anthocyanin Biosynthesis in Purple Pericarp of Wheat. *Journal of Experimental Botany* doi: 10.1093/jxb/ery101.

Jisha V, Dampanaboina L, Vadassery J, Mithofer A, Kappara S, Ramanan R. 2015. Overexpression of an AP2/ERF Type Transcription Factor OsEREBP1 Confers Biotic and Abiotic Stress Tolerance in Rice. *PLoS One* 10: e0127831.

Katiyar A, Smita S, Lenka SK, Rajwanshi R, Chinnusamy V, Bansal KC. 2012. Genome-wide classification and expression analysis of MYB transcription factor families in rice and Arabidopsis. *BMC Genomics* 13: 544.

Kim SH, Oikawa T, Kyozuka J, Wong HL, Umemura K, Kishi-Kaboshi M, Takahashi A, Kawano Y, Kawasaki T, Shimamoto K. 2012. The bHLHRac Immunity1 (RAI1) Is Activated by OsRac1 via OsMAPK3 and OsMAPK6 in Rice Immunity. *Plant and Cell Physiology* 53:740-754.

Kitomi Y, Ito H, Hobo T, Aya K, Kitano H, Inukai Y. 2011. The auxin responsive AP2/ERF transcription factor CROWN ROOTLESS5 is involved in crown root initiation in rice through the induction of OsRR1, a type-A response regulator of cytokinin signaling. *The Plant Journal* 67: 472-484.

Ko SS, Li MJ, Lin YJ, Hsing HX, Yang TT, Chen TK, Jhong CM, Ku MS. 2017. Tightly Controlled Expression of bHLH142 Is Essential for Timely Tapetal Programmed Cell Death and Pollen Development in Rice. *Frontiers in Plant Science* 8: 1258.

Ko SS, Li MJ, Sun-Ben Ku M, Ho YC, Lin YJ, Chuang MH, Hsing HX, Lien YC, Yang HT, Chang HC, Chan MT. 2014. The bHLH142 Transcription Factor Coordinates with TDR1 to Modulate the Expression of EAT1 and Regulate Pollen Development in Rice. *Plant Cell* 26: 2486-2504.

Kou X1, Liu C2, Han L2, Wang S2, Xue Z3. 2016. NAC transcription factors play an important role in ethylene biosynthesis, reception and signaling of tomato fruit ripening. *Molecular Genetics and Genomics* 291: 1205-1217.

Krieger EK, Allen E, Gilbertson LA, Roberts JK, Hiatt W, Sanders RA. 2008. The FlavrSavr Tomato, an early example of RNAi technology. *Horticultural Science* 43: 962–964.

Lee DK, Chung PJ, Jeong JS, Jang G, Bang SW, Jung H, Kim YS, Ha SH, Choi YD, Kim JK. 2017. The rice OsNAC6 transcription factor orchestrates multiple molecular mechanisms involving root structural adaptions and nicotianamine biosynthesis for drought tolerance. *Plant Biotechnology Journal* 15: 754-764.

Lee DK, Yoon S, Kim YS, Kim JK. 2017. Rice OsERF71-mediated root modification affects shoot drought tolerance. *Plant Signaling & Behavior* 12: e1268311

Lee H, Cha J, Choi C, Choi N, Ji HS, Park SR, Lee S, Hwang DJ. 2018. Rice WRKY11 Plays a Role in Pathogen Defense and Drought Tolerance. *Rice (N Y)* 11: 5.

Li CW, Su RC, Cheng CP, Sanjaya You SJ, Hsieh TH, Chao TC, Chan MT. 2011. Tomato RAV transcription factor is a pivotal modulator involved in the AP2/EREBP-mediated defense pathway. *Plant Physiology* 156: 213-227.

Li F, Li M, Wang P, Cox KL Jr, Duan L, Dever JK, Shan L, Li Z, He P. 2017. Regulation of cotton (Gossypiumhirsutum) drought responses by mitogen-activated protein (MAP) kinase cascade-mediated phosphorylation of GhWRKY59. *New Phytologist* 215: 1462-1475.

Li J, Guo X, Zhang M, Wang X, Zhao Y, Yin Z, Zhang Z, Wang Y, Xiong H, Zhang H, Todorovska E, Li Z. 2018. OsERF71 confers drought tolerance via modulating ABA signaling and proline biosynthesis. *Plant Science* 270: 131-139.

Liu B, Hong YB, Zhang YF, Li XH, Huang L, Zhang HJ, Li DY, Song FM. 2014. Tomato WRKY transcriptional factor SlDRW1 is required for disease resistance against Botrytis cinerea and tolerance to oxidative stress. *Plant Science* 227: 145-156.

Liu B, Ouyang Z, Zhang Y, Li X, Hong Y, Huang L, Liu S, Zhang H, Li D, Song F. 2014. Tomato NAC transcription factor SlSRN1 positively regulates defense response against biotic stress but negatively regulates abiotic stress response. *PLoS One* 9: e102067.

Liu B, Zhu Y, Zhang T. 2015. The R3-MYB gene GhCPC negatively regulates cotton fiber elongation. *PLoS One* 10: e0116272.

Liu G, Li X, Jin S, Liu X, Zhu L, Nie Y, Zhang X. 2014. Overexpression of rice NAC gene SNAC1 improves drought and salt tolerance by enhancing root development and reducing transpiration rate in transgenic cotton. *PLoS One* 9: e86895.

Liu J, Chen X, Liang X, Zhou X, Yang F, Liu J, He SY, Guo Z. 2016. Alternative Splicing of Rice WRKY62 and WRKY76 Transcription Factor Genes in Pathogen Defense. *Plant Physiology* 171:1427-1442.

Liu M, Gomes BL, Mila I, Purgatto E, Peres LE, Frasse P, Maza E, Zouine M, Roustan JP, Bouzayen M, Pirrello J. 2016. Comprehensive Profiling of Ethylene Response Factor Expression Identifies Ripening-Associated ERF Genes and Their Link to Key Regulators of Fruit Ripening in Tomato. *Plant Physiology* 170: 1732-1744.

Lv Y, Yang M, Hu D, Yang Z, Ma S, Li X, Xiong L. 2017. The OsMYB30 Transcription Factor Suppresses Cold Tolerance by Interacting with a JAZ Protein and Suppressing â-Amylase Expression. *Plant Physiology* 173:1475-1491.

Ma N, Feng H, Meng X, Li D, Yang D, Wu C, Meng Q. 2014. Overexpression of tomato SlNAC1 transcription factor alters fruit pigmentation and softening. *BMC Plant Biology* 14: 351.

Ma NN, Zuo YQ, Liang XQ, Yin B, Wang GD, Meng QW. 2013. The multiple stress-responsive transcription factor SlNAC1 improves the chilling tolerance of tomato. *Physiologia Plantarum* 149: 474-486.

Makhloufi E, Yousfi FE, Marande W, Mila I, Hanana M, Bergès H, Mzid R, Bouzayen M. 2014. Isolation and molecular characterization of ERF1, an ethylene response factor gene from durum wheat (Triticumturgidum L. subsp. durum), potentially involved in salt-stress responses. *Journal of Experimental Botany* 65: 6359-6371.

Makhloufi E, Yousfi FE, Pirrello J, Bernadac A, Ghorbel A, Bouzayen M. 2015. TdERF1, an ethylene response factor associated with dehydration responses in durum wheat (Triticumturgidum L. subsp. durum). *Plant Signaling & Behavior* 10: e1065366.

Mao C, Ding J, Zhang B, Xi D, Ming F. 2018. OsNAC2 positively affects salt-induced cell death and binds to the OsAP37 and OsCOX11 promoters. *The Plant Journal* doi: 10.1111/tpj.13867.

Mao C, Lu S, Lv B, Zhang B, Shen J, He J, Luo L, Xi D, Chen X, Ming F. 2017. A Rice NAC Transcription Factor Promotes Leaf Senescence via ABA Biosynthesis. *Plant Physiology* 174: 1747-1763.

Mathew IE, Das S, Mahto A, Agarwal P. 2016. Three Rice NAC Transcription Factors Heteromerize and Are Associated with Seed Size. *Frontiers in Plant Science* 7: 1638.

Miao M, Niu X, Kud J, Du X, Avila J, Devarenne TP, Kuhl JC, Liu Y, Xiao F. 2016. The ubiquitin ligase SEVEN IN ABSENTIA (SINA) ubiquitinates a defense-related NAC transcription factor and is involved in defense signaling. *New Phytologist* 211: 138-148.

Nakano T, Fujisawa M, Shima Y, Ito Y. 2014. The AP2/ERF transcription factor SlERF52 functions in flower pedicel abscission in tomato. *Journal of Experimental Botany* 65: 3111-3119.

Nakano T, Suzuki K, Fujimura T, Shinshi H. 2006. Genome-Wide Analysis of the ERF Gene Family in Arabidopsis and Rice[W]. *Plant Physiology* 140: 411–432.

Nakayasu M, Shioya N, Shikata M, Thagun C, Abdelkareem A, Okabe Y, Ariizumi T, Arimura GI, Mizutani M, Ezura H, Hashimoto T, Shoji T. 2018. JRE4 is a master transcriptional regulator of defense-related steroidal glycoalkaloids in tomato. *Plant Journal* doi: 10.1111/tpj.13911.

Nuruzzaman M, Sharoni AM, Kikuchi S. 2013. Roles of NAC transcription factors in the regulation of biotic and abiotic stress responses in plants. *Front Microbiology* 4: 248.

Ogawa S, Kawahara-Miki R, Miyamoto K, Yamane H, Nojiri H, Tsujii Y, Okada K. 2017. OsMYC2 mediates numerous defence-related transcriptional changes via jasmonic acid signalling in rice. *Biochemical and Biophysical Research Communications* 486: 796-803.

Ogo Y, Itai RN, Nakanishi H, Kobayashi T, Takahashi M, Mori S, Nishizawa NK. 2007. The rice bHLH protein OsIRO2 is an essential regulator of the genes involved in Fe uptake under Fe-deficient conditions. *Plant Journal* 51: 366-377.

Oh SJ, Kim YS, Kwon CW, Park HY, Jeong JS, Kim JK. 2009. Overexpression of the transcription factor AP37 in rice improves grain yield under drought conditions. *Plant Physiology* 150: 1368–1379.

Ouyang Z1, Liu S2, Huang L2, Hong Y2, Li X2, Huang L2, Zhang Y2, Zhang H2, Li D2, Song F2. 2016. Tomato SlERF.A1, SlERF.B4, SlERF.C3 and SlERF.A3, Members of B3 Group of ERF Family, are required for resistance to Botrytis cinerea. *Frontiers in Plant Science* 7: 1964.

Pan Y, Seymour GB, Lu C, Hu Z, Chen X, Chen G. 2012. An ethylene response factor (ERF5) promoting adaptation to drought and salt tolerance in tomato. *Plant Cell Reports* 31: 349-360.

Peng X, Wang H, Jang JC, Xiao T, He H, Jiang D, Tang X. 2016. OsWRKY80-OsWRKY4 Module as a positive regulatory circuit in rice resistance against *Rhizoctonia solani*. *Rice (N Y)* 9:63.

Phukan UJ, Jeena GS, Shukla RK. 2016. WRKY Transcription Factors: Molecular regulation and stress responses in plants. *Frontiers in Plant Science* 7: 760.

Phukan UJ, Jeena GS, Tripathi V, Shukla RK. 2017. Regulation of Apetala2/Ethylene response factors in plants. *Frontiers in Plant Science* 8: 150.

Phukan UJ, Jeena GS, Tripathi V, Shukla RK. 2018. MaRAP2-4, a waterlogging-responsive ERF from Mentha, regulates bidirectional sugar transporter AtSWEET10 to modulate stress response in Arabidopsis. *Plant Biotechnology Journal* 16 :221-233.

Phukan UJ, Mishra S, Timbre K, Luqman S, Shukla RK. 2014. Mentha arvensis exhibit better adaptive characters in contrast to Mentha piperita when subjugated to sustained waterlogging stress. *Protoplasma* 251: 603-14.

Qi W, Sun F, Wang Q, Chen M, Huang Y, Feng YQ, Luo X, Yang J. 2011. Rice ethylene-response AP2/ERF factor OsEATB restricts internode elongation by down-regulating a gibberellin biosynthetic gene. *Plant Physiology* 157: 216-228.

Qiu Z, Wang X, Gao J, Guo Y, Huang Z, Du Y. 2016. The Tomato Hoffman's Anthocyaninless Gene Encodes a bHLH Transcription Factor Involved in Anthocyanin Biosynthesis That Is Developmentally Regulated and Induced by Low Temperatures. *PLoS One* 11: e0151067.

Raineri J, Wang S, Peleg Z, Blumwald E, Chan RL. 2015. The rice transcription factor OsWRKY47 is a positive regulator of the response to water deficit stress. *Plant Molecular Biology* 88: 401-413.

Ranjan R, Khurana R, Malik N, Badoni S, Parida SK, Kapoor S, Tyagi AK. 2017. bHLH142 regulates various metabolic pathway-related genes to affect pollen development and anther dehiscence in rice. *Scientific Reports* 7: 43397.

Ren D, Li Y, Zhao F, Sang X, Shi J, Wang N, Guo S, Ling Y, Zhang C, Yang Z, He G. 2013. MULTI-FLORET SPIKELET1, which encodes an AP2/ERF protein, determines spikelet meristem fate and sterile lemma identity in rice. *Plant Physiology* 162: 872-884.

Rong W, Qi L, Wang A, Ye X, Du L, Liang H, Xin Z, Zhang Z. 2014. The ERF transcription factor TaERF3 promotes tolerance to salt and drought stresses in wheat. *Plant Biotechnology Journal* 12: 468-479.

Ruan W, Guo M, Xu L, Wang X, Zhao H, Wang J, Yi K. 2018. An SPX-RLI1 module regulates leaf inclination in response to phosphate availability in rice. *Plant Cell* doi: 10.1105/tpc.17.00738.

Rushton PJ, Somssich IE, Ringler P, Shen QJ. 2010. WRKY transcription factors. *Trends in Plant Science* 15: 247-258.

Sakuma Y, Liu Q, Dubouzet JG, Abe H, Shinozaki K, Yamaguchi-Shinozaki K. 2002. DNA-binding specificity of the ERF/AP2 domain of Arabidopsis DREBs, transcription factors involved in dehydration- and cold-inducible gene expression. *Biochemical and Biophysical Research Communications* 290: 998-1009.

Sakuraba Y, Piao W, Lim JH, Han SH, Kim YS, An G, Paek NC. 2015. Rice ONAC106 inhibits leaf senescence and increases salt tolerance and tiller angle. *Plant and Cell Physiology* 56: 2325-2339.

Samota MK, Sasi M, Awana M, Yadav OP, AmithaMithra SV, Tyagi A, Kumar S, Singh A. 2017. Elicitor-induced biochemical and molecular manifestations to improve drought tolerance in rice (*Oryza sativa* L.) through Seed-Priming. *Frontiers in Plant Science* 8: 934.

Shen H, Liu C, Zhang Y, Meng X, Zhou X, Chu C, Wang X. 2012. OsWRKY30 is activated by MAP kinases to confer drought tolerance in rice. *Plant Molecular Biology* 80: 241-253.

Shen J, Lv B, Luo L, He J, Mao C, Xi D, Ming F. 2017. The NAC-type transcription factor OsNAC2 regulates ABA-dependent genes and abiotic stress tolerance in rice. *Scientific Reports* 7: 40641.

Shim JS, Oh N, Chung PJ, Kim YS, Choi YD, Kim JK. 2018. Overexpression of OsNAC14 improves drought tolerance in rice. *Frontiers in Plant Science* 9: 310.

Shoeva OY. 2018. Complex regulation of the TaMyc1 gene expression in wheat grain synthesizing anthocyanin pigments. *Molecular Biology Reports* doi: 10.1007/s11033-018-4165-0.

Sun W, Gao D, Xiong Y, Tang X, Xiao X, Wang C, Yu S. 2017. Hairy Leaf 6, an AP2/ERF Transcription Factor, Interacts with OsWOX3B and regulates trichome formation in rice. *Molecular Plant* 10: 1417-1433.

Tan J, Wang M, Shi Z, Miao X. 2018. OsEXPA10 mediates the balance between growth and resistance to biotic stress in rice. *Plant Cell Reports* doi: 10.1007/s00299-018-2284-7.

Thagun C, Imanishi S, Kudo T, Nakabayashi R, Ohyama K, Mori T, Kawamoto K, Nakamura Y, et al. 2016. Jasmonate-Responsive ERF Transcription Factors Regulate Steroidal Glycoalkaloid Biosynthesis in Tomato. *Plant and Cell Physiology* 57: 961-975.

Thirumalaikumar VP, Devkar V, Mehterov N, Ali S, Ozgur R, Turkan I, Mueller-Roeber B, Balazadeh S. 2018. NAC transcription factor JUNGBRUNNEN1 enhances drought tolerance in tomato. *Plant Biotechnol Journal* 16: 354-366.

Ullah A, Sun H, Hakim, Yang X, Zhang X. 2018. A novel cotton WRKY gene, GhWRKY6-like, improves salt tolerance by activating the ABA signaling pathway and scavenging of reactive oxygen species. *Physiologia Plantarum* 162: 439-454.

Vo KTX, Kim CY, Hoang TV, Lee SK, Shirsekar G, Seo YS, Lee SW, Wang GL, Jeon JS. 2018. OsWRKY67 Plays a Positive Role in Basal and XA21-Mediated Resistance in Rice. *Frontiers in Plant Science* 8: 2220.

Wang B, Wei J, Song N, Wang N, Zhao J, Kang Z. 2017. A novel wheat NAC transcription factor, TaNAC30, negatively regulates resistance of wheat to stripe rust. *Journal of Integrative Plant Biology* doi: 10.1111/jipb.12627.

Wang F, Wang C, Liu P, Lei C, Hao W, Gao Y, Liu YG, Zhao K. 2016. Enhanced Rice Blast Resistance by CRISPR/Cas9-Targeted Mutagenesis of the ERF Transcription Factor Gene OsERF922. *PLoS One* 11: e0154027.

Wang FZ, Chen MX, Yu LJ, Xie LJ, Yuan LB, Qi H, Xiao M, Guo W, Chen Z, Yi K, Zhang J, Qiu R, Shu W, Xiao S, Chen QF. 2017. OsARM1, an R2R3 MYB Transcription Factor, Is Involved in Regulation of the Response to Arsenic Stress in Rice. *Frontiers in Plant Science* 8: 1868.

Wang H, Hao J, Chen X, et al. 2007. Overexpression of rice WRKY89 enhances ultraviolet B tolerance and disease resistance in rice plants. *Plant Molecular Biology* 65: 799-815.

Wang H, Meng J, Peng X, Tang X, Zhou P, Xiang J, Deng X. 2015. Rice WRKY4 acts as a transcriptional activator mediating defense responses toward Rhizoctoniasolani, the causing agent of rice sheath blight. *Plant Molecular Biology* 89: 157-171.

Wang J, Tao F, Tian W, Guo Z, Chen X, Xu X, Shang H, Hu X. 2017. The wheat WRKY transcription factors TaWRKY49 and TaWRKY62 confer differential high-temperature seedling-plant resistance to Pucciniastriiformis f. sp. tritici. *PLoS One* 12: e0181963.

Wang L, Ying Y, Narsai R, Ye L, Zheng L, Tian J, Whelan J, Shou H. 2013. Identification of OsbHLH133 as a regulator of iron distribution between roots and shoots in Oryza sativa. *Plant, Cell & Environment* 36: 224-236.

Wang L, Zhang XL, Wang L, Tian Y, Jia N, Chen S, Shi NB, Huang X, Zhou C, Yu Y, Zhang ZQ, Pang XQ. 2017. Regulation of ethylene-responsive SlWRKYs involved in color change during tomato fruit ripening. *Scientific Reports* 7: 16674.

Wang Z, Tian X, Zhao Q, Liu Z, Li X, Ren Y, Tang J, Fang J, Xu Q, Bu Q. 2018. The E3 ligase drought hypersensitive negatively regulates cuticular wax biosynthesis by promoting the degradation of transcription factor ROC4 in rice. *Plant Cell* 30: 228-244.

Wu C, Avila CA, Goggin FL. 2015. The ethylene response factor Pti5 contributes to potato aphid resistance in tomato independent of ethylene signalling. *Journal of Experimental Botany* 66: 559-570.

Wu X, Shiroto Y, Kishitani S, Ito Y, Toriyama K. 2009. Enhanced heat and drought tolerance in transgenic rice seedlings overexpressing OsWRKY11 under the control of HSP101 promoter. *Plant Cell Reports* 28: 21–30.

Xie Z, Zhang ZL, Zou X, Huang J, Ruas P, Thompson D, Shen QJ. 2005. Annotations and functional analyses of the rice WRKY gene superfamily reveal positive and negative regulators of abscisic acid signaling in aleurone cells. *Plant Physiology* 5: 176–189.

Xiong W, Wang P, Yan T, Cao B, Xu J, Liu D, Luo M. 2018. The rice "fruit-weight 2.2-like" gene family member OsFWL4 is involved in the translocation of cadmium from roots to shoots. *Planta* 247: 1247-1260.

Xu JJ, Zhang XF, Xue HW. 2016. Rice aleurone layer specific OsNF-YB1 regulates grain filling and endosperm development by interacting with an ERF transcription factor. *Journal of Experimental Botany* 67:6399-6411.

Xue GP, Way HM, Richardson T, Drenth J, Joyce PA, McIntyre CL. 2011. Overexpression of TaNAC69 leads to enhanced transcript levels of stress up-regulated genes and dehydration tolerance in bread wheat. *Molecular Plant* 4: 697-712.

Yamamura C, Mizutani E, Okada K, Nakagawa H, Fukushima S, Tanaka A, Maeda S, Kamakura T, Yamane H, Takatsuji H, Mori M. 2015. Diterpenoidphytoalexin factor, a bHLH transcription factor, plays a central role in the biosynthesis of diterpenoidphytoalexins in rice. *The Plant Journal* 84: 1100-1113.

Yang T, Hao L, Yao S, Zhao Y, Lu W, Xiao K. 2016. TabHLH1, a bHLH-type transcription factor gene in wheat, improves plant tolerance to Pi and N deprivation via regulation of nutrient transporter gene transcription and ROS homeostasis. *Plant Physiology and Biochemistry* 104: 99-113

Yang WT, Baek D, Yun DJ, Lee KS, Hong SY, Bae KD, Chung YS, Kwon YS, Kim DH, Jung KH, Kim DH. 2018. Rice OsMYB5P improves plant phosphate acquisition by regulation of phosphate transporter. *PLoS One* 13: e0194628.

Ye Y, Liu B, Zhao M, Wu K, Cheng W, Chen X, Liu Q, Liu Z, Fu X, Wu Y. 2015. CEF1/OsMYB103L is involved in GA-mediated regulation of secondary wall biosynthesis in rice. *Plant Molecular Biology* 89: 385-401.

Yin X, Cui Y, Wang M, Xia X. 2017. Overexpression of a novel MYB-related transcription factor, OsMYBR1, confers improved drought tolerance and decreased ABA sensitivity in rice. *Biochemical and Biophysical Research Communications* 490:1355-1361.

Yokotani N, Sato Y, Tanabe S, et al. 2013. WRKY76 is a rice transcriptional repressor playing opposite roles in blast disease resistance and cold stress tolerance. *Journal of Experimental Botany* 64: 5085-5097.

Yu Y, Yang D, Zhou S, Gu J, Wang F, Dong J, Huang R. 2017. The ethylene response factor OsERF109 negatively affects ethylene biosynthesis and drought tolerance in rice. *Protoplasma* 254: 401-408.

Zhan Y, Sun X, Rong G, Hou C, Huang Y, Jiang D, Weng X. 2017. Identification of two transcription factors activating the expression of OsXIP in rice defence response. *BMC Biotechnology* 17: 26.

Zhang Z, Huang R. 2010. Enhanced tolerance to freezing in tobacco and tomato overexpressing transcription factor TERF2/LeERF2 is modulated by ethylene biosynthesis. *Plant Molecular Biology* 73: 241-249.

Zhao D, Derkx AP, Liu DC, Buchner P, Hawkesford MJ. 2015. Overexpression of a NAC transcription factor delays leaf senescence and increases grain nitrogen concentration in wheat. *Plant Biology (Stuttg)* 17: 904-913.

Zhao Y, Cheng S, Song Y, Huang Y, Zhou S, Liu X, Zhou DX. 2015. The interaction between rice ERF3 and WOX11 promotes crown root development by regulating gene expression involved in cytokinin signaling. *Plant Cell* 27: 2469-2483.

Zheng L, Ying Y, Wang L, Wang F, Whelan J, Shou H. 2010. Identification of a novel iron regulated basic helix-loop-helix protein involved in Fe homeostasis in *Oryza sativa*. *BMC Plant Biology* 10: 166.

Zhu M, Chen G, Zhang J, Zhang Y, Xie Q, Zhao Z, Pan Y, Hu Z. 2014. The abiotic stress-responsive NAC-type transcription factor SlNAC4 regulates salt and drought tolerance and stress-related genes in tomato (*Solanum lycopersicum*). *Plant Cell Reports* 33: 1851-1863.

Zhu M, Chen G, Zhou S, Tu Y, Wang Y, Dong T, Hu Z. 2014. A new tomato NAC (NAM/ATAF1/2/CUC2) transcription factor, SlNAC4, functions as a positive regulator of fruit ripening and carotenoid accumulation. *Plant Cell Physiology* 55: 119-135.

Zhu N, Cheng S, Liu X, Du H, Dai M, Zhou DX, Yang W, Zhao Y. 2015. The R2R3-type MYB gene OsMYB91 has a function in coordinating plant growth and salt stress tolerance in rice. *Plant Science* 236: 146-156.

Zhu X, Qi L, Liu X, Cai S, Xu H, Huang R, Li J, Wei X, Zhang Z. 2014. The wheat ethylene response factor transcription factor pathogen-induced ERF1 mediates host responses to both the necrotrophic pathogen Rhizoctonia cerealis and freezing stresses. *Plant Physiology* 164: 1499-1514.

Zhu Z, Chen G, Guo X, Yin W, Yu X, Hu J, Hu Z. 2017. Overexpression of SlPRE2, an atypical bHLH transcription factor, affects plant morphology and fruit pigment accumulation in tomato. *Scientific Reports* 7: 5786.

Zong Y, Xi X, Li S, Chen W, Zhang B, Liu D, Liu B, Wang D, Zhang H. 2017. Allelic variation and transcriptional isoforms of wheat TaMYC1 gene regulating anthocyanin synthesis in pericarp. *Frontiers in Plant Science* 8: 1645.

15

Mucuna pruriens (L.) DC: Ensuring Sustainability and Societal Enabling Through Agricultural Practices

Susheel Kumar Singh* and Sunita Singh Dhawan

Biotechnology Division, CSIR-Central Institute of Medicinal and Aromatic Plants P.O. CIMAP, Lucknow-226015, Uttar Pradesh, INDIA
*Email: susheelbt05@gmail.com

ABSTRACT

Green revolution provided sufficient amount of foods but the way resources were utilized causes the degradation of natural resources in various aspect. The adverse effect on the ecological system can be visualized in the form of various diseases in human and loss of biodiversity. The population burst is the root cause of imbalanced ecological system because agricultural land decreases drastically. Therefore we must aim to develop and improve the agri-economic growth of farmers by introducing elite variety of *Mucuna pruriens* (L.) DC. *M. pruriens,* a member of Fabaceae family, is an adaptogen and known for its tonifying, strengthening and all around beneficial properties from every part of the plant. *M. pruriens* has range of medicinal applications, because of unusual amino-acid L-DOPA (L3, 4dihydroxyphenylalanine) a precursor of neurotransmitter dopamine (Anti Parkinson's) and involved in various secondary metabolite pathways. It acts like aphrodisiac, anti-snake venom, anti-depressant, antidiabetic, anti-microbial. It has a nutritional quality comparable to soyabeans. *M. pruriens* exhibits a tolerance for various kinds of environmental stresses such as drought, low soil fertility, and high soil acidity. *M. pruriens* suppresses weeds through indeterminate growth and large leaves reduce the amount of light for weeds. It shows allelopathy and effective in lowering nematodes. *M. pruriens* is used as forage, fallow, green manure prevents the soil erosion. It has a longer time span to be used in rotations for management of various pathogens.

Keywords: Cover crop, Green manure, *Mucuna pruriens*, Sustainable development

INTRODUCTION

Mucuna pruriens (L.) DC. belongs to Fabaceae family, is commonly known as Kewanch, Velvet Bean, Cowhage, Cowitch, Atmagupta, Kapikachu. The species is indigenous and endemic to the tropical climates of Asia. *M. pruriens* is an annual climbing herb, which has pinnately trifoliate leaves with purple or white colored flowers and usually self-pollinated. Fruits are called pods or legumes, which are 9-12 cm long and linear with pubescent hairs, and its seeds are uniseriate with white or black colored or mottled seed coat (Fig.1). The genus *Mucuna* covers approximately 100 species of annual and perennial plants (Buckles 1995, Pugalenthi et al. 2005).

M. pruriens has a wide range of medicinal applications containing amino-acid L-DOPA (L-3, 4 dihydroxyphenylalanine) in its seed, which converts into dopamine, an important neurotransmitter involved in the mood, sexuality and movement. L-DOPA is a precursor of neurotransmitter dopamine mainly used for treatment of Parkinson's disease and mental disorders (Lieu et al. 2012) & dopamine itself involve in various secondary metabolites pathway (Singh et al. 2016). In another way it act like aphrodisiac (Lampariello et al. 2012), anti-snake venom (Kumar et al. 2016), anti-depressant properties, anti-diabetic (Eze and Ndukwe 2011), anti-microbial and anti-epileptics, improves the fertility of sperm (Singh et al. 2013), anti-cancerous, anti-oxidant properties (Agbafor et al. 2011), arthritis, dysentery, and cardiovascular diseases etc.

Mucuna pruriens, medicinal values and applied aspects

Fig. 1: Habit representing pods of *Mucuna pruriens*

Mucuna pruriens as a boon to Parkinsonism

The dried seeds (endocarp) powder of the *M. pruriens* bean is used to treat parkinsonism because it contains L-DOPA which is a catecholic amino acids responsible for generation of important neurotransmitter dopamine and other hormones (Barbeau 1969, Manyam 1990). Dopamine is responsible for the movement of body and maintenance of its posture. PD patients treated with *M. pruriens* do not develop drug-induced dyskinesias (Lieu et al. 2012). Therefore, *M. pruriens* extract is better than the artificial L-DOPA. The most probable reason for the potential activity of *M. pruriens* endocarp may contain unidentified antiparkinsonian compounds in addition to L-DOPA, or it may have adjuvant that enhances the efficacy of L-DOPA (Hussian and Manyam 1997, Manyam 1990; Collins-Praino et al. 2011, Goetz 2011).

Mucuna pruriens as an aphrodisiac

Though there are various reasons for male infertility but, low sperm count and motility were commonly seen in an infertile patient. Spermatogenic loss in men could involve several factors such as age, hypertension, oxidative stress, life style, pathological complications, nutritional deficiency, toxicity etc. *M. pruriens* widely used since long time for the treatment of sexual disorders as cited in Charak Samhita, Bhawprakash etc. Dopamine increases the production of testosterone which could further lead to improvement vigor and vitality.

M. pruriens proved effective anti-oxidant, improve the defense system of infertile men under stress along with the quality of their sperm (Shukla et al. 2010). *Mucuna* also contains pruieninin which may help to slow the heart rate, decrease blood pressure and stimulate the intestines. Roots of *M. pruriens* help in regulating menstrual cycle. Treatment with *M. pruriens* significantly improved T, LH, dopamine, adrenaline, and noradrenalin levels in infertile men and reduced levels of FSH and PRL. Sperm count and motility were significantly recovered in infertile men after treatment. Treatment with *M. pruriens* regulates steroidogenesis and improves semen quality in infertile men (Hussian and Manyam 1997; Shukla et al. 2010, Singh et al. 2013)

Mucuna pruriens as an alternative for diabetes

Diabetes mellitus (DM) hampers spermatogenesis, sperm quality, maturation and mitochondrial functioning. Hyperglycaemia-induced free radicals and mitochondrial reactive oxygen species can be a key player in the sperm damage (Padrón et al. 1984; Brownlee 2001). The diabetic rats supplemented with *M. pruriens* extract showed a significant recovery in antioxidant levels and reduced lipid peroxidation (Suresh and Prakash 2012).

Agricultural Practices: Soil Improvement and Green Manuring

Population pressures, intensive cultivation, enhanced the used of pesticide and herbicide   have deteriorated the soil fertility. The encroachment of noxious weeds that another major threat to sustainable agricultural production in developing countries. *M. pruriens* could be used for reclamation of unfertile soil (Duke 1981, Buckles 1995). *M. pruriens* have better adoptability to various kind of environmental and soil condition (Pugalenthi et al. 2005). Application of green manure, crop rotation was better strategy to replenish the soil nutrient that had been used up by single cash crop and to provide the humus or organic material to established the favorable environment for soil micro-biota for sustainable agriculture (Mulvaney et al. 2009). The broader leaf area reduces the soil erosion drastically.

Mucuna is indeed a legume cover crop that is an efficient, cost effective source of nitrogen with additional property to improve soil fertility in intensified cropping regimen (Buckles 1995, Lal and Cummings 1979, Buckles 1995, Weber 1996).

Mucuna pruriens are herbaceous forage and food legumes which has a longer growth period (about 3 to 12 months), found widespread usage as rotation crops for management of various pests and pathogens, as well as in soil improvement and weed control (Buckles 1995). *M. pruriens* could also be used as a cover crop as it is a type of efficient climber, and fast growth, higher herbage yield and large size of leaf. *M. pruriens* controls various kinds of diseases and pests. They are called green manure because they provide nutrients to the soil much like manure does and as living mulches cover crops prevent soil erosion. Once grown, cover crops are usually mowed and then tilled into the soil (Tian et al. 2001).

Mucuna pruriens for weed control

It is also efficient in controlling noxious weeds such as *Imperata cylindrical* and nut grass (*Cyperus rotundus*), two of the most difficult weeds to control in the tropics (Tian et al. 2001, Lawson et al. 2006). There are reports that *Mucuna* can be used for reducing nematode populations (Vargas-Ayala et al. 2000). In the dry season, *Mucuna* ends its life cycle, leaving thick mulch free of weeds. This allows for a subsequent maize crop during the major rainy season with little or no land preparation or weeding. *Mucuna* does not require additional weeding as compared to other crop thus reduces the investment by the farmer. *M. pruriens* were grown in many parts of various countries such as South Africa, Ghana, India, Nigeria etc. for soil fertility and weeds control. *Mucuna* is a vigorous grower and better competes for sunlight and nutrient.

Mucuna pruriens as intercropping

M. pruriens intercropping was used in Ghana and also in some African country. *Mucuna* can also be intercropped with maize and with other crop having strong and thin stem as *Mucuna* require support for proper growth and development. *Mucuna* could be planted in a relay cropping system with food crops such as maize. Maize is planted first, and *Mucuna* seed are sown 40 to 45 days later (Blanchart et al. 2006).

Allelopathic effects of *Mucuna pruriens*

M. pruriens was reported to naturally reduce the weed, and resist against insect, nematode infestation. Allelopathy caused by volatile compounds was studied with velvet bean. Root growth of *Lactuca sativa* L. seedlings was inhibited by volatile gas from velvet bean seedlings planted in Agripot (Fujii et al. 1991). The L-DOPA and dopamine was found to contributed the Allelopathic effect of *M. pruriens* (Fujii et al. 1991, Nishihara et al. 2005).

Limitation of *Mucuna pruriens*

Despite of tremendous property *Mucuna* have some demerits which discourage its cultivation. *Mucuna* is highly sensitive to frost and show poor growth in cold and wet soils. Certain species of *Mucuna* causes severe itching due to its stinging trichome (Singh and Dhawan 2017). Proper stacking or support is required for its optimum yield which require major investment (Singh et al. 2016). *Mucuna* is susceptible for viral disease for instance *Velvet bean Severe Mosaic Virus* (Zaim et al. 2011), but more research input is required to revel the role of viruses with respect to primary and secondary metabolites production. The market of *Mucuna* is not properly developed. Seed of *Mucuna* contains some ant-nutritional factor such as tannins, L-DOPA, lectins, protease inhibitor etc (Pugalenthi et al. 2005). The availability or selection of elite genotype reduces above mentioned problem.

Benefits using *Mucuna pruriens*

Mucuna, have scientifically proven potential to treat various kinds of disorder. Thus, global demand constantly is increasing. More than 200 herbal formulation of *Mucuna* is available in market to treat various kinds of diseases. The cost of *Mucuna* in Indian market was ranged from 50-500 rs/kg. Income of famer could be enhanced by using *Mucuna* with intercropping. The other benefits of *M. pruriens* are biocontrol agent for weed and disease, nitrogen fixation or enrichment of soil nutrient, preservation of soil microbiota, protection of soil from erosion, increase the yield of maize cultivar etc.

Future Perspectives: Development of Genotypes with Better Agronomic Traits

The selection of better adoptable genotype is highly advisable. ã-irradiation was frequently used for various type of crop improvement program as shown in FAO/IAEA Mutant Varieties Database (http://mvgs.iaea.org/Search.aspx). ã-irradiation was also used for the improvement of *M. pruriens* var. *CIM-Ajar*. The new mutant was reported with improved and stable traits such as yield and no stinging genotype of *M. pruriens* (Singh et al. 2016, Singh and Dhawan 2017). The mutants with early maturity, seeds yield and high L-DOPA content were selected and further analysis is continued for better agronomic traits as well as the suitability of crop for the different cropping regimes (Singh et al. 2016). However, proper L-DOPA pathway is still highly desirable.

UPCOMING CHALLENGES

The major problems are smaller land holdings combined with large family size and higher and regularly increasing input costs. The unreliable weather conditions have resulted into reduced net returns from conventional agricultural crops. Therefore, agriculture has been gradually proving to be an uneconomical option leading to grower's inclination towards urbanization. Therefore the current requirement is to develop and improve the agri-economic growth of farmers by introducing some specialty economically viable crops. Only directed scientific efforts would be useful for improving the agricultural productivity for making a significant improvement in rural economy.

CONCLUSION

Mucuna pruriens was found suitable for sustainable agricultural practices and enhancing the socio-economic condition of farmer. The demerits associated with *Mucuna* could be eliminated with use of suitable genotype for instance non stinging, high yielding genotype of *M. pruriens* is advisable for commercial cultivation. Therefore the crops should be developed as per rural demand and their dependence on agriculture which may further provide a support in the form of natural products as drugs in industrial partnership and further developed as an agro-business with the support of technology. Which would then enable the society by ensuring sustainability through generation of employment, by increasing income which provide higher living standards and nutritional security and education through the input of research and development.

Abbreviations

L-DOPA (L-3,4 dihydroxyphenylalanine); *M. pruriens (Mucuna pruriens)*

Conflict of Interest

Authors declare no conflict of interests.

ACKNOWLEDGEMENT

We gratefully acknowledge the Director CSIR-CIMAP, Lucknow, India and senior research fellowship from ICMR (Indian Council Medical Research), New Delhi, India.

REFERENCES CITED

Agbafor KN, Nwachukwu N, Agbafor KN, Nwachukwu N. 2011. Phytochemical analysis and antioxidant property of leaf extracts of *Vitex doniana* and *Mucuna pruriens*. *Biochem Res Int* 2011:e459839. doi: 10.1155/2011/459839,

Lawson IYD, DzomekuIKD, Asempa R, Benson S. 2006. Weed control in maize using *Mucuna* and *Canavalia* as intercrops in the Northern Guinea Savanna Zone of Ghana. *J Agron* 5:621–625. doi: 10.3923/ja.2006.621.625

Barbeau A. 1969. L-DOPA therapy in Parkinson's disease. *Can Med Assoc J* 101:59–68

Blanchart E, Villenave C, Viallatoux A, et al. 2006. Long-term effect of a legume cover crop (*Mucuna pruriens* var. utilis) on the communities of soil macrofauna and nematofauna, under maize cultivation, in southern Benin. *Eur J Soil Biol* 42:S136–S144. doi: 10.1016/j.ejsobi.2006.07.018

Brownlee M. 2001. Biochemistry and molecular cell biology of diabetic complications. *Nature* 414:813–820. doi: 10.1038/414813a

Buckles D. 1995. Velvetbean: A "new" plant with a history. *Econ Bot* 49:13–25

Duke JA. 1981. Handbook of legumes of world economic importance. Springer US, Boston, MA

Eze JI, Ndukwe S. 2011. Effect of methanolic extract of *Mucuna pruriens* seed on the immune response of mice. *Comp Clin Pathol* 21:1343–1347. doi: 10.1007/s00580-011-1294-4

Fujii Y, Shibuya T, Yasuda T. 1991. L-3,4-dihydroxyphenylalanine as an allelochemical candidate from *Mucuna pruriens* (L.) DC. var. utilis. *Agric Biol Chem* 55:617–618. doi: 10.1080/00021369.1991.10870627

Hussian G, Manyam BV. 1997. *Mucuna pruriens* proves more effective than L-DOPA in Parkinson's disease animal model. *Phytother Res* 11:419–423.

Kumar A, Gupta C, Nair DT, Salunke DM. 2016. MP-4 contributes to snake venom neutralization by *Mucuna pruriens* seeds through an indirect antibody-mediated mechanism. *J Biol Chem* 291:11373–11384. doi: 10.1074/jbc.M115.699173

Lampariello LR, Cortelazzo A, Guerranti R, et al. 2012. The magic velvet bean of *Mucuna pruriens*. *J Tradit Complement Med* 2:331–339

Lieu CA, Venkiteswaran K, Gilmour TP, et al. 2012. The antiparkinsonian and antidyskinetic mechanisms of *Mucuna pruriens* in the MPTP-treated nonhuman primate. *Evid Based Complement Alternat Med* 2012:e840247. doi: 10.1155/2012/840247

Manyam BV. 1990. Paralysis agitans and levodopa in/ Ayurveda: Ancient Indian medical treatise. *Mov Disord* 5:47–48. doi: 10.1002/mds.870050112

Mulvaney RL, Khan SA, Ellsworth TR. 2009. Synthetic nitrogen fertilizers deplete soil nitrogen: a global dilemma for sustainable cereal production. *J Environ Qual* 38:2295. doi: 10.2134/jeq2008.0527

Nishihara E, Parvez MM, Araya H, et al. 2005. L-3-(3,4-Dihydroxyphenyl)alanine (L-DOPA), an allelochemical exuded from velvetbean (*Mucuna pruriens*) roots. *Plant Growth Regul* 45:113–120. doi: 10.1007/s10725-005-0610-x

Padrón RS, Dambay A, Suárez R, Más J. 1984. Semen analyses in adolescent diabetic patients. *Acta Diabetol Lat* 21:115–121. doi: 10.1007/BF02591100

Pugalenthi M, Vadivel V, Siddhuraju P. 2005. Alternative food/feed perspectives of an underutilized legume *Mucuna pruriens* var. utilis—a review. *Plant Foods Hum Nutr Dordr Neth* 60:201–218. doi: 10.1007/s11130-005-8620-4

Shukla KK, Mahdi AA, Ahmad MK, et al. 2010. *Mucuna pruriens* reduces stress and improves the quality of semen in infertile men. *Evid-Based Complement Altern Med* 7:137–144. doi: 10.1093/ecam/nem171.

Singh AP, Sarkar S, Tripathi M, Rajender S. 2013. *Mucuna pruriens* and its major constituent L-DOPA recover spermatogenic loss by combating ros, loss of mitochondrial membrane potential and apoptosis. *PLOS ONE* 8:e54655. doi: 10.1371/journal.pone.0054655.

Singh SK, Dhawan SS. 2017. Analyzing trichomes and spatio-temporal expression of a cysteine protease gene Mucunain in *Mucuna pruriens* L. (DC). *Protoplasma*. doi: 10.1007/s00709-017-1164-2.

Singh SK, Yadav D, Lal RK, et al. 2016. Inducing mutations through ã-irradiation in seeds of *Mucuna pruriens* for developing high L-DOPA-yielding genotypes. Int J Radiat Biol 0:1–10. doi: 10.1080/09553002.2016.1254832.

Suresh S, Prakash S. 2012. Effect of *Mucuna pruriens* (Linn.) on sexual behavior and sperm parameters in streptozotocin induced diabetic male rat. *J Sex Med* 9:3066–3078. doi: 10.1111/j.1743-6109.2010.01831.x.

Tian G, Ishida F, Keatinge D, et al. 2001. *Mucuna* cover crop fallow systems: potential and limitations. In: SSSA Special Publication. Soil Science Society of America and American Society of Agronomy.

Vargas-Ayala R, Rodrýìguez-Kábana R, Morgan-Jones G et al. 2000. shifts in soil microflora induced by velvetbean (*Mucuna deeringiana*) in cropping systems to control root-knot nematodes. *Biol Control* 17:11–22. doi: 10.1006/bcon.1999.0769.

Zaim M, Kumar Y, Hallan V, Zaidi AA. 2011. Velvet bean severe mosaic virus: a distinct begomovirus species causing severe mosaic in *Mucuna pruriens* (L.) DC. *Virus Genes* 43:138–146. doi: 10.1007/s11262-011-0610-z.

16

Taxonomy, Anatomy and Phytochemical Screening of *Eulaliopsis binata* (Retz.) C.E. Hubb. (Family-Poaceae)

Shilpa Dinda[1], Debashree Ghosh[2], Souradut Ray[3] and Amal Kumar Mondal[4]

Department of Biological Sciences, Midnapore City College, Kuturia Bhadutata Midnapore, Paschim Medinipure-721129, West Bengal, INDIA
[2,3,4]Plant Taxonomy, Biosystematics and Molecular Taxonomy Laboratory UGC-DRS-SAP-SAP-II & DBT-BOOST-WB, Department of Botany and Forestry Vidyasagar University Midnapore-721102, West Bengal, INDIA
Email: shilpadindabotvu@gmail.com; souradutray@gmail.com amalcaebotvu@gmail.com

ABSTRACT

Eulaliopsis binata (Retz.) C.E. Hubb. is a valuable grass growing in Midnapore district of West Bengal known for their commercial purposes. Livelihood of and many of peoples are depends on the production of this grass. Besides this we try to present their phytochemical constituents mainly alkaloids, flavonoids and phenolic compounds for medicinal study. In this investigation, we studied the morpho-taxonomy and anatomical characterization as well as phytochemical screening of different chemicals. Besides this we make an idea about presence of sugar molecules by HPLC study and chemical bonding properties with the help of FTIR. The detailed study gives an idea about *E. binata* and its medicinal properties.

Keywords: *Eulaliopsis binata*; Morpho-taxonomy; Anatomy; Phytochemical screening; HPLC; FTIR.

INTRODUCTION

India is a big mega diversity country and known for biodiversity rich sites such as Western Ghats, The Sundarland Himalaya, Indo-Myanmar and Himalayas (Mayers et al. 1990). The major parts of Indian vegetation are covered by forests of various compositions. Grass constitute one of the major components of any forests and NTFP derived from forests product which are used for daily needs as well as for economic purposes by the forest dwellers.

Grasses are the natural widely distributed homogeneous group of plants belonging to the family Poaceae *(nom. alt.* Gramineae). The value of grasses to the mankind is well recognized. As the members of this group are present in all the conceivable habitats suitable for the plant growth they have become natural associate of man. In nutshell, the people not only depend on forest for food, clothes, shelter but also depend on them for their medical needs.

In spite of that many of the usages of grasses are still unrecorded. Tribal people who mainly depend on plants for their day to day needs and livelihood are the custodians of the traditional knowledge systems of utilization of plants. In the present study, an attempt has been made to record the utilization of grasses by different tribal groups to discover their knowledge bank, before it gets lost forever under the pressure of fast pace of westernization of different tribal communities disconnecting them from the past (Mitra and Mukherjee 2009).

Here we intend to discuss about a particular grass *Eulaliopsis binata* (Retz.) C.E.Hubb. (Syn: *Pollinidium angustifolium* Haines) locally known as Sabai grass and its ayurvedic name is Balvaja (Khare 2007). Sabai grass is perennial plant which grows in different parts of Asian countries like India, China, Nepal, Bhutan, Bangladesh, Pakistan, and Myanmar (Jayaswal 2003). In India, this species is cultivated in West Bengal, Bihar, Jharkhand, Odisha, Punjab, Haryana and Jammu and Kashmir. In West Bengal, it is mostly cultivated in the district of Paschim Midnapore, Bankura and Purulia. It has thin and long leaves producing high quality fibre, making it suitable as a major raw material for paper industry Basu et al (2006). The tribal peoples such as Lodha, Oraon, Sabar, Santal, Munda and Lohara uses this grass for making hardy rope. They also use this grass as ethnic medicines to cure fever, bronchitis and diuretics (Kirtikar and Basu 1935).

Sabai is considered as "The Money Plant" which ensures cash receipt throughout the year. The cottage industry based on sabai grass is associated with the activities of raising production of grass and processing of consumer goods such as ropes, mats, carpets, sofa sets, wall hangings and other sophisticated fashionable articles Hathy (Sahu and Satpathy 2010).

The traditional practice of making ropes from this grass by hands is a laborious, time consuming and slow. *Eulaliopsis binata* is widely collected for paper-making, strings, ropes and mats. It frequents found in hot dry areas and its extensive underground root-system enables it to survive forest fires. It is not eaten by cattle except in times of hardship or in winter season.

Plant Sample Collection

For the present study, fresh leaves, stem, root from each plants collected from in natural forest of south West Bengal is characterized by lateritic soils and dry deciduous forests like in Midnapore district. This survey was conducted in some natural forest areas like Kankra jhor (22°34'N, 86°55'E) Kushboni forest (22°49'N, 86°97'E to 22°51'N,86°94'E), Hoomgarh forest (22°82'N, 87°23'E to 22°83'N, 87°25'E), Bulanpur forest (22°73'N, 87°10'E to 22°74'N, 87°11'E). In Purulia District we covered Ajodhya hills (23°13'N 86°06'E to 23°27'N, 86°19'E), Panchakot hills (23°59'N, 86°75'E to 23°64'N, 86°77'E), and in Bankura district, we visited Sarenga jungle (22°76'N, 87°01'E to 22°77'N, 87°03'E) by Garmin.

Morphological Studies

The detailed morphological study was performed with Stereozoom Microscope (Leica S8APO) and identifies silica bodies with the help of Leica compound microscope (DM1000).

Anatomical Characters

For anatomical study plant materials are preserved into 70% ethanol. Leaves cross section are made at1/4 leaf width and at the main vein, using a freezing microtome. Leaves cross section are made by using microtome. The photos of sections are taken by using a microscope Leica DM 1000 LED).

Plant Extract Preparation for Chemical Analysis

The entire plant was separated and dried in Hot Air Oven. The dried plant materials were ground to course powdered. The powdered plant was extract with Acetone, Methanol and Water then it preserved in freez for further chemical test Ghai (2004).

Phyto-Chemical Tests

Qualitative Tests for Carbohydrate and Protein

Carbohydrate

Molisch's Test: Add a two drops of Molisch's reagent (5% 1-napthol in alcohol) to about 2ml of test solution and mix well. Incline the tube and add about 1ml of concentrated sulphuric acid along the sides of the tube. Observe the colour at the junction of two liquids.

Fehling's test: To 1ml of Fehling's solution 'A', add 1ml of Fehling's solution 'B' and a few drops of the test solution. Boil for a few minutes.

Benedict's test: To 2ml of Benedict's reagent add five drops of the test solution. Boil for five minutes in a water bath. Cool the solution.

Protein

Biuret test: To 2ml of the test solution add 2ml of 10% NaOH. Mix. Add two drops of 0.1% $CuSO_4$ solution.

Ninhydrin test : To 4 ml of the solution which should be at neutral p^H add 1 ml of 0.1% freshly prepared ninhydrin solution. Mix the contents and boil for a couple of minutes. Allow to cool.

Glyoxylic Reaction for Tryptpophan

Hopkins-Cole test: Add 2ml of glacial acetic acid to 2ml of the test solution. Then add about 2ml of conc. H_2SO_4 carefully down the sides of the test tube. Observe the colour, change at the junction of the two liquids.

Qualitative Tests for the Identification of Tannin, Starch, Alkaloids and Oil

Preparation of the extract: Take 0.2 gm of the powdered drug in test tube, add 2ml of the solvent and shake it for 1minute. Filter through cotton wool plug over a glass funnel. If required the solvent with the powder may be heated on a water bath for 2 min before filtration.

Tanins: Ferric chloride test: To the drug extract add 1% ferric chloride solution. Blue, green or brownish color indicates the presence of tannins.

Iodine: Add a few drops of iodine solution to about 1ml of the test solution. Appearance of blue colour is due to the formation of starch iodine complex.

Glycosides: Take 0.5 ml of alcoholic extract of the drug add 1ml of water and 1ml of sodium hydroxide solution. A yellow color indicates the presence of glycosides.

Saponins: Weight 0.1 gm of the drug and place it in a clean dry test tube. Add 10ml of D/W put a stopper and vigorously shake for 30 minutes allow the tube to stand in a vertical position and observe for a period of 30 minutes if a honeycomb forth more than 3cm above the surface of the liquid persists for 30 minutes, the drug is presumed to contain saponins.

Oils: To the thin section of the drug add alcoholic solution of Sudan III; Red color indicates presence of volatile oil.

Flavonoids: Add 5ml Lead acetate to about 100mg of plant extract.

Alkaloids: Add 10 ml HCl to the 250 plant extract and then the solution filtered. 2ml of filtrate material add with the 2ml of picric acid.

Quantitative Tests

Extraction and Estimation of the Carteniod Content

Total carotenoid content (mainly β-carotenoid) was extracted and estimated according to the method of Kuhu and Grundmann (2007).1g plant tissue was extracted with 5ml of acetone reagent. Then centrifuged at 10,000rpm for 10min, supernatant was collected. Then aliquote 4ml taken in a test tube for experiment after incubation 30 min in room temperature, the colour intensity was measured by spectrophotometer at 474nm.

Extraction and Estimation of Protein by Lowry's Method

1gm nodule was extracted with 5ml of 50Mm phosphate buffer (pH- 7.5) centrifuged at 10,000g for 10 min. To the supernatant known as buffer soluble protein equal volume of 20% TCA was added and kept for 30 min as 4^0C.The precipitate obtained after centrifugation collected. To the precipitate dissolve in 1(M) NaOH solution was considered as TCA insoluble fraction and the supernatant was considered as TCA soluble fraction. The known amount of protein extract from buffer soluble, TCA insoluble and TCA soluble fractions were mixed with 5ml of Lowry reagent. After 10 min 0.5 ml of Folin phenol reagent was added and intensity of color developed was measured in a spectrophotometer at 650nm after 30 min incubation.

Extraction and Estimation of the Sugar Content

1g nodules were extracted with 5ml of 50mM K-phosphate buffer (pH-8) at 4000g for 15min. Then supernatant was taken and 5 volumes of 80% ethanol was added, kept at 4°C for 4days. Aquous phase was taken, 2ml of aliquote was mixed with 4ml of chilled anthrone reagent and heated at 90°C for 15min in water bath. Then the tubes were allowed to stand at room temperature and the absorbancy was taken at 655nm using spectrophotometer (Lo and Garceau, 1975).

Extraction and Estimation of the Polyphenol Content

1g tissue grinds with 5ml of 80% ethanol and centrifuged at 1000g for 20 min. The supernatant was taken and kept. Again re-extract the residue with same reagent, spin and pool the supernatant as above. Then dissolve the residue in 2ml distilled water, take 1ml in a test tube for experiment. In 1ml aliquote add 2ml 2% Na_2CO_3 after 2min incubation add 1ml of Folin-ciocalteau reagent. Then after incubation 30min in room temperature and the absorbancy was taken it 720nm (Malik and Singh, 1980).

Catalase activity

1gm tissue crush with 5ml 100mM k-phosphate buffer (pH-8) and then spin with 10,000rpm for 10 min. The supernatant was collected as a enzyme extract. Then the reaction mixtures are containing enzyme Blank -800µl-2400µl Phosphate buffer +800µl EDTA. Substrate blank -400µ Lenzyme extract +2800µl Phosphate buffer+800µlEDTA. Experimental Set-400µ lEnz. extract+800µL H_2O_2 +2000µl Phosphate buffer+800µlEDTA. Then take O.D. value at 240nm by spectrophotometer (Luck 1974).

Extraction and Estimation of Ascorbic Acid

Weight 42 mg sodium bicarbonate into a small volume of distilled water. Dissolve 52mg 2,6 dichloro phenol indophenols in it and make up to 200mL with distilled water. Dilute 10mL of the stock solution to 100ML with4% oxalic acid used as a standard. The concentration of working standard is 100µg/mL. Pipette out 5mL of the working standard solution into a 100mL conical flask. Add 10mL of 4% oxalic acid and titrate against the dye (V_1ml). End point is the appearance of pink colour which persists for a few mimutes. The amount of the dye consumed is equivalent to the amount of ascorbic acid. Extract the sample (0.05-5g dependending on the sample) in 4% oxalic acid and make up to a known volume (100ml) and centrifuge. Pipette out 5mLof this supernatant, add 10mL of 4% oxalic acid and titrate against the dye (V_2ml) (Miller et al. 1995).

Calculation

Amount of ascorbic acid mg/100g sample = 0.5/V_2ml*V2/5ml*100ml/wt. of the sample*100]

Extraction and Estimation of Flavonoid Content

Aluminium chloride colorimetric method was used for flavonoid content determination (Ghai 2004). Each extract (0.5 ml of 1: 10 g/ml) in methanol was mixed with 1.5 ml of methanol, 0.1 ml 10 % aluminium chloride, 0.1 ml of 1 M potassium acetate and 2.8 ml of distilled water. It remained at room temperature for 30 min; the absorbance of the reaction mixture was measured at 415 nm. The calibration curve was prepared by preparing gallic acid solutions at concentrations 12.5 to 100 µg/ml in methanol.

HPLC ANALYSIS

Extraction and Purification Procedures for Carbohydrate

100-300 mg plant tissue were extracted overnight in 10-30 ml 75% EtOH (adjusted to pH 7) at room temperature and the homogenate filtered through

Whatman 42 paper. Extraction steps are repeated 3 times (Victoria and Bonta, 2008). The 30-90 ml solution was then evaporated to dryness at room temperature under reduced pressure (Rotavapor Buchi R-144,Switzerland) and diluted with 1 ml, pH 7,water (Milli Q grade, Millipore Italia). Analytical method was carried out using a binary LC pump 250 (Agilent) equipped with an automatic injection system. A waters column heater module controlled by a Temper olive tissues, was used as internal standard and added to the crushed material .The recovery was estimated for each carbohydrate .Thus 0.25, 0.50, 0.75 and 1.0 ml 1 mg ml^{-1} carbohydrate solutions were fractioned and analysed as described for samples. Recovery ranged from 92 to 99 %.Calibration curves were performed for glucose, fructose, sucrose, galactose, mannose, and sorbitol, xylose etc. The recovery and calibration curves for different sugars were constructed after preparative HPLC using *Eulaliopsis binata* leaf, stem and root extracts.

PSA ANALYSIS

BSA (Sigma Aldrich) was dissolved in 1x Phosphate Buffer Saline (137 mM NaCl, 2.7 mM KCl, 10 mM Na_2HPO_4, and 1.8 mM KH_2PO_4) at a concentration of 2 mg/mL; brief bath sonication was applied before use (Mitrano, et.al. 2014). BSA was filtered through a 0.1 micron filter (Whatman); 50 ìl of filtered BSA was pipetted into a previously calibrated Delsa Max CORE quartz cuvette, which was immediately capped and placed in the cuvette. The quartz cuvette had been calibrated with toluene at 20°C; additional calibration of the water solvent at 20°C and 90°C was necessary as well, in order to make DLS-based on molecular weight measurements (MW-S). Knowledge of the sample dn/dc value typically 185 mL/g for proteins is required, and knowledge of the second virial coefficient (A2) helps to improve accuracy (0.00014 mol * L/cm for BSA). An event schedule was programmed to allow the BSA and cuvette to initially come to temperature equilibrium at 37°C followed by molecular weight measurements using the combined SLS and DLS. After the measurement, the temperature incrementally rose by 5°C, followed by a molecular weight measurement; this process continued until the temperature reached 82° plant extract samples were also used for molecular weight analysis at both 0.5 mg/mL and 1 mg/mL concentrations. For BSA measurements, the quartz cuvette was calibrated at 25°C and90°C. For the plant extracts measurements, the quartz cuvette was calibrated at 25°C (Mitrano et al. 2014).

Fourier Transform Infrared Spectrophotometer (FTIR) Analaysis

Preparation of Leaf Extract

The shade dried leaves of each plant (at 20°C) were powdered in mechanical grinder. 20 grams of leaf powder (of each plant) was weighed, 150 ml of solvent was added and kept for 3 days. The extract was filtered using Whatman No.1 filter paper and the supernatant was collected. The residue was again extracted two times (with 3 days of interval for each extraction) and supernatants were collected. The supernatants were pooled and evaporated (at room temperature, 28 ± 1 C) until the volume was reduced to 150 ml. Extracts of the leaf powder of the four plants with 2 different solvents such as water (WA) and methanol (ME) were prepared and stored in air tight bottles for further analysis.

Estimation

Liquid samples of different solvent extracts of each plant materials were used for FTIR analysis. The liquid sample of each plant specimen was loaded in FTIR spectroscope (Shimadzu), with a Scan range from 400 to 4000 cm 1 with a resolution of 4 cm.

RESULTS AND DISCUSSION

Morphological Characterization

Perenials. Culms tufted, erect, wooly at base, nodes glabrous. Leaf sheath terete, basal sheaths woolly. Leaf blade flat, linear, hairy on upper side, finely acuminate at apex. Racemes 2-4 on axillary and terminal peduncles. Sessile spikelet, narrowly obovate, villous yellowish brown hairs. Lower glume membranous, narrowly ovate, densely hairy in the lower half, margins ciliate hairy. Upper glume membranous, densely hairy on dorsal side, 1-keeled, densely hairy below the middle, 5-nerved, awned at apex. Lower lemma hyaline, ovate elliptic, margins ciliate on upper side, nerveless, 3 lobed at apex. Palea hyaline, narrowly ovate, margins inflexed, apex acute-acuminate. Lodicules-2. Stamens 3. Upper lemma membranous, narrowly ovate, margins hairy, 1-nerved, awned at apex. Palea hyaline, obovate, margins inflexed, apex three lobed, ciliate hairy. Lodicules 2. Stamens 3. Pedicelled spikelet similar to sessile spikelet.

Flowering and Fruting Period – December to February

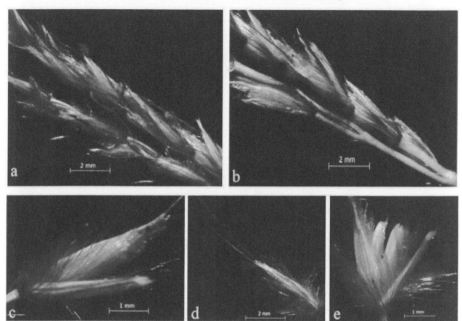

Fig. 1: Morphological study of *Eulaliopsis binata* under
Stereozoome Microscope (Leica)

Micro-morphological Study

Leaf epidermis of *Eulaliopsis binata* contain silica bodies, stomata and trichomes. Grass is the major source of silica bodies found in the soil which are deposited in the cells of the dermal, vascular and fundamental tissues of the plant. *Eulaliopsis binata* deposit two types of silica bodies viz bilobate short bar with squared sphere shaped and narrow elliptical rectangular shaped. Bilobate short bar with squared sphere silica bodies are found on one or two rows over the vascular bundle, they fit against the cell in same rows. Narrow elliptical rectangular shaped silica bodies are found in different position of the epidermal cells of the leaf. These two types of silica bodies are observed under Light, Polarized and Fluroscence Microscope (Fig. 3). Diacytic stomata and unicellular trichome are also present on leaf epidermis. Stomata are located in two or three rows between the vascular bundle and the stomata are distantly arranged in the same row. The stomata are formed by the four cells. Two interior cells i.e the guard cells which are on either side of the stomata and the two exterior cells i.e the subsidiary cells. The trichomes are Sharp pointed and unicellular. They are found over the vascular bundle. The prickle hairs may be hook shaped. The trichomes are pointed towards the apex of the plant parts.

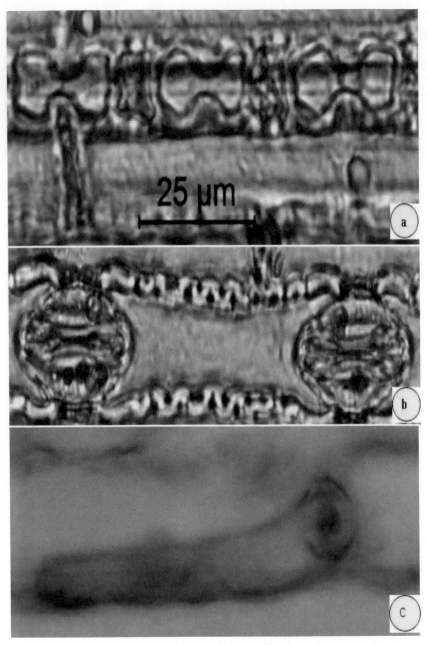

Fig. 2: Silica bodies (a), stomata (b) and trichome (c) view of *Eulaliopsis binata* under compound Microscope (Leica DM-1000)

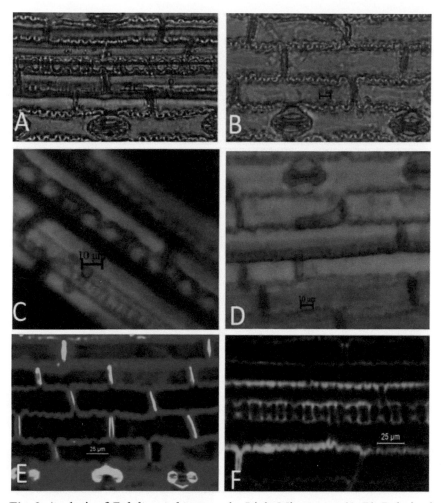

Fig. 3: Analysis of *Eulaliopsis binata* under Light Microscope (A, B), Polarized Microscope (C, D) and Fluroscence Microscope (E, F).

ANATOMICAL STUDY

Leaf Anatomy: The leaf is isobilateral, the upper and the lower epidermises of leaf are similar and consist of single layer of cells. The cells of the epidermis is cuticularized, mesophyll not differentiated into palisade and spongy tissues. The cells are spherical and covered the whole surface of the leaf; no intercellular spaces. Many large and small vascular bundles are present in the leaf; vascular bundles are conjoint, collateral and closed.

Root Anatomy: It is the outermost layer of cells with large number of unicellular cells, called epiblema. Cortex lies below the epiblema, upto endodermis, multilayered, composed of circular parenchymatous cells with intercellular space, casperian bands present in some cells. Pericycle one layer thick composed of sclerenchymatous tissues, present below the endodermis. Vascular bundles are polyarch, radial and exarch. Pith present in centre and composed of loosely arranged parenchymatous cells.

Stem anatomy: Epidermis is the outermost layer of the stem and single cell thick. The outer wall of the cells is cuticularized. Hypodermis made up of sclerenchymatous cells and is 2-4 layers thick. Below the hypodermis parenchymatous ground tissues are present and vascular bundles are embedded within it. Vascular bundles are scattered throughout the ground tissue. Vascular bundles are conjoint, collateral, endarch and closed.

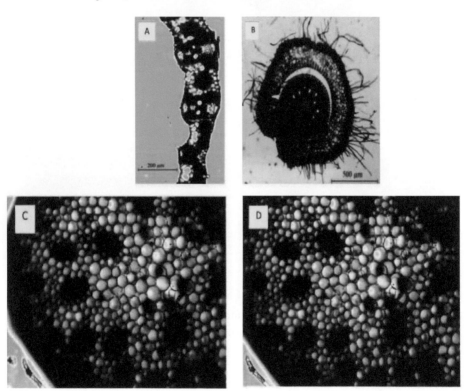

Fig. 4: Leaf anatomy (A), root anatomy (B) and stem anatomy (C, D) of *Eulaliopsis binata*.

PHYTOCHEMICAL CHARACTERS AND VARIATIONS

Qualitative Analysis

The qualitative analysis of different phytochemicals shows the presence of carbohydrate, proteins, alkaloides, flavonoid and phenolics. The water and methanolic extract of different plant parts shows positive result in carbohydrate test. In case of protein test results is positive. In case of tannin and glycosides water extract shows negative result but positive result in methanolic extract. In case of saponins, alkaloid, phenolic and flavonoid shows always positive result in water extract and methanolic extract. In case of iodine and saponins shows intermediate response in water extract and methanolic extract.

Table 1: Phytochemical screening of leaves powder of *Eulaliopsis binata*

Phytochemical tests	Results					
	Water extract			Methanol extract		
	Leaf	Stem	Root	Leaf	Stem	Root
1. Carbohydrate	+	+	+	+	+	+
a. Felling's test	+	+	+	+	+	+
b. Benedict test	+	+	+	+	+	+
c. Molish's test	+	+	+	+	+	+
d. Seliwanoff's test	+	+	+	+	+	+
e. Barfoed's test	+	+	+	+	+	+
2. Protein						
a.Biuret test	+	+	+	+	+	+
b.Ninhydrin test	+	+	+	+	+	+
c.Glyoxylic reaction for Tryptophan	+	+	+	+	+	+
3. Tanin test: Ferric chloride test	-	-	-	+	+	+
4. Glycoside test	-	-	-	+	+	+
5. Iodine	+	-	-	+	+	+
6. Saponins	+	+	+	+	+	+
7. Alkaloid: Hanger's test	+	+	+	+	+	+
8. Flavonoid	+	+	+	+	+	+
9. Volatile oil	+	+	+	-	-	-
10. Phenolic	+	+	+	+	+	+

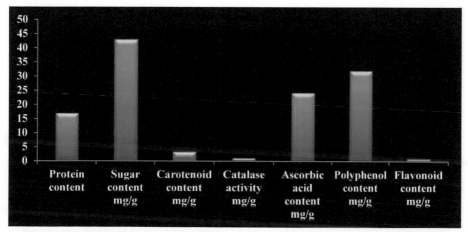

Fig. 5: Graphical representation of phytochemical content of fresh leaf of tissue.

After qualitative test, we perform quantitative test for quantification of these phytochemicals. We quantify the nutritional components as well as secondary metabolites in leaf extract. In case nutritional components the leaf extract shows highest sugar content (43.05 mg/g) and low level of protein content (17mg/g). The other ascorbic acid (24.24mg/g), carotenoid content (3.382mg/g) and catalase activity (1.165mg/g) is very low. In case of polyphenol content (32mg/g) the leaf extract of *Eulaliopsis binata* shows moderately high value than other tests. The leaf contain very low amount of secondary metabolites like flavonoid content (0.923mg/g).

Table 2: Phytochemical analysis of leaves powder of *Eulaliopsis binata*

Sl No.	Phytochemical test	Amount (mg/g)
1.	Protein content	17
2.	Sugar content mg/g	43.05
3.	Carotenoid content mg/g	3.382
4.	Catalase activity mg/g	1.165
5.	Ascorbic acid content mg/g	24.24
6.	Polyphenol content mg/g	32
7.	Flavonoid content mg/g	0.923

PSA Analysis

The analysis of molecular size of protein molecule is performed by the PSA analysis in form of Z-average size determination of protein molecule. From these data we calculate the protein size in kilo-dalton form. The exact molecular weight of protein molecule is 28777KD.

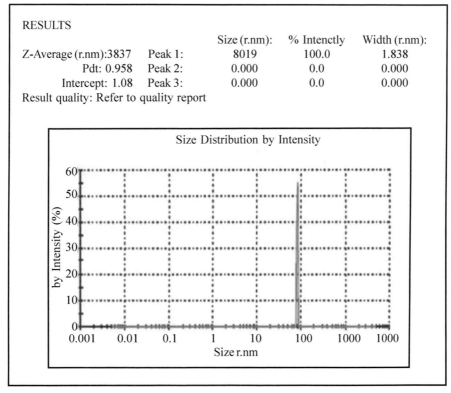

Fig. 6: Molecular Size of the Protein molecule of *Eulaliopsis binata*

HPLC Analysis

After quantification of carbohydrate molecule by quantitative test, we also perform HPLC analysis of leaf extract for proper determination of presence of individual sugar molecule. HPLC data shows presence of glucose and galactose molecule present in leaf extract of *Eulaliopsis binata*.

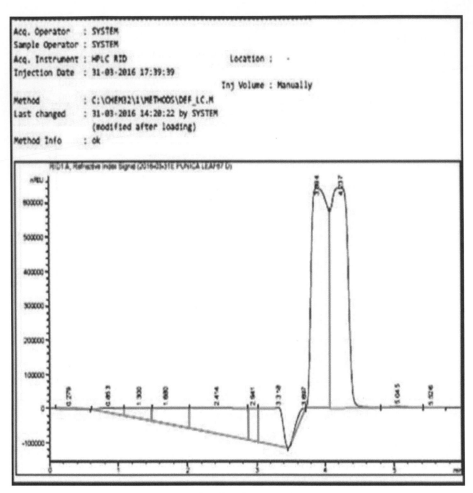

```
Acq. Operator   : SYSTEM
Sample Operator : SYSTEM
Acq. Instrument : HPLC RID                    Location :   ·
Injection Date  : 31-03-2016 17:39:39
                                             Inj Volume : Manually
Method          : C:\CHEM32\1\METHODS\DEF_LC.M
Last changed    : 31-03-2016 14:20:22 by SYSTEM
                  (modified after loading)
Method Info     : ok
```

Fig. 7: HPLC analysis of *Eulaliopsis binata* leaf extract.

FTIR Spectroscopy

In FTIR assay we confirmed that the presence of different chemical bond stretching groups, that denotes the presence of alkenes, aromatic, ketone, phenols, acids etc. By this graphical output, we confirmed that the leaf extracts of *Eulaliopsis binata* have N-H stretching bond (3400-3500) for protein molecule, C=C stretching (1450-1600) for aromatic compounds, C=S stretching for any type of sulphur containing compounds and C-H stretching for alkane group. From this data we confirmed that the leaf extract of *Eulaliopsis binata* have so many chemical compounds and have medicinal importance.

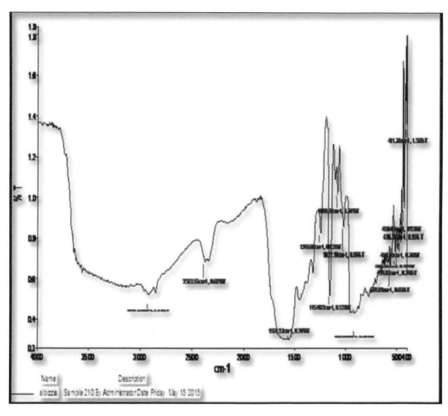

Fig. 8: FTIR Sectroscopy result of *Eulaliopsis binata.*

Economic usages of *Eulaliopsis binnata*

Sabai grass is one of the ornamental plant in the family Poaceae. Sabai grass is not only the soil binder plant but also provides raw material for paper industry. The maximum rural peoples are depend in sabai grass for this income and livelihood. The sabai grass can be used to manufacture house-holds articles such as ropes, mats, sofa, carpets, basket, wall hanging and sophisticated fashionable things. Sabai grass is collected every year in the moth of November to December. Ladies and children are involved in collection of plant samples. Since there is no alternative employment, families engaged in rope making to provide supplemental income (income is very small). Some Government society and NGO are also helping for increasing the marketing potencial of sabai grass product.

Fig. 9: Cultivation of sabai grass and making procedure of rope by the rural people.

CONCLUSION

Herb is an integral part of nature containing natural substances that can promote human health. Natural products especially those derived from plant sources is gaining much interest for therapeutic use than that of the conventional ones. This is due to development of resistance and unwanted side effects. The isolation of different phytochemical compounds and characterization is a initial leads obtained from the traditional system of medicine. *E. binata* is traditional herbs in Midnapore districts and many of the peoples depend on its high yielding and commercialization purposes. Explanation of chemical constituents of plants and phytochemical screening may provide us the basis for the developing the leads for the development of novel agents that cure many of the diseases. The investigation carried out on *E. binata* upon the literature survey; try to established a idea about total plants containing phytochemical compounds. Thus it could be concluded that it has a commercial properties as well as medicinal properties because possesses various bioactive compounds that can be used for treatment of various diseases basesd upon the phytochemical screening in our studies.

Abbreviations

LM : Light Microscope

PM : Polarized Microscope

SM : Stereozoome Microscope

FTIR : Fourier Transform Infrared Spectroscopy

HPLC : High Performance Liquid Chromatography

PSA : Partical size Analyzer

Conflict of Interest

There is no conflict of interest

ACKNOWLEDGEMENT

We would like to acknowledge UGC for their financial assistance in the form of DRS-SAP (Phase-I) [2011-2016]. I also like to thanks all of the research scholars in the Plant Taxonomy, Systematic and Molecular Taxonomy Laboratory; UGC-DRS-SAP Department, Department of Botany and Forestry, Vidyasagar University, Midnapore-721102, West Bengal, India.

REFERENCES CITED

Basu M, Mahapatra SC, Bhadoria PBS. 2006. Performance of Sabai grass (*Eulaliopsis binate* (Retz.) C.E. Hubb) under different level of organic and inorganic fertilizer in acid soil; American-Eurasia*Journal of Agriculture & Environmental Sciences*. 1(2):102-105.

Ghai CL. 2004. *In: A Text Book of Practical Physiology*. New Delhi : Jaypee Brothers Medical Publishers. 6th ed., 130-175.

Khare CP. 2007. *Indian Medicinal Plants: An illustrated Dictionary*. New Delhi : Springer Reference.

Kirtikar KR, Basu BD. 1935. *Indian Medicinal Plants* (Vol. I- IV). Allahabad.

Luck H. 1974. *In: methods in enzymatic analysis.*, New York : Academic Press.

LO SN. Garceau II. 1975. A spectrophotometric method for quantitative analysis of sugar mixtures containing known sugars. *The Canadian Journal of Chemical Engineering.* 53:112.

Malik CP, Singh MB. 1980. *Plant enzymology and Histro-enzymology*, New Delhi :Kalyani Publishers, 278.

Myers, N, Mittermeier, R A, Mittermeier C G; da Fonseca, Gustavo A. B.; Kent, J. (2000). Biodiversity hotspots for conservation priorities. *Nature.* 403 (6772): 853–858.

Miller NJ, Diplock AT, Rice-Evans CA. 1995. Evalution of the total antioxidant as a marker of the deterioration of apple juice on storage. *Journal of Agriculture and Food Chemistry* 43: 1794-1801.

Mitra S, Mukherjee S. 2009. Ethnobotany of some Grasses of West Bengal (India), *Advances in Plant Biology* 5. India.

Mitrano D, Ranville JF, Stephan C. 2014. Quantitative Evaluation of Nanoparticles Dissolution Kinetics using single particle ICP-MS. *A case study with silver Nanoparticle* : Parkin Elmer Application Note ,USA.

Ramamurthy N, Kennan S. 2007. Fourier transforms infrared spectroscopic analysis of plant (*Calotropis gigantean* Lill.) from industrial village, Cuddalore Dt, Tamilnadu, India. *Romanian Journal of Biophysics* 17(4): 269-276.

Tee ES, Lim CL. 1991.The analysis of carotenoids and retinoids:a review. *Food Chemistry* 41: 147-19.

Victoria L, Bonta A. 2008. High Performance Liquid Chromatographic analysis of sugar in Transylvanian Honey. *Bulletin UASVN Animal Science and Biotechnology* 65:1-2.

17

Indian Rhododendrons and their Value Addition: Revision of Species Diversity and Review with Reference to Product Development

Sumit Singh and Bikarma Singh*

[1]Plant Sciences (Biodiversity and Applied Botany Division), CSIR-Indian Institute of Integrative Medicine, Jammu 180001, Jammu and Kashmir, INDIA
[2]Academy of Scientific and Innovative Research, New Delhi 110001, INDIA
*Email: drbikarma@iiim.res.in; drbikarma@iiim.ac.in

ABSTRACT

Evolution theory postulates that the collision between the Indian and the Eurasian plates in early Eocene and Miocene epoch leads to formation of the Himalayas, and currently recognized across the globe as repository of life. Some species of this regions is categorized under endangered and endemic category. Rhododendrons placed under family Ericaceae is an important genus of flowering plants. It is one of the multiferous keystone member of plants, mostly endemic to Southeast Asia. Wide distribution of rhododendrons concentrated in Himalayas. The Eastern Himalaya, especially Arunachal Pradesh and Sikkim states of India is the main centre of species distribution and provide wide range of habitat for ecological studies. Several species exhibits many nutritional, medicinal and aromatic properties as evident from tribal knowledge and published research sciences. Different parts of plant such as flowers and leaves used in folklore herbal medicine and for preparation of local wines. Chemically, rhododendrons are repository of several bioactive molecules in the form of taraxerol, hyperoside, betulinic acid, quercitin, arbutin, rutin, coumaric acid, and several other fine molecules in minor quantities. In the present communication, Indian rhododendrons species diversity of different regions are revised and updated the current list for India. Besides, its value addition and future prospective is also presented for Indian economy. The checklist, chemistry, traditional knowledge, ecological requirement, keystone taxa, potential value added products for future, threat and future perspectives of rhododendrons for conservation is also discussed and presented in different subheads. First time checklist of all Indian rhododendrons are prepared and provided here in this communication.

Keywords: Ericaceae, Future Perspective, Indian Himalaya, Rhododendrons, Checklist, Phytochemistry, Value Addition.

INTRODUCTION

Plants are the basic need upon which life of all other organisms depends. Current loss of biodiversity is one of the greatest concerns for ecologist and environmentalist of 21st Century. Climate change is causing species shifts along altitudinal gradients. Plants in mountaineous and hilly regions are sensitive to climate change, and therefore, mountains and other similar habitats are the most suitable sites for studying climate-induced biological response to species composition. Himalaya represents an important globally recognized biodiversity hotspot. Plant adaptation to harsh climate involves modification of certain morphological and eco-physiological traits, niche shifts and production of secondary metabolites.

India is placed as 12th mega biodiversity zone due to percentage of threats, presence of four globally recognized hotspots, *viz*., Western Ghats, Sundarland Himalayas and Indo-Myanmar (former Indo-Burma), based on species rarity and endemism (Myers et al. 2000). Climate in hotspot regions provide ecosystem for species requirement and there exists continuous process of evolution and species extinction. Forests and vegetations components varies from tropical rainforests to alpine meadows. Estimated 8,000 species of flowering plants abode in Himalayan regions of India, 3,160 species recorded as endemic and endangered (Singh and Hajra 1996). Evolution theory postulates that a species either adapts to climate change through eco-physiological, morphological, biochemical and genetic modifications, or migrates to new climate suitable for their survival and reproduction. In this paper, an attempt have been made to gather information on distribution and uses of rhododendrons, provide first time checklist of Indian rhododendrons, their traditional uses, commercial values and future strategies for their conservation.

Discovery and History of Rhododendrons

The genus *Rhododendron* is an important commercial flowering plants of Asia first described and published by Carl von Linnaeus in 'Species Plantarum' in 1753 and placed under family Ericaceae. The study on rhododendrons started in India at the end of 18th Century, with the establishment of Botanical Garden at Calcutta by British Government in 1792 (de millevile 2002).

Rhododendron arboreum Sm.was the first species discovered from Southest Asia and identified by Captain Thomas Hardwicke from the Siwalik Hills near Srinagar (formely called Sireenagur) of Jammu and Kashmir in 1796 (Hooker 1849) however, the first rhododendron classified was *Rhododendron hirsutum* L. which was discovered by Charles L' Ecluse in 16th Centuary. According to Mao (2010), Sir Joseph Dalton Hooker (1817-1911) visited to the Sikkim Himalaya between 1848-1850, collected and described the rhododendrons for Northeastern India. From Sikkim, he described 34 new species in his monograph entitled '*Rhododendrons of Sikkim Himalaya*'. Afterwards C.B.Clarke (1882) published 'Indian rhododendron' which has 46 species. Since then many new species have been described and recorded from India by various workers (Calder et al. 1926, Razi 1959, Nayar and Ramamurthy 1973, Cullen 1980, Nayar and Karthikeyan 1981, Chamberlain 1982, Pradhan 1985-1986, Ghosh and Samaddar 1989, Pradhan and Lachunga 1990, Naithani 1990, Mao et al. 2002, Sastry and Hajra 2010, Pradhan 2010). Arunachal Pradesh is recognized as Centre of Rhododendron Diversity for India and for Southeast Asia as a whole.

Diversity and Distribution of Indian Rhododendrons

Mountains, hills and valleys ecosystem of India represents wide distribution of rhododendrons recorded from different geographic locations, and species composition varies due to wide variations in climatic conditions and elevation differences. Indian rhododendrons grows in a unique climate and reported mostly from Himalayas, Indo-Myanmar and Western Ghats. As per current data, the species of rhododendrons were not been reported from Siwaliks, Indian Deserts and plains of Central India, but repository centre of rhododendrons diversity include Eastern Himalaya (Maximum in Arunachal Pradesh and Sikkim), Himachal Pradesh, Uttarakhand, Jammu and Kashmir of Western Himalayas, Nilgiris, Anamalai and Palani Hills of Western Ghats. The Greater Himalayas also provides an ideal ecosystem for rhododendrons. TPL (2018) recorded family Ericaceae represented by 151 genera and 3,554 species distributed all across the world, and majority of this group members has medicinal as well as ornamental importance. The habits of plants under this group are trees, shrubs and epiphytes, but no parasitic species of rhododendron reported till to date. Figure 1,2 and 3 provide photoplates of Indian rhododendrons distributed in Himalayas and Indo-Myanmar hotspots.

Rhododendron arboreum   Rhododendron anthopogon   Rhododendron baileyi

Rhododendron campanulatum   Rhododendron barbatum   Rhododendron cephalanthum

Fig. 1: Rhodoendrons of Himalayas and Indo-Myanmar Hotspots in India

Currently, the genus *Rhododendron* L. comprised of 641 species (TPL, 2018), distributed in different climate zones across the globe, however, most of the species reported is confined to Northern hemisphere especially in Sino-Himalayas. Literature reveals wide distribution of this genus is is found in Afghanistan, Australia, China, Indonesia, Japan, Myanmar, Malaysia, Philippines, Pakistan, Thailand, southern Europe and northern America. After revision of Rhododendron species in India, authors prepared a checklist of 116 species, 15 subspecies and 10 varieties (Table1).

Table1: Altitudinal range, habit and distribution of Indian Rhododendrons

S.No.	Botanical name	State/ Global Distribution	Habit/ range(m)
1.	*Rhododendron anthopogon* D.Don*		
	R. anthopogon subsp. *hypenanthum* (Balf. f.) J.Cullen*	Arunachal Pradesh, Himachal Pradesh, Sikkim, Uttarakhand, West Bengal, J&K/Bhutan, China, Nepal, Tibet	Shrub/3350-5000
2.	*Rhododendron arboreum* Sm.		
	R. arboreum subsp. *cinnamomeum* (Wall. ex G.Don) Tagg.*	Sikkim, West Bengal/Nepal	Tree/1500-3000
	R. arboreum var. *roseum* Lindl.	Sikkim, West Bengal/ Bhutan, China, Myanmar	Tree/2500-3600
	R. arboreum subsp. *delavayi* (Franch.) D.F.Chamb.	Arunachal Pradesh, Manipur, Meghalaya/ China, Myanmar, Thailand	Tree/2500-3200
	R. arboreum subsp. *nilagiricum* (Zanker) Tagg	Endemic to Western Ghat	Tree/1500-2000
	R. arboreum var. *peramoenum* (Balf.f. & Forrest) D.F. Chamb.	Endemic to Arunachal Pradesh	Tree/3000-3200
3.	*Rhododendron argipeplum* Balf.f. & R.E. Cooper	Arunachal Pradesh, Sikkim/ Bhutan, China	Shrub/2750-3500
4.	*Rhododendron arizelum* Balf.f. & Forrest	Arunachal Pradesh/China, Myanmar, Tibet	Shrub/2500-4000
5.	*Rhododendron arunachalense* D.F. Chamb & Rae	Endemic to Arunachal Pradesh	Shrub/2000-2100
6.	*Rhododendron assamicum* Kingdon-Ward	Endemic to Arunachal Pradesh (former Assam)	Shrub/2000-3000
7.	*Rhododendron baileyi* Balf.f.*	Sikkim/Bhutan, Tibet	Shrub/3000-4000
8.	*Rhododendron barbatum* Wall. ex G. Don*	Arunachal Pradesh, Sikkim, Uttarakhand, West Bengal/ Bhutan, China, Nepal	Tree/2500-3700
9.	*Rhododendron beanianum* Cowan	Arunachal Pradesh/ Myanmar	Shrub/3000-3350
10.	*Rhododendron bhutanense* D.G. Long & Bowes Lyon	Arunachal Pradesh/ Bhutan, China	Shrub/2100-3800
11.	*Rhododendron boothii* Nutt.	Arunachal Pradesh/Bhutan, China	Shrub/1800-2500
12.	*Rhododendron bulu* Hutch.	Arunachal Pradesh/China, Tibet	Shrub/3000-3800
13.	*Rhododendron callimorphum* Balf. & W.W.Sm.	Arunachal Pradesh/China	Shrub/3000-4000
14.	*Rhododendron calostrotum* Balf. f. & Kingdon-Ward		
	R. calostrotum subsp. *riparium* (King-don-Ward) J. Cullen	Arunachal Pradesh/ China, Myanmar	Shrub/3000-3500

Contd.

15.	*Rhododendron camelliaeflorum* Hook.f.	Arunachal Pradesh, Sikkim/China, Nepal, Tibet	Shrub/2700-4000
16.	*Rhododendron campanulatum* D.Don*		
	R. campanulatum subsp. *aeruginosum* (Hook.f.) D.F.Chamb.	Sikkim/Bhutan, Nepal	Shrub/4500-5000
	R. campanulatum var. *wallichii* Hook.f.	Arunachal Pradesh, Sikkim/ Bhutan, Nepal	Shrub/3500-4000
17.	*Rhododendron campylocarpum* Hook.f.	Arunachal Pradesh, Sikkim/ Bhutan, Nepal	Shrub/3000-4000
18.	*Rhododendron campylogynum* Franch.*	Arunachal Pradesh/China, Myanmar	Shrub/2700-4000
19.	*Rhododendron candelabrum* Hook.f.	Endemic to Sikkim/ Arunachal Pradesh	Shrub/3500-4000
20.	*Rhododendron cephalanthum* Franch.	Arunachal Pradesh/ China, Myanmar	Shrub/3000-4500
21.	*Rhododendron cerasinum* Tagg	Arunachal Pradesh/ China, Myanmar, Tibet	Shrub/3000-3500
22.	*Rhododendron chamaethomsonii* (Tagg) Cowan & Davidian	Endemic to Arunachal Pradesh	Shrub/3600-5000
23.	*Rhododendron charitopes* subsp. *tsangpoense* (Kingdon-Ward) Cullin	Arunachal Pradesh/China	Shrub/2500-4000
24.	*Rhododendron ciliatum* Hook.f.*	Arunachal Pradesh, Sikkim/ Bhutan, China, Nepal	Shrub/2700-3400
25.	*Rhododendron cinnabarinum* Hook.f.		
	R. cinnabarinum subsp. *xanthocodon* (Hutch.) J.Cullen*	Arunachal Pradesh/Bhutan, China	Shrub/3000-4000
26.	*Rhododendron concinnoides* Hutch. & Kingdon-Ward	Endemic to Arunachal Pradesh	Shrub/2500-3500
27.	*Rhododendron coxianum* Davidian	Endemic to Arunachal Pradesh	Shrub/1200-1800
28.	*Rhododendron crinigerum* Franch.	Arunachal Pradesh/	Shrub/3000-4000
29.	*Rhododendron dalhousiae* Hook.f.	China, Myanmar	
	R. dalhousiae Hook.f. var. *rhabdotum* (Balf.f. & Cooper) J.Cullen*	Arunachal Pradesh/ Bhutan, China	Shrub/1500-2000
	Rhododendron dalhousiae var. *tashii* U.C.Pradhan & S.T. Lachungpa	Endemic to Sikkim	Shrub/2500-3500
30.	*Rhododendron decipiens* Lacait.	Endemic to Sikkim (former West Bengal)	Shrub/2500-3300
31.	*Rhododendron dendricola* Hutch.	Arunachal Pradesh/ China, Myanmar	Shrub/1000-1500
32.	*Rhododendron edgeworthii* Hook.f.*	Arunachal Pradesh, Sikkim/ Bhutan, China, Myanmar	Shrub/2100-3000

Contd.

33.	*Rhododendron elliottii* Watt ex Brandis	Endemic to Manipur and Nagaland	Tree/2700-3000
34.	*Rhododendron eudoxum* Balf.f. & Forrest		
	R. eudoxum subsp. *tamenium* (Balf.f. & Forrest) Tagg	Arunachal Pradesh/ Bhutan, China	Shrub/3500-4000
35.	*Rhododendron exasperatum* Tagg	Arunachal Pradesh/ China, Myanmar	Tree/3000-4000
36.	*Rhododendron falconeri* Hook.f.		
	R. falconeri subsp. *eximium* (Nattau) D.F. Chamb.*	Endemic to Arunachal Pradesh	Shrub/3000-3500
37.	*Rhododendron faucium* D.F. Chamb.	Arunachal Pradesh/ China	Shrub/2600-3400
38.	*Rhododendron flinckii* Davidian	Arunachal Pradesh/ Bhutan	Shrub/2000-3000
39.	*Rhododendron formosum* Kingdon-Ward		
	R. formosum var. *inaequale* (Hutch.) J.Cullen	Endemic to Meghalaya	Shrub/1500-1800
40.	*Rhododendron fragariflorum* Kindon -Ward	Arunachal Pradesh/ Bhutan, China	Shrub/2500-3500
41.	*Rhododendron fulgens* Hook. f.	Arunachal Pradesh, Sikkim, West Bengal/ Bhutan, China, Nepal, Tibet	Shrub/3000-4300
42.	*Rhododendron fulvum* Balf.f. & W.W.Sm.	Arunachal Pradesh/ China, Myanmar, Tibet	Tree/2400-3300
43.	*Rhododendron glaucophyllum* Rehder		
	R. glaucophyllum var. *tubiforme* Cowan & Davidian	Arunachal Pradesh/ Bhutan, China, Myanmar	Shrub/2800-3100
44.	*Rhododendron glischrum* B.Balf. & W.W.Sm.	Arunachal Pradesh/ China, Myanmar	Shrub/2400-3600
45.	*Rhododendron grande* Wight*	Arunachal Pradesh, Sikkim, West Bengal/ Bhutan, China, Napel, Tibet	Tree/2200-3400
46.	*Rhododendron griffithianum* Wight	Arunachal Pradesh, Sikkim, West Bengal/ Bhutan, Napal, Tibet	Tree/2200-2800
47.	*Rhododendron haematodes* subsp. *chaetomallum* (Balf.f. & Forrest) D.F.Camb	Arunachal Pradesh/ China	Shrub/3100-4000
48.	*Rhododendron hodgsonii* Hook.f.	Arunachal Pradesh, Sikkim, West Bengal/ Bhutan, Napal, Tibet	Tree/3100-3700
49.	*Rhododendron hookeri* Nutt.	Arunachal Pradesh/Bhutan	Tree/2500-3700
50.	*Rhododendron hylaeum* Balf.f. & Farrer	Arunachal Pradesh/ China, Myanmar	Shrub/2400-3000
51.	*Rhododendron imberbe* Hutch.	Endemic to Arunachal Pradesh	Shrub/2700-2800

Contd.

52.	*Rhododendron imperator* Kingdon-Ward	Arunachal Pradesh/ China, Myanmar	Shrub/2500-3500
53.	*Rhododendron johnstoneanum* Watt ex Hutch.	Endemic to Arunachal Pradesh, Manipur, Mizoram	Shrub/1200-3200
54.	*Rhododendron kasoense* Hutch. & Kingdon-Ward	Arunachal Pradesh/China	Shrub/2500-2700
55.	*Rhododendron kendrickii* Nutt.	Arunachal Pradesh/ Bhutan, China	Tree/2300-2800
56.	*Rhododendron kesangiae* D.G.Long & Rushforth		
	Rhododendron kesangiae var. *album* Namgyal & D.G.Long	Arunachal Pradesh/Bhutan	Shrub/3000-3500
57.	*Rhododendron keysii* Nutt.	Arunachal Pradesh, Sikkim/ Bhutan, China, Tibet	Shrub/2500-3600
58.	*Rhododendron lacteum* Franch.	Arunachal Pradesh/ China	Shrub/3000-4100
59.	*Rhododendron lanatum* Hook. f.*	Arunachal Pradesh, Sikkim/ Bhutan, China, Tibet	Shrub/3100-4000
60.	*Rhododendron lanigerum* Tagg	Arunachal Pradesh/ China, Tibet	Tree/3000-3500
61.	*Rhododendron lepidotum* Wall. ex D.Don	Arunachal Pradesh, Sikkim, J&K, Himachal Pardesh, Uttarakhand/ Bhutan, China, Myanmar, Napal, Pakistan	Shrub/2100-4500
62.	*Rhododendron leptocarpum* Nutt.	Arunachal Pradesh, Sikkim/ Bhutan, China, Myanmar, Tibet	Shrub/2300-4000
63.	*Rhododendron lindleyi* T. Moore	Arunachal Pradesh, Sikkim, West Bengal, Manipur/ Bhutan, China, Myanmar, Nepal, Tibet	Shrub/1800-3100
64.	*Rhododendron ludlowii* Cowan	Arunachal Pradesh/China	Shrub/3900-4200
65.	*Rhododendron macabeanum* Watt ex Balf.f.	Endemic to Manipur and Nagaland	Tree/2500-3000
66.	*Rhododendron maddenii* Hook.f. *R. maddenii* subsp. *Crassum* (Franch.) J.Cullen	Arunachal Pradesh, Manipur, Nagaland, Meghalaya/ Bhutan, China, Myanmar, Vietnam	Tree/2200-3000
67.	*Rhododendron mechukae* A.A.Mao & A.Paul	Endemic to Arunachal Pradesh	Tree/2400-3000
68.	*Rhododendron megacalyx* Balf.f. & Kingdon-Ward	Arunachal Pradesh, Sikkim/ Bhutan, China	Tree/2400-3600
69.	*Rhododendron megeratum* Balf.f. & Forrest*	Arunachal Pradesh/ China, Myanmar, Tibet	Shrub/3100-4100
70.	*Rhododendron mekongense* Franch. *R. mekongense* var. *rubrolineatum* (Balf.f. & Forr.) J.Cullen	Arunachal Pradesh, Sikkim /China	Shrub/3300-4200

Contd.

71.	*Rhododendron moulmainense* Hook.	Arunachal Pradesh/China, Japan, Myanmar, Taiwan	Shrub/700-1500
72.	*Rhododendron nayarii* G.D.Pal	Endemic to Arunachal Pradesh	Shrub/2500-3000
73.	*Rhododendron neriiflorum* Franch.		
	R. neriiflorum subsp. *phaedropum* (Balf.f. & Farrer) Tagg	Arunachal Pradesh/ Bhutan, China, Myanmar	Shrub/2900-3000
74.	*Rhododendron nivale* Hook.f.*	Sikkim, Uttarakhand/ Butan, China, Nepal	Shrub/4000-5500
75.	*Rhododendron niveum* Hook.f.	Endemic to Arunachal Pradesh and Sikkim	Tree/3100-3700
76.	*Rhododendron nuttallii* Booth ex Nutt.	Arunachal Pradesh/Bhutan	Tree/1200-3000
77.	*Rhododendron obtusum* Hort. ex Wats	Arunachal Pradesh/ Bhutan, China, Myanmar	Shrub/1200-1500
78.	*Rhododendron papillatum* Balf.f. & R.E.Copper	Arunachal Pardesh, Sikkim/ Bhutan, Nepal	Shrub/1800-3300
79.	*Rhododendron pemakoense* Kingdon-Ward	Arunachal Pradesh/China	Shrub/2400-3000
80.	*Rhododendron pendulum* Hook.f.*	Arunachal Pradesh, Sikkim/ Bhutan, China, Nepal, Tibet	Shrub/2200-3600
81.	*Rhododendron piercei* Davidian	Arunachal Pradesh/China	Shrub/2000-4000
82.	*Rhododendron pocophorum* Balf.f. ex Tagg	Arunachal Pradesh/China	Shrub/3500-4000
83.	*Rhododendron protistum* Balf.f. & Forrest	Arunachal Pradesh, Myanmar, China	Tree/ 2400-4200
84.	*Rhododendron pruniflorum* Hutch. & Kingdon-Ward	Arunachal Pradesh/Myanmar	Shrub/3000-4000
85.	*Rhododendron pumilum* Hook.f.	Arunachal Pradesh, Sikkim/ Bhutan, China, Myanmar,	Shrub/3500-4500
86.	*Rhododendron rawatii* D.Rai & B.S.Adhikari	Endemic to Uttarakhand	Shrub/2200-3000
87.	*Rhododendron rex* H.Leveille.	Arunachal Pradesh, Myanmar	Tree/2300-4000
	Rhododendron rex Levl. subsp. *Arizelum* (Balf.f. & Forr.) D.F.Chamb.	Arunachal Pradesh/China, Myanmar	Tree/3000-4000
88.	*Rhododendron santapaui* Sastry, Kataki, P.Cox & P.Hutch.	Endemic to Arunachal Pradesh	Shrub/2200-2300
89.	*Rhododendron setosum* D.Don*	Arunachal Pradesh, Sikkim, West Bengal/ Bhutan, China, Nepal, Tibet	Shrub/2100-5000
90.	*Rhododendron sherrifii* Cowan	Arunachal Pradesh/ China	Shrub/
91.	*Rhododendron sidereum* Balf.f.	Arunachal Pradesh/ China, Myanmar	Tree/2700-3000
92.	*Rhododendron sikkimense* U.C.Pradhan & S.T.Lachungpa	Endemic to Arunachal Pradesh, Sikkim	Shrub/3500-3700
93.	*Rhododendron sinogrande* Balf.f.	Arunachal Pradesh/ China, Myanmar, Tibet	Tree/3000-4300

Contd.

94.	*Rhododendron smithii* Nutt.	Arunachal Pradesh, Sikkim/ Bhutan, China	Shrub/2100-3700
95.	*Rhododendron stenaulum* Balf.f. & W.W.Smith	Arunachal Pradesh/ China	Shrub/2700-2800
96.	*Rhododendron stewartianum* Diels*	Arunachal Pradesh/ China, Myanmar, Tibet	Shrub/3000-4300
97.	*Rhododendron subansiriense* D.F.Chamb. & Cox	Endemic to Arunachal Pradesh	Shrub/2600-2800
98.	*Rhododendron succothii* Davidian	Arunachal Pradesh/ Bhutan	Shrub/3400-4200
99.	*Rhododendron taggianum* Hutch.	Arunachal Pradesh/China, Myanmar	Shrub/2160-3390
100.	*Rhododendron tanastylum* Balf.f. & Kingdon-Ward	Arunachal Pradesh/ China, Myanmar	Shrub/1800-3350
101.	*Rhododendron tephropeplum* Balf.f. & Farrer	Arunachal Pradesh/ China, Myanmar	Shrub/2450-4300
102.	*Rhododendron thomsonii* Hook.f.	Arunachal Pradesh, Nagaland, Sikkim, West Bengal/ Bhutan, Nepal, Tibet	Shrub/3390-4000
103.	*Rhododendron titapuriense* Mao, Cox & Chamb.	Endemic to Arunachal Pradesh	Tree/2340-3000
104.	*Rhododendron triflorum* Hook.f.		
	R. triflorum var. *bauhiniiflorum* (Watt ex Hutch.) J.Cullen	Endemic to Manipur	Shrub/2470-3080
105.	*Rhododendron trilectorum* Cowan	Arunachal Pradesh/China	Shrub/3600-4300
106.	*Rhododendron tsariense* Cowan	Arunachal Pradesh/Bhutan, China	Shrub/2000-3000
107.	*Rhododendron uvarifolium* Diels	Arunachal Pradesh/ China	Shrub/2160-2470
108.	*Rhododendron vaccinioides* Hook.f.	Arunachal Pradesh, Sikkim, West Bengal/ Bhutan, Myanmar, Nepal, Tibet	Shrub/1850-3700
109.	*Rhododendron virgatum* Hook.f.		
	R. virgatum subsp. *oleifolium* (Franchet) J.Cullen	Arunachal Pradesh/ China, Tibet	Shrub/2200-3000
110.	*Rhododendron veitchianum* Hook.	Mizoram/ Laos, Myanmar, Thailand	Shrub/1230-1700
111.	*Rhododendron wallichii* Hook. f.	Sikkim, West Bengal/ Bhutan, China, Nepal	Shrub/4000-4500
112.	*Rhododendron walongense* Kingdon-Ward	Arunachal Pradesh/ China	Shrub/1500-2150
113.	*Rhododendron wattii* Cowan	Endemic to Arunachal Pradesh, Manipur	Shrub/2700-
114.	*Rhododendron wightii* Hook.f.	Arunachal Pradesh, Sikkim/ Bhutan, China, Myanmar, Nepal, Tibet	Shrub/3050-4310
115.	*Rhododendron xanthocodon* Hutchinson	Arunachal Pradesh, Manipur/ Bhutan	Shrub/2900-4100
116.	*Rhododendron xanthostephanum* Merrill	Arunachal Pradesh, Sikkim/ China, Myanmar	Shrub/1500-3000

* Species provided with photo (Figures 1,2,9)

While undertaking field survey tours in different parts of India, it has been observed that Eastern parts of Himalayas are repository of Indian rhododendrons, and the maximum diversity can be recorded in Himalayan biodiversity hot spot region. State of Arunachal Pradesh is a hub of rhododendrons (106 species) followed by Sikkim (47 species) and West Bengal (18 species). Details under this aspect is given in Table 2. In Western Himalaya, *Rhododendron arboreum, R. anthopogon* subsp. *hypenanthum, R. barbatum, R. campanulatum, R. nivale* and *R. lepidotum* were commonly grows between 2000-5000 m above mean sea level. Only *Rhododendron arboreum* var. *nilagiricum* described as a new variety from Western Ghats of India.

Increasing anthropogenic pressure in last few decades were seen as the major concern for this pristine landscape. According to a report, 46 species of rhododendron classified as rare and threatened category in the Eastern Himalayas of India (Paul et al. 2005). Preliminary enumerations and inventories of the genus were made by botanists *viz.* Pradhan (1986), Gosh and Samaddar (1989), Sastry and Hajra (1983), Pradhan and Lachungpa (1990), Mao et al. (2002), Singh et al. *(*2003), Yumnam (2008) and Mao et al. (2010). This scientists made a significant contribution on the discovery of different species of rhododendrons, and also provided status of rare and endemic rhododendrons grows and found in specific location of India.

Table 2: State-wise distribution of Indian rhododendrons and their endemic states

State	Total no. of species	Rare and endemic Rhododendrons
Arunachal Pradesh	95	*R. arboreum* var. *peramoenum, R. arunachalense, R. assamicum R. beanianum, R. boothii, R. bulu, R. calostrotum* subsp. *riparium, R. campylogynum, R. cephalanthum, R. cerasinum, R. chamaethomsonii, R. cinnabarinum, R. cinnabarinum* subsp. *xanthocodon, R. concinnoides, R. coxianum, R. crinigerum, R. dalhousiae* var. *rhabdotum, R. eudoxum* subsp. *tamenium, R. falconeri* subsp. *eximium, R. fulvum, R. glaucophyllum* var. *tubiforme, R. hookeri, R. imberbe, R. johnstoneanum, R. kasoense, R. kendrickii, R. lanigerum, R. megacalyx, R. megeratum, R. mekongense* var. *rubrolineatum, , R. nayarii, R. neriiflorum* subsp. *Phaedropum R. nuttallii, R. obtusum, R. pemakoense, R. pocophorum, R. pruniflorum, R. rex* subsp. *Arizelum, , R. santapaui, R. sidereum R. sinogrande, R. stenaulum, R. stewartianum, R. subansireiense, R. titapuriense, R. wattii.*
Himachal Pradesh	5	-
Jammu and Kashmir	4	-

Contd.

Manipur	8	*R. elliottii, R. macabeanum, R. triflorum* var. *bauhiniflorum.*
Meghalaya	5	*R. formosum*
Mizoram	3	-
Nagaland	6	*R. macabeanum, R. wattii*
Sikkim	46	*R. baileyi, R. campanulatum* subsp. *aeruginosum, R. candelabrum, R. ciliatum, R. glaucophyllum, R. niveum, R. decipiens, R. sikkimense,*
Western Ghats	1	*R. arboreum* var. *nilagiricum*
Uttarakhand	7	*R. rawatii*
West Bengal	18	-

Himalaya is one of the most fragile mega-ecosystem and holds an enormous repository of bio-diversity, and most of the these diversity is under pressure due to anthropogenic activities induced by humans and some are natural occurrence.

Rhododendron ciliatum   *Rhododendron dolhousiae*   *Rhododendron falconeri*

Rhododendron cinnabarinum   *Rhododendron grande*   *Rhododendron edgeworthii*

Fig. 2: Rhodendrons of Himalayas and Indo-Myanmar Hotspots in India

Different state Govt. has recognized flower of rhododendron as State plant, for examples, *Rhododendron campanulatum* is a state flower of Himachal Pradesh, *Rhododendron arboreum* is a state flower of Nagaland. *Rhododendron niveum* is recognized as state tree of Sikkim and *Rhododendron arboreum* as State tree of Uttarakhand (Kant 2004, Joshi and Sharma 2005). Therefore, there is a need to investigate the ecological requirement for preservation and conservation of rhododendrons, and other similar species of plants for future perspectives.

Rare and Endangered Rhododendron Species of Indian Himalaya

There are nine species of rhododendron considered as rare and endangered viz. *Rhododendron dalhousiae* var. *rhabdotum, R. edgeworthii, R. hookeri R. kendrickii, R. pendulum, R. maddenii, R. megaratum* and *R. tanastylum* (Paul et al. 2010). Table 1 explains the distributional area and distributional ranges for all rhododendron species of India.

Ecological Requirement of Rhododendrons

Slopes, Terraine and Elevation

Mountaineous slopes and valleys are important ecological factors responsible for the distribution of many high altitude plant species, and rhododendrons are one of them. Several species such as *Rhododendron falconeri, Rhododendron grande, Rhododendron hodgsonii, Rhododendron maddenii* and *Rhododendron niveum* prefers to grow along ground, while *Rhododendron formosum* and *Rhododendron johnstoneanum* prefers adapting in ravines and on steep slopes. Similar factors are directly responsible for the water content of the soil and other ecological requirements for growth of the species. As most species prefer well-drained mountaineous soils, therefore such group of plant communities are slope loving. Maximum species diversity occurs in south facing slopes, and reason for its occurrence is availability of light and moisture for their phenological growth and reproduction. Therefore, slopes, terraive and elevation of hills play a major role in distribution of high altitude species.

Elevation and Altitudinal Gradients

The ability of the *rhododendron* species to cope up with the environment is directly relates to the altitude and ecosystem in which they prefer to live. In general, any tall tree species of comparatively lower elevations may get dwarfed when exposed to unfavourable climatic conditions of higher ridges. Elevation coupled with exposure to light and temperature is responsible for the growth of species and it is also a case for rhododendrons and other wild plant species. It has been observed that high elevations and wind velocity make rhododendrons dwarf, which normally attains much taller height in absence of these parameters. Species growing higher altitudes often characterized by thick indumentum than those occurring at lower elevations. However, the thickness of indumentum occasionally varies within the same species growing at different altitudes in hilly regions. Scientific evidence reveals that plants at higher altitude usually take less time for fruiting and seed ripening in comparison to those growing at lower altitudes.

Edaphic Parameters

Rhododendrons grows well in acidic soils found in mountaineous ecosystem. Soils of this nature are rich in organic matter and low inorganic contents. Most species grow on slopes where there is no water logging. *Rhododendron falconeri* prefers to grow in swampy soils, and this species usually left behind fibrous remnants of old leaves and fallen branches on the ground and this take time for decomposition. Epilithic and epiphytic rhododendrons grow either on heavily loaded mossy rocks or pendulous from cliffs or tree trunks rich in humus content. In alpine vegetation, *Rhododendron anthopogon, Rhododendron setosum, Rhododendron lepidotum* and *Rhododendron nivale* grows on soil with limited water and nutrients. During winter, alpine zones species get buried under the snow and when summer arrives, snow melts and the soil gets exposed to for a period of 3-4 months. During this season, plant starts sprouting and flowering and complete their life cycle. In temperate regions, soil used to get exposed earlier in comparison to alpine zones and sprouting starts at lower elevation and gradually proceed towards upward. *Rhododendron arboreum* grows well in the subtropical regions of Himalayas and Western Ghats, and soils free from ice and snow.

Humidity and Rainfall

Eastern belt of Himalaya hotspot receives annually 200-500 cm rainfall and provide a suitable climate for growth of rhododendrons, and this is the reason which makes Arunachal Pradesh and Sikkim in Northeastern states of India as one of the mega diversity centre for Indian rhododendrons, and other species suitable to similar climatic conditions. Some species such as *Rhododendron anthopogon* and *R. lepidotum* grows at high altitude in Jammu and Kashmir, Uttarakhand and Himachal Pradesh where annual average rainfall varies from 20-30 mm every year. *Rhododendron nivale* occurs at 5000 m above mean sea level or more, where habitat and ecosystem looks dry with incessant cold rain or some kind of sleet in summer.

Intensity of Sunlight

Rhododendrons of alpine meadows are dwarf and usually have stunted growth. Phenological time is highly effected by duration of light due to formation of fog and cloud which usually comes in flowering season. *Rhododendron falconeri, Rhododendron grande, Rhododendron hodgsonii and Rhododendron sinograde* were recorded growing in cloud-affected areas of high altitude mountaineous region. *Rhododendron barbatum* and *Rhododendron campylocarpum* prefers to grow in subtropical broad-leaved forests and were characterized by small thick leaves. Flower colour varies with climate and different slopes on which they grows. Plants with dark shade of flowers observed growing in sun-exposed slopes, becomes pale and faded when forcefully grown in shade condition.

Temperature

Phenological growth parameters and flowering stages of different species of Rhododendron is directly related to temperature and allied climatic variables. Dwarf rhododendrons species remain dormant under snow during winter seasons in alpine zones. *Rhododendron arboreum* from Lachung (temperature recorded was 7°C-10°C during day, -2°C to + 4°C during night in April), and same species in the nursery of Indian Botanic Garden, Howrah grows where average temperature ranges from 25°-30°C in summer. *Rhododendron falconeri* in Rhododendron Sanctuary of Sikkim (2900 m above sea level, temperature recorded was 12°C-15°C during day, 1°C-5°C during night in April) was introduced in new climate at the Garden of Sikkim Himalayan Circle (1800 m above sea level, average temperature ranges from 15°C-23°C), Botanical Survey of India, Gangtok and there growth rate of this species recorded were very slow. Hence, temperature plays an important role for the growth of species in mountaineous ecosystem.

Importance of Rhododendron as Keystone and Umbrella Species

Indian rhododendrons plays the role of keystone species in Himalaya regions specially in Eastern Himalaya. Different species under the genus plays role in landscaping, accent and woodland planting. This leads to support a wide range

Rhododendron stewartianum *Rhododendron nivale* *Rhododendron megeratum*

Rhododendron setosum *Rhododendron pendulum* *Rhododendron lanatum*

Fig. 3: Rhododendron of Himalayas and Indo-Myanmar Hotspots in India

of plant, birds and animal diversity. Alpine zone are timberline areas and observed to be the most fragile ecosystem in hilly belt of Himalaya. The rhododendrons growing in similar habitat is responsible for doing ecosystem services. Most of the rhododendrons species prefers to flourish in areas of high rainfall, high humidity and acidic soils, and this type of climatic condition is not suitable for many other angiosperms for their survival and reproduction. Therefore, we can say that rhododendrons play a pronounced role of keystone species and provides an ecological stability to the vegetation communities and associated niches.

Potential and Values of Rhododendrons

Aesthetic and Sacred Values

Since ancient times, Rhododendrons were one of the most popular plants for their beautiful flower, foliage, shape and usage in local daily preparation. Species and hybrids were used extensively for ornamental purpose in landscaping ecology of many species of the world. Emergence of eco-tourism in recent years as an engine of economic growth and rhododendrons and other similar species have tremendous importance. Beautiful and magnificent flowers of rhododendrons have attracted enthusiastic, botanists and researchers of the world. In current scenario, 50% of rhododendron species is under cultivation across the globe and there is report that 5000 to 6000 hybrids of rhododendrons available for marketing as horticulture plant (Sally and Greer 1986). Such hybrid plants species are mainly grown in gardens, parks, temples and other important tourist places for their attractive flowers.

Traditional Usages

Flowers of Rhododendron species occuring in Western Himalaya is considered as efficacious (Biswas and Chopra 1982). Fresh flowers used as local medicine in curing diarrhoea and dysentery. According to Bhattacharjee (1998), dried flowers of *Rhododendron arboreum* were edible after frying with ghee, and this help to cure blood dysentery. In hilly ecosystem, the flowers of *Rhododendron arboreum* along with sugar used in making of squash, jams, jellies and local brew. Local people believes that such drinks and edible products are pleasant and also used to cure and prevent high altitude sickness. Fresh petals of flowers of *R. arboreum* are used to prepare local chutney known as *Barah Ki Chutney*. It is of record that flowers when consumed in excess is known to cause intoxication (Anonymous 1972). Young twigs and tender leaves reported to be used as vegetables (Nayar et al. 1994). In Uttarakhand, leaves are used as poultice in high fever and headache. Juice extracted from leaves are helpful to get rid of lice. Many hybrid species of rhododendrons grown in gardens of Europe and America for their magnificent flowers and foliage.

Rhododendrons were also used as fuel wood and for medicinal purposes (Paul et al. 2010). Tribal people believes that woods of rhododendrons such as *R. arboreum* and *R. grande* are the best quality and used as fuel, and useful in making of cups, spoons, boxes and saddles (Paul et al. 2005). Woods are used to make charcoal by using local techniques. Flowers and leaves can be fitted in long ropes and tied around the houses and the temples to get ride off ghost or evil spirit (Chauhan 1999). The leaves of *Rhododendron cinnabarinum* is poisonous to livestock due to its toxic composition and is not given along with fodder. The smoke of leaves and wood causes inflammation of eyes and, therefore, childrens are kept away from this smoke. Sain (1974) reported that the petals are used for making local jams by Lamas and Tibetan tribes in Ladakh, Nepal and Arunachal Pradesh. A mixture of powdered dried leaves is used as snuff to cure hemicarnia and colds (Chopra et al. 1958). Table 3 depicts traditional usages of different species of *Rhododendron*.

Table 3: Ethnomedicinal uses of Indian rhododendrons

Species	Medicinal uses	Parts used
R. anthopogon	In treatment of catarrh; also in curing cold, cough, chronic bronchitis and asthma; administered to produce sneezing.	Decoction of leaves
	In treating indigestion and lung infection and mixed with oil and used in massage in post-delivery complications	Decoction of leaves and flowers
R. anthopogon	Eaten with butter in Leucorrhoea and Gonorrhoea	Powdered dried leaves and young shoots
R. lepidotum	Drinks made are supposed to be purgative.	Bark
R. arboreum	In treatment of hill diarrhoea and dysentery	Fresh flowers
	Taken with ghee after frying to check blood dysentery	Dried flowers
	Used as poultice in high fever and headache	Leaves
R. campanulatum	Used in treating chronic rheumatism, syphilis and sciatica	Leaf decoction
	Used as snuff after mixing with tobacco leaves to cure Hemicarnia, cough and colds	Powdered dried leaves
	In treatment of chronic fevers	Dried twigs and wood in powdered form.
R. assamicum	Used in treating chronic rheumatism, syphilis	Crushed leaves.
R. falconeri	Used to cure cold, fever.	Leaves and flowers.
R. lepidotum	Drinks made are supposed to be purgative.	Flowers

In India, rhododendrons were mostly utilized by the Bhutias, Lepchas and Nepalis dominant inhabitants of the Indo-Himalayan biogeographic locations located in high altitude regions. Many rhododendrons such as *R. arboreum, R. barbatum, R. campanulatum, R. falconeri* and *R. hodgsonii* were used as firewood by local people. *Rhododendron niveum* is used for indoor decoration. Leaves of

Rhododendron campanulatum from Western Himalaya reported to be used in treating of chronic rheumatism and syphilis. Rhododendrons possess anti-tubercular properties. Besides, it has been reported that rhododendron causes poisonous effect or toxication due to the presence of toxic molecules and compounds such as andromedotoxin, asebotoxin and rhodotoxin. Twigs of *Rhododendron anthopogon* is in good demand as raw material used in manufacturing of herbal botanicals. Decoction of leaves of *Rhododendron anthopogon* used to cure cold, cough, chronic bronchitis, asthma and mucus formation in the nose and throat (Kumar and Srivastava 2002). According to Balodi and Singh (1997), leaves and flowers of *Rhododendron anthopogon* used to cure indigestion and lung infection in hilly tribe such as Tibetans and Lamas. Watt (1892) reported medicinal uses of *Rhododendron lepidotum* similar to those of *Rhododendron anthopogon*. Herbal tea made from bark of *Rhododendron lepidotum* can be used as purgative.

CHEMISTRY OF RHODODENDRONS

Genus Rhododendron contain several varieties of chemical compounds that are required for their biological functioning in harsh environment. Few chemical constituents in rhododendrons are unique to one species, while others occur in two or more in the same genus. Various compound and secondary metabolites have been isolated, characterized and studied from *Rhododendrons*. The major

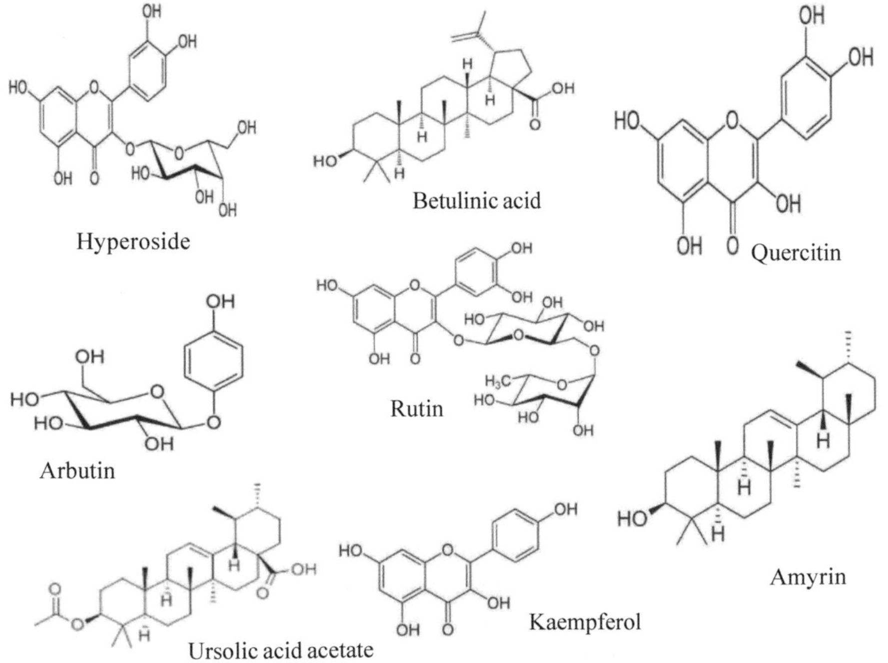

Fig. 4: Major chemical constituents of Rhododendrons

chemical constituents isolated from rhododendrons are taraxerol, hyperoside, quercitin, arbutin, rutin, coumaric acid, amyrin and epifridilen. Chemical structures of some important compounds isolated and characterized from rhododendrons is provided in Fig. 4.

Major Essential Oil Components

Rhododendron anthopogon is studied for essential oil contents by Innocenti et al. (2010). Major components includes α-pinene (37.4%), β-pinene (16.0%), limonene (13.3%), γ-cadinene (9.1%) (Fig. 5). Leaves could be distilled for aromatic oils with possible uses in perfumery and cosmetics (Pradhan and Lachungpa 1990). Young et al. (2011) reported 37 different components from essential oils of *Rhododendron anthopogonoides*. Some of the major constituents includes 4-Phenyl-2-butanone (27.22%), nerolidol (8.08%), 1,4 cineole (7.85%), caryophyllene (7.63%), γ-elemene (6.10%), α-tarnesene (4.4%) and spathulenol (4.19%). The quantity and quality of essential oils also varies depending on climate, soil and geographic location, plant parts used, time of collection and analytical method applied for quantification. Predominant ones were 4-thujene, 5-(1-methylethyl)-bicyclo [3.1.0] hex-3-en-2-one and (-)-4-terpineol. 4-Phenyl-2-butanone of *R. anthopogonoides* possess fumigant toxicity against *Sitophilus zeamais*. Another species, *Rhododendron tomentosum* has essential oils and according to Zhao et al. (2016), it contains 57 constituents.

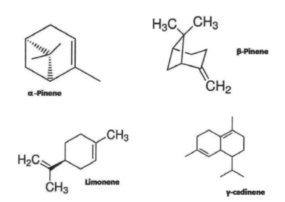

Fig. 5: Major essential oil constituents in rhododendrons

PHARMACOLOGY ASPECTS

Anti-microbial Activity

Flower extracts of *Rhododendron* species contain multiple organic components including flavonoids, saponins and alkaloids. These group of compounds known to possess anti-bacterial effects (Ani et al. 2011). Extract prepared rich in phenolic compounds indicate the high anti-microbial properties.

Anti-diabetic activity

Flowers of *Rhododendrons* species shows anti-diabetic properties, which would be helpful in development of drugs, nutraceuticals and functional food for diabetes (Bhandary and Kuwabata 2008).

Anti-diarrhoeal activity

Ethyl acetate fraction of rhododendron flowers shows potent anti-diarrhoeal activity. The ethyl acetate fractions of rhododendron flowers are found to reduce magnesium sulfate induced diarrhea significantly which could be due to increased absorption of water and electrolytes (Verma et al. 2011).

Hepato-protective activity

Ethyl acetate fraction of leaves and flowers of rhododendrons exhibit significant hepato-protective potential against carbon tetrachloride (CCL_4) induced liver damage in preventive and curative models. Fraction at a dose of 100, 200 and 400 mg/kg applied orally in CCL_4 treated models. There is also report that, the ethyl acetate fraction significantly prevented the elevation of hepatic malondialdehyde formation and depletion of glutathione content in CCL_4 induced liver (Verma et al. 2011).

Anti-oxidant property

Ethanolic extract of *Rhododendrons* show significant anti-oxidant property as by mitigating the effect of acute and chronic stress induced biochemical and physiological perturbation. Flavonoids isolated from the leaves of *R. arboreum* possess potent anti-oxidant property (Dhan et al. 2007).

Anti-cancerous activities

The genus *Rhododendron* is also a rich source of phenolic compounds, especially flavonoids, essential oils, chromones, terpenoids and steroids. Leaf extract of *Rhododendron* species exhibit selective cytoxicity against colon and liver cancer cells compared to normal fibroblast cells, while this selective cytotoxicity was not observed in breast cells (Srivastava 2012).

Anti-inflammatory

Ethanolic herb extract of *Rhododendrons* leaves contain bioactive compounds with anti-inflammatory activities and support ethnomedical claim of usages of plant in the management of inflammatory conditions.

Commercial Aspects and Value Added Products from Rhododendrons

Rhododendron species with high levels of phenols, promising anti-oxidant and free radical scavenging activities may be utilized in the development of health care products. Juice from fresh flowers of several species were extracted and sold in regional village markets. Anti-oxidants play a key role to scavenge free radicals and are associated with reduced risk of cancer and cardiovascular diseases (Willcox et al. 2004) and is considered as the best as it reduces the oxidation process in the body (Krishanaih et al. 2007). NGOs and Government Food Processing Centres are fully harnessing the potential of bioprospecting of *Rhododendron* species for economic benefits, thereby, preparing, the value-added products such as sauce and squash.

In hilly regions of Himalayas, when flower blooms in January and March, tribal local people consume burans chutney to prevent setnes from sickness caused by changes in weather. Chutney is prepared by grinding together burans petals with soaked tamarind, chillies and garlic. In Sikkim and Darjeeling, local wine called 'Gurans Ko Raksi' is traditionally made from flowers of *Rhododendron arboreum* and people in Sikkim consumed it as herbal medicine. It taste slightly sour and has leafy flavor of aroma of the petals. Local wine amidst tribals from chilling weather and cold climate of the mountains and hills.

THREAT ASPECTS

It is estimated that approximately 97% of Rhododendron species in Himalaya is lossing their identity due to indiscriminate felling and loss of habitat. This is causing rhododendron flowering plants vulnerable in natural habitat and slowely but steadily leads to extinction of species. In the recent few decades Indian Himalayan is greatly affected by various threats imposed by the nature and other anthropogenic activities. Increasing human population and associated activities through direct and indirect involvement is heavily loading pressure on forests and naturing wild life population. The primary forests are degraded into scrub lands by rapid deforestation and desertification. Rhododendrons routinely cut for firewood by local people in summer season, and thus threatening the survival of many species. According to Menon et al. (2012) uncontrolled, indiscriminate and unsustainable harvesting for firewood has resulted several rhododendron species under rare, endangered and threatened categories. Change in climatic variables are one of the important factors affecting the surviving population of rhododendrons. Indiscriminent grazing and jhum cultivation have threatened the natural habitat of Rhododendron species up to a large extent. Roadways and hydel power projects is also imposing equal pressure on the natural habitat and population survival for rhododendron species.

CONSERVATION AND FUTURE PERSPECTIVES

Looking into the economic potential and rhododendrons as keystone species, this group of plants community requires several urgent conservation measures as *in-situ* and *ex-situ* conservation mode. *In-situ* conservation can be undertaken about by establishing rhododendron sanctuaries, parks and national parks. Presently Sikkim is the only state in India where *in-situ* conservation have been taken up for Rhododendron species. *Ex-situ* conservation effected by cultivating rhododendron species in gardens and parks under a suitable climate or by using plant tissue culture techniques. Tissue culture studies of Indian rhododendrons are recently initiated and only few species especially *Rhododendron maddeni* has been successfully propagated through plant tissue culture methods (Singh and Gurung 2009). Endemic rhododendrons and similar species requires urgent measures in terms of conservation and preservation as natural reporting. The *Rhododendron arboreum* can be propagated through cutting.

CONCLUSION

As rhododendrons possess anti-inflammatory, hepato-protective, anti-diarrhoeal, anti-diabetic and anti-oxidant pharmacological properties due to the presence of flavonoids, saponins, tannins and various other phytochemicals. There is urgent need to investigate chemicals isolated from different species. Well-drained, well-aerated, acidic soil rich in humus and low in elements, high rainfall, humid atmosphere, low temperature, high elevations and sun-exposed slopes are optimal for luxuriant growth of rhododendrons, and such ecosystem need to be conserved for future research. Agrotechnology development for captive cultivation is required for proper growth and conservations of rhododendrons. There is need of more value added products from rhododendrons species that will help in growth of economy, and will be a great contribution towards the global health.

ACKNOWLEDGEMENT

Authors would like to thank Director IIIM Jammu for facilities and also would like to extend gratitude to scientist of plant sciences for encouragement, and AcSIR for allowing the first author to work for his Ph.D. thesis. This article represent institutional publication no. IIIM|2233|2018.

LITERATURES CITED

Ani AE, Diarr B, Dahle UR, Lekuk C, Yetunde F, Somboro AM. 2011. Identification of Mycobacteria and other acid fast organisms associated with pulmonary disease. *Asian Pacific Journal of Tropical Medicine* 4: 259-262.

Anonymous. 1972. The Wealth of India. Publication and Information Directorate, CSIR, New Delhi, India.

Balodi B, Singh DK. 1997. Medico-ethnobotany of Ladakh. *Annals of Forest Research* 5: 189-197.

Bhandary MR, Kuwabata J. 2008. Antidiabetic activity of Laligurans (*Rhododendron arboreum* Sm.) flower. *Journal of Food Science and Technology* 4: 61-63.

Bhattacharjee SK. 1998. Handbook of Medicinal Plants. Pointer Publishers, Jaipur, India.

Biswas K, Chopra RN. 1982. Common medicinal plants of Darjeeling and the Sikkim Himalayas. Periodical Experts Book Agency, Delhi, India.

Calder C, Narayanaswami B, Ramaswami MS. 1926. List of species and genera of Indian phanerogams not included in Sir JD Hooker's *Flora of British India*. Record of Botanical survey of India, Calcutta, India.

Chamberlain DF. 1982. A revision of *Rhododendron* 1. Subgenus *Rhododendron*, Section *Rhododendron* and *Pogonanthum*. *Notes Royal Botanic Garden*. Edinburgh, London.

Chauhan NS. 1999. Medicinal and aromatic plants of Himachal Pradesh. Indus Publishing Company, New Delhi, India.

Chopra RN, Chopra IC, Handa KL, Kapur LD. 1958. Indigenous Drugs of India. UN Dhar & Sons Private Limited, Calcutta, India.

De Milleville R. 2002. The Rhododendrons of Nepal. Himal books, Kathmandu, Nepal.

Dhan P, Garima U, Singh BN, Ruchi D, Sandeep K, Singh KK. 2007. Free radical scavenging activities of Himalayan Rhododendrons. *Current Science* 92: 526-32.

Gosh RB, Samaddar UP. 1989. The Rhododendrons of the North-East India. *Journal of Economic and Taxonomic Botany* 1: 205-220.

Hajra PK, Singh DK. 1996. Floristic Diversity in: G. S. Gujral and V. Sharma, Eds., Changing Perspectives of Biodiversity Status in the Himalaya, British Council, New Delhi, India.

Hooker JD. 1854. *Himalayan Journals*. Ward Lock, Bowden Co. London, Melbourne, New York.

Hutchinson J. 1947. The Distribution of Rhododendrons. The Rhododendron Year Book 1947: 87-98. The Royal Horticultural Society, London.

Innocenti G, Dall'Acqua S, Scialino G, Banfi E, Sosa S, Gurung K, Barbera M, Carrara M. 2010. Chemical composition and biological properties of *Rhododendron anthopogon* essential oil. *Molecules* 15: 2326-2338.

Joshi AP, Sharma N. 2005. Flower power. Rhododendron is a health freak's delight. Science and Environment fortnightly. *Down to Earth* 3: 52.

Joshi PC, Joshi N. 2004. *Biodiversity and conservation*, A.P.H Publishing Corporation, New Delhi, India.

Kant A. 2004. Wildlife (the northeast travel special). *Sanctuary Asia* 2: 67-84.

Kingdon-Ward F. 1934. A Plant Hunter in Tibet. Jonathan Cape, London.

Krishnaih D, Sarbatly R, Bono A. 2007. Phytochemical antioxidants for health and medicine-a move towards nature. *Biotechnology and Molecular Biology Reviews* 2: 97-104.

Kumar S, Srivastava N. 2002. Herbal research in Garhwal Himalayas: Retrospects & Prospect. *Annals of Forest Research* 10 (1): 99-118.

Mao AA. 2010. The genus Rhododendron in Northeast India. *Journal of Plant Science,* 7:26-34.

Mao AA, Singh KP, Hajra PK. 2002. "Rhododendrons," In: N.P. Singh and D.K. Singh, Eds., Floristic Diversity and conservation Strategies in India, BSI, Calcutta. 2167-2202.

Menon S, Khan ML, Paul A, Peterson AT. 2012. Rhododendron species in the Indian Eastern Himalaya: new approaches to understanding rare plant species distribution. *Journal of American Rhododendron Society* 66: 78-84.

Myers N, Muttermeier RA, Muttermeier CA, da Fonseca AB, Kent J. 2000. Biodiversity hotspots for conservation priorities. *Nature* 853-858.

Naithani HB. 1990. Flowering plants of India, Nepal and Bhutan (not recorded in Sir JD Hooker's Flora of British India). Surya Publications, Dehradun, India.

Nayar MP, Karthikeyan S. 1981. Fourth list of species of genera of Indian phanerogams not included in Sir JD Hooker's *Flora of British India*. Record of Botanical survey of India, Calcutta, India.

Nayar MP, Ramamurthy K, Agarwal VS. 1994. Economic plants of India. Botanical Survey of India, Kolkata, India

Paul A, Khan ML, Arunachalam A, Arunachalam K. 2005. Biodiversity and conservation of Rhododendrons in Arunachal Pradesh in the Indo-Burma biodiversity hotspot. *Current Science* 89(4): 623-634.

Paul A, Khan ML, Das AK. 2010. Utilization of rhododendrons by Monpas in western Arunachal Pradesh, India. *Journal American Rhododendron Society* 64(2): 81-84.

Pradhan KC. 2010. The Rhododendrons of Sikkim. Sikkim adventure, Sikkim, India.

Pradhan UC, Lachungpa ST. 1990. Sikkim-Himalayan Rhododendrons. Primulaceae Books, Kalimpong.

Pradhan UC. 1986. A preliminary Enumeration of Rhododendrons of the Indian region- part 2. *Himalayan Plant Journal* 11: 73-76.

Razi BA. 1959. A second list of species and genera of Indian Phanerogams not included in JD Hooker's Flora of British India. Record of Botanical survey of India, Calcutta, India.

Sain M. 1974. Rhododendrons of Darjeeling and Sikkim Himalaya. Seattle.

Sally HE, Greer HE. 1986. Rhododendrons Hybrids- A guide to the origins. B.T. Batsford Ltd, London.

Sastry ARK, Hajra PK. 1983. "Rare and endemic species of Rhododendrons in India- A preliminary Study, In: Jain, SK and Rao, RR Eds., An assessment of Threatened Plants of India BSI, Calcutta. 222-231.

Sastry ARK, PK. Hajra 2010. Rhododendrons in India- Floral and Johar splendor of the Himalayan flora. BS Publications, Hyderabad, India.

Singh DK, Hajra PK. 1996. Floristic Diversity. In: Gujral, G. S. and Sharma, V. Eds., changing perspectives of Biodiversity Status in Himalaya, British council, New Delhi, India.

Singh KK, Gurung B. 2009. In vitro Propagation of R. maddeni Hook. f. an Endangered Rhododendron Species of Sikkim Himalaya. *Notulae Botanicae Horti Agrobotanici Cluj-Napoca* 37(1): 79-83.

Singh KK, Kumar SL, Rai K, Krishna AP 2003. Rhododendron Conservation in Sikkim Himalaya. *Current Science* 85(2): 602-606.

Srivastava P, 2012. *Rhododendron arboreum*: An overview. *Journal of Applied Pharmaceutical Science* 2(1): 158-162.

TPL. 2018. Plant list 2018, Version 1.1. Published on the internet (http://www.theplantlist.org)

Verma N, Singh AP, Amresh G, Sahu PK, Rao CV. 2011. Protective effect of ethyl acetate fraction of Rhododendron arboreum flowers against carbon tetrachloride-induced hepatotoxicity in experimental models. *Indian Journal of Pharmacology* 43(3): 291-295.

Watt G. 1893. Dictionary of the economic products of India. Printed by Superintendent of Govt. Printing, Harvard.

Willcox JK, Ash SL, Catignani GL. 2004. Antioxidants and prevention of chronic disease. *Critical Review of Food Science and Nutrition* 44: 275-295.

Young K, Zhou YX, Wang CF, Du SS, Deng ZW, Liv QZ, Liv ZL (2011). Toxicity of *Rhododendron anthopogonoides* essential oil and its constituents compounds towards *Sitophilus zeamais*. *Molecules* 16: 7320-7330.

Yumnam JY.2008. Rhododendrons in Meghalaya Need Attention. *Current Science* 95(7): 817-818.

Zhao Q, Ding Q, Yuan G, Li B, Wang J, Ouyang J. 2016. Comparison of the essential oil composition of wild *Rhododendron tomentosum* stem, leaves and flowers in bloom and non bloom China. *Journal of Essential Oil* 19 (5): 1216-1223.

18

Cyperus pangorei Rottb. (Cyperaceae) an Economically Important Mat Grass for Rural Prosperity in India: Cultivation, Processing and Marketing

Anup Kumar Bhunia[1] and Amal Kumar Mondal[2]

[1,2]*Plant Taxonomy, Biosystematics and Molecular Taxonomy Laboratory*
UGC-DRS-SAP and DBT-BOOST-WB Funded Departmen,
Department of Botany & Forestry, Vidyasagar University, Midnapore-721102
West Bengal, INDIA
Email: anupbhunia1988@gmail.com, akmondal@mail.vidyasagar.ac.in

ABSTRACT

Cultivation of mat grass (*Cyperus pangorei* Rottb.) is valuable and expensive products can plays an important role in rural areas such as Sabong and Pingla in Paschim Medinipur of West Bengal in India. Mat grass plays an act to fill the empty employment field to the resource of poor farming community to support and secured their livelihoods and maintain the economical balance of their area. This grass usually cultivated by the poor and marginal farmers. The mat sticks are used to make various household utility items, such as floor and bed mats and other attractive goods like bags, purse, TV covers. Now a days the products prepared from this grass is exported to different countries. Mat sticks are attractive durable and is eco-friendly substances which easily gets decomposed by organisms, does not create any environmental hazards and gets more profit as compared to growing rice per annum. Even an old man or woman in rural areas can earn a net income of about Rs.120 to Rs150 per day. Major focus of the study is to establish the scientific principles behind processing, separation of clum strand, texture and strength properties of the mat fibers and increase mat production which leads to improve the economic condition of rural people.

Keywords: *Cyperus pangorei,* Mat industry, Cultivation, Processing, Marketing, Eco-friendly, Rural Prosperity

INTRODUCTION

Cultivation of mat-grass (*Cyperus pangorei* Rottb.) and its expensive products and production of mat-reed of different quality can play an important role in economic upliftment of rural community. In West Bengal, about 100,000 farm families are associated and solely or partly dependent on mat grass cultivation for their secured livelihoods. Mat industry is the indigenous art of mother of Sabong in Paschim Medinipur District and is one of the noticeable example with respect to hand craft collection of West Bengal. Naturally mat grass grows near river side and seashore but at present artificially it is cultivated by farmers in their own interest and their personal requirements.

The flood prone Kaleghai river basin and its tributaries are often in undated by flood and water logging is the natural habitat for the Madur grass. Madur sticks are used to make various household utility items, such as floor and bed mats and other attractive goods such as bags, purse, TV covers etc. Those products are exported to different countries. As over a period of time, the demand of Sabong mat grew, other surrounding areas took up the cultivation of this grass. Though the major occupation is agriculture, mat weaving forms the next largest source of income of women, children and even the aged can be involved in production of madur sticks. The weaving of mat is almost exclusively women's work, employing as much as 80% women.

Different NGO are engaged in mat industry and produce different products from it with the help of local villagers. They export the different goods produce from madur sticks in the different countries. Many villages, families take the mat industry as their daily profession and help to increase their economic stability. Madur grass is also cultivated now at Narayanghar, Pingla of Paschim Medinipur and at Barasat of 24-Paragana District, but the quality of madur stick is not good as Sabong.

Study Area

Paschim Medinipur district is located in south West Bengal, bordered by district Bankura on the north, Hugli on the northeast, Jhargram on the west, and Bay of Bengal in the south. The district has three sub-divisions, namely, Kharagpur, Ghatal and Medinipur Sadar one of the country's most backward and one of the largest districts in the state covering a total land area of 9368 square kilometers.

Sabong block of West Bengal is located in Kharagpur sub-division of Paschim Medinipur district and lies between 22°25'N to 87°65'E. Mat is the indigenous small scale industry of villages and is mainly restricted to the state of West Bengal in India. In West Bengal, this is mainly found at the Block Sabong and partly found it at adjoining Blocks such as Pingla, Narayanghar of Paschim Medinipur district. But now madur grass is also cultivated in some parts of Howrah, Burdwan, Paschim Dinajpur and 24-Paragana as District (Ghosh1993). But the quality of MS is not so good as of Sabong (Samat 2008). Midnapur remains the most important mat weaving district in West Bengal, the major centers being the block of sabang, Egra, Ramnagar, Narayangarh, Panskura and Patashpur.

Data Collection

Cyperus pangorei Rottb. are collected from in Sabong of Paschim Medinipur district. The survey was conducted in Sabang (20°10'34"N, 87° 36' 04"E) and adjoining areas like Pingla (22° 16'04"N, 87° 35' 08"E) and Naryangarh (22° 09'05"N, 87° 23' 34"E). Through a long drawn investigation, it is convinced that mat grass is a unique raw material, on which the entire mat industry is based. Therefore, preliminary steps have been taken to contact with the cultivators producing mat grass. Comprising more than 200 (two hundred) fields, large and small, were keenly supervised, including personal contact with the producers. The data concerning soil, manure, saplings with roots, irrigation and insecticides was to be collected. The study was carried out during 2007 to 2018.Every data has been collected through personal contact. Analysis of data has been made with the help of manually operated calculators. Not only that, at a difference of time, harvesting method, that is cutting, slicing, drying and classifying was also speculated step by step. Every information has been collected, with a view to further specific development of the process, in future. Morphological Studies: The detailed morphological studies were performed with Stereozoom microscope (Leica S8APO).

Taxonomical Characters

Cyperus Pangorei is a perennial grass, creeping with rhizomatous stem. Culms 50-90 cm tall, stout, obtusely 3-angled, smooth. Leaves originate from the base of stem, apically bladeless or with a short blade. Bracts 3 to 5, leaf like, longer than inflorescence, lenceolate. Inflorescence decompounds or compound anthela, rays 5-7 cm, unequal, is with 3 to 8 raylets. Spikes broadly ovoid, 1×2 cm, with 4-15 spikelates. Spikelates laxly arranged, lenier, 8-20cm, slightly compressed, obliquely spreading, 6-30 flowered; rachilla wings reddish brown. Glumes reddish brown on both surface but middle green, oblong, 2-3mm, 3-5 veined, margin slightly revolute at maturity, apex obtuse to rounded. Stamen-3; anther linear; connective prominent beyond anther. Style of medium length; stigmas-3. Nutlet dark brown, obovoid oblong (Figure 1).

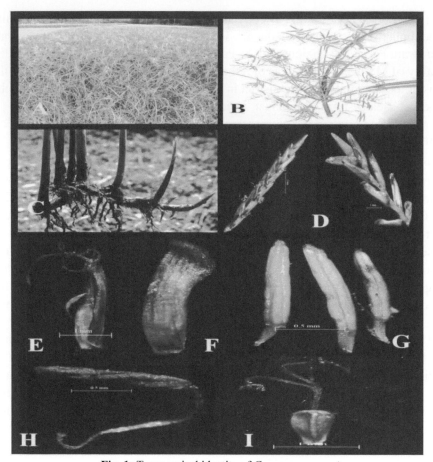

Fig. 1: Taxonomical identity of *Cyperus pangorei*

Phenology Season

The flowering season of *Cyperus pangorei* is November to January

Cultivation Technique

The flood prone river basin has become a natural habitat for the madur grass at Sabong. It is cultivate in Kaleghai river basin and its tributaries. Soil used to get prepared in the month of March to April. At first the soil is spaded in to 2 - 3 deep and dried up to avoid the weeds in production. As a result, the weeds which are present on the soil are dried. The land of the madur plantation is elevated up to one it height from the soil of neighboring area for well drainage of rain water. The dried spaded soil is flooded with water then up to 5 to 6 days later the soil is spaded to become granular soil . In the month of June the soil is spaded to prepare straight line which is 8 to 10 inches apart to each other and

2 to 3 inches deep. Then the rhizome if madur grass is planted at about 6- 8 inches distance. Fertilizer is also used before the plantation of rhizomes on the deep spaded lines mainly the fertilizer which are used are Bio-fertilizer green manure, farmyard manure, bone fertilizer etc. is about 10 kg. / Bigha of the mixture of this Bio-fertilizer some times chemical fertilizer N.P.K. is used 10 kg./ Bigha (in the ratio of 2:1:1). Rhizome is covered with the soil by spaded. Irrigation need is not necessary as during this period, the rain water is sufficient for it. If rain water is not sufficient then watering is done regularly. After 15 days the young plants comes out through the soil from the rhizomes. After 2-3 months, the grass is matured & sticks are harvested after maturation of flower. The sticks are collected in 2 times in a year such September - October and April - May. The sticks are collected up to 10 years from one time plantation. Every times after harvesting the crops, the most promising weed *Cyprus kelinga* is removed from the field and remaining rhizomes are covered with clay soil for again new crop production.

Processing of Mat Sticks and its Preservation

After the harvesting of flowering sticks from the mat grass field, the small leaves and flowers are removed by a sharp knife (Fig. 2).

Then sticks are split into 4 - 8 longitudinal pieces, after cleaning the weeds. But before this, the green sticks are classified according to different heights. Then these are spread under the Sun, in equal lines and height for about 2 hours. In this way the sticks become soft. After splitting the sticks with a knife, the white portion inside them are cleaned. The split filaments are then soaked overnight in water to destroy the fungal cysts and making the filaments fungal resistant. After this the sticks are sundried for two days. Thereafter the sticks are bundled into 'khones.' These are called 'seasoned sticks'. Thus the sticks are protected from fungus or insects. In this process, the color of sticks or mats remains unchanged. At the eve of weaving the dried madur sticks should be kept at least 12 hours in a tank or Pond. Purifying madur sticks are naturally be cream or ass cloured. Such mat lasted for along. No infection is made even in Nainy season. Its colour remains unchanged. After purifying the madur stick, it is bundled first and then kept in the sheet one by one. It shall be covered with polithin to keep the same free from cold and rain.

Fig. 2: Processing of Mat Sticks, A-Mat Sticks are hervested; B-Clearing the weeds; C-Leaves and flower removed by a sharp knife from the sticks; D-Sticks are classified according to different height; E-Split the sticks in to 4-8 longitudinal pieces; F-Sticks are sundried for two days; G-Sun dried sticks are bundles.

Processing of Masland Madur Sticks

A very fine and valuable mat is called 'Masland.' An artisan splits the mat sticks with the help of his teeth, according to the density of the threads and fineness of the 'Masland.' Now-a-days, another modern process has been invented; now the filaments are splitted with a fine needle (Fig. 3).

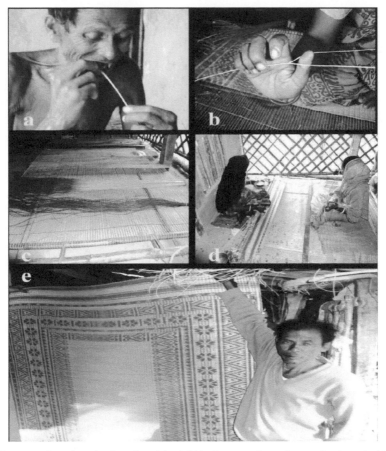

Fig. 3: Processing of masland madur sticks (a&b) ; Process of weaving masland mat (c,d&e)

Process of Weaving Masland Mat

In the process of weaving masland mat two artisans are required. Two persons sit facing each other and weave a masland. The first artisan begins with a mat stick and then the second one begins in front of him with another mat sticks. It should be reminded that threads are spread out, before hand, longitudinally within a wooden frame. Then only the weaving begins. The first one begins from the right side, takes the head of the stick under alternative threads and draws it to the left end. But the second one begins from the opposite side and draws the mat sticks to the right end. Thus it will go on till the masland (or mat) is completed (Samat 2008).

When various designs are depicted in a masland - such flowers or creepers the process, described above, are changed slightly in specific ways. Then the mat stick is drawn sometimes below 2 threads or sometimes below 3 threads alternately or in any pattern, as required for a special design.

In modern ages, especially after 1960, the artisans have been investing novelty in the field of designs, namely (a) zigzag, (b) beehive, (c) peacock, (d) garnish, (e) 'Kadam' flower, (f) fountain (water-fall) etc. loveliness has come into the designs, imitating the boarder of 'Baluchari' sari or hand-loom made saris of Bengal. With the help of various process of weaving - different names, portraits of great men, peacocks, other birds and animals, natural sceneries, butterfly, Asoka-pillar, map of India, Srimati Indira Gandhi, the father of the Nation Mahatma Gandhi, the child Jesus with mother Mary and so many other things are being depicted on the beautiful and valuable maslands or mats.

Modern Process of Mat Sticks Dissection

From 1990 there is a remarkable change in the process. The sticks are sliced with needle-combs. On the eve of weaving the sticks are dipped in water for about 3 to 4 hours. The sticks will be puffed up then and the soft portion inside will easily be removed with a knife. Then the fine sticks are divided longitudinally into 5 to 10 finer sticks. After making bundles, the weaving begins.

Modern Technology of Mat Weaving

The use of 'Kedi loom' is an example of modern technology. The word 'Kedi', perhaps is originated from the word 'caddy', a small box for holding tea. However, Kedi loom is an uncommon loom, made of wood and operated by hands. In it mat sticks are woven with the help of a shuttle (maku). There are 3 kinds of Kedilooms, namely (i) Kediloom with 2 springs (Jhap), (ii) Kediloom with 3 springs and (iii) Kediloom with 4 springs. This modern machine has no definite measurement. Generally, it is 403 in length and 403 in breadth. But it may be made up to 503-753 in length, if required. The machine is very complex one with at least 30 little parts and a auxiliary parts.

ECONOMIC ASPECTS AND COMMERCIAL VALUES

Madur grass is an important plant. It is also used for different purposes.

Fodder

Young and green madur grass used as fodder for domestic animals such as Cows, Goats, Sheep and Buffaloes.

Medicinal Uses

The extract of rhizomes are used as a tonic and stimulating medicine.It is used medicinally as astringent,diaphoretic and diuretic. Rhizome are extemsively used by the tribal people as febrifuge, digestive, laxative, ulcer and anthelmintic (Chatterjee et al. 2011).

Dye

Rhizomes are certain dye preparation to impart a perfume to the fabric and fragrance in hair lotions. Mat sticks, made of mat grass are widely used in our daily life, in various ways. In Bengal, when a great or a relative comes to a family, a mat is spread out for him either on the ground or on a cot. Especially in mid-summer noon, a cold mat on a bed, brings refreshment in the tired body and mind. Mat sticks are used not only for making mats, but they have other uses also.

A variety of products have been made by using by madur sticks such as maslanda, simple mat, door and window curtain with ring, TV and computer cover, shoulder bag, ladies bag, ladies purse, table mat with tassel, lamp shade, roller blind, spectacle case, file folder, asan mat, wall organizer, mobile cover and painting wall etc. Madur sticks and leaves are used as fuel in the villages. The madur sticks are used as a binding materials for fencing. It is use to bind the jute sticks in the 'Pan Borose' and to bind the paddy seedlings before plantation. Madur sticks are also used for decoration of different puja pandel.

Socio-Economic Importance

Mostly the madur sticks from which the different daily uses household utility items, are developed by many poor people, especially women and children are involved for the production of such materials and earn some money and increase the economic stability of their families. A common mat generally costs Rs.175 to Rs.250 while a designed and embroidered mat costs Rs.500 to Rs.8000.The profit under good management condition may be much higher than a farmer can get by growing two crops of rice per annum. Mat making is very easy. The spitted and dried mat stick may be used for making mat or may be kept for use in the off time of the season, when there is no other farming operation. Even an older man or woman is rural areas can earn a net income of about Rs.120 to Rs150 per day.

Luxury Materials from Mat

In modern times, summerous luxury materials are made up by mat grass or matsticks. Worth mentioning of them is Masalande. It is very fine and costly also. Besides, door and window curtain with ring, T.V. and Computer cover, shoulder bag, ladies bag, ladies purse, table mat with tassel, lamp shade, roller blind, spectacle case, file - folder, as a mat, wall organizer, mobile cover - and so many other useful things are prepared with madur sticks. In wealthy families, cushion-boxes, bed covers, foldable floor mats, sofa covers are being used also, and all of them are made with mat sticks.

TYPES OF MATS

Generally there are four types of mats, *viz*., (a) Single mat, (b) Double mat, (c) Masland mat and (d) Folding mat. Besides these, there is another kind of mat, which is called 'Chala' mat; this is just like double mat, but with more density of the sticks (Singha 1958) Now a short description is given below, about 4 kinds, mentioned above.

i) Single Mat

For weaving a single mat (SM), single threads are spread out longitudinally through little holes of the wooden shovel (Hata). Comparatively less quantity of fine mat-sticks are required for it. The sticks are woven plainly though alternative threads. This kind of mat is prepared throughout the whole area of Sabong. In most cases, a single artisan weaves a single mat.

ii) Double Mat

To make DM, pair (2) of threads are spread out through the holes of wooden shovel. More quantity of less fine sticks are used in it. Mat sticks are drawn through pair of threads (2threads) alternatively. This kind of double mat is prepared in area of Balpai in Sabong.

iii) Masland Mat

The most fine, famous and valuable mat is *'Masalanda'* or *'Mataranji'*. There are are about 12 kinds of wooden shovel for preparing masalanda mat (MM). The shovels are named after according to the density of bores or holes in the shovels. The weaving of masalanda mat has been described before.

iv) Folding Mat

When mats are prepared with advanced appliances, operated by hands using fibres of matgrass, they may be folded easily. These FM may be rolled just like a map also. To discuss about modern process and moder technology in mat kaking, we must remember the names of Balai Das and Abhiram Jana. Puspa Rani Jana is the wife of Abhiram Babu. In 1966 Balai Das engaged himself in research of modern technology in mat making. The Jana family tried deeply to develop the process of 'making fine mats i.e. *masalanda* or *mataranji*.

MARKETING

Structure of marketing shows that mat products are sold to its users through some specific channels. A multiple number of middlemen play their respective roles in these distributive channels. Among the middlemen in the market, mahajan paikars are the most influential ones (Maity 1996). Sales of finished mats are affected through some specialized village markets, conventionally known as hats. There are as many as 12 such markets in Sabong, *viz.*,

(i)	Sabong Bazar (Hat),	(ii)	Dashagram Bazar,
(iii)	Belki Bazar,	(iv)	Bural Bazar,
(v)	Mohar Bazar,	(vi)	Jhikuria Bazar,
(vii)	Balpai Bazar,	(viii)	Mohini Bazar,
(ix)	BishnupurBazar,	(x)	Pernua Bazar,
(xi)	HarnanBazar,	(xii)	Khagrageria Bazar.

These markets are situated within a distance of 3 - 5 kilometers from the mat - producing units. Variation in the value of transactions is from 1.5 Lakhs to 5 thousand in a market day. Sales and purchases are performed in cash. Prices of mats are determined by paikars (Table 1, Graph 1).

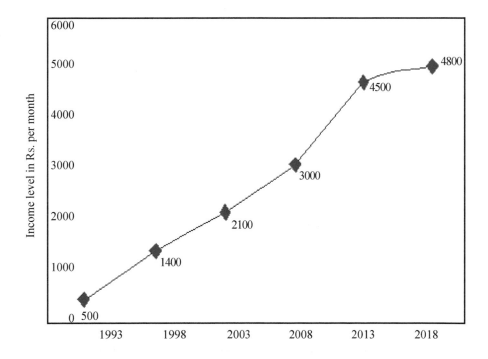

Graph 1: Income level of diversified mat weaver artisan per month

Table 1: Yearly income from Mat in Sabang

Yearly Income from Mat in Sabong	
Financial Year	Yearly Income
2007-2008	4,44,12,425/-
2008-2009	5,05,01,905/-
2009-2010	3,57,17,200/-
2010-2011	5,12,54,169/-
2011-2012	5,17,23,312/-
2012-2013	6,96,86,223/-
2013-2014	7,96,56,125/-
2014-2015	7,36,96,963/-
2015-2016	8,02,13,584/-
2016-2017	8,69,56,204/-

FUTURE PERSPECTIVES

Before starting implementation of every valuable thing we always think about their future, if future makes any positive side then we can start to put extra emphasis on these things. Here mats plant has some valuable things and it has a lot of future prospective. Future prospectives are given below:

1. Mat cultivation is the primary cultivation of Sabong areas' people and we all know about the environment of Sabong, which is very much moderate temperature as well as good soil texture ultimately helps to better cultivation of mat. If we could be increased mat production automatically economical prospects will be increased which will be also helpful entire Sabong people.

2. Different parts of mat are very useful for developing different areas such as for making perfume. For making our daily necessary things like bag, TV covers etc. More or less if we start our Research and development with mat the rural people use it for making different things as an alternate economic source.

3. Now a day's fashion industry is growing and all of the people want to be very sophisticated, by using mat we can make beautiful design by using mat.

4. Increase the dependency of the mat industry among the people and both central and state govt should focus on mat as an alternate economy.

5. Major focus of the study is to establish the scientific principles behind processing, separation of clum strands, texture and strength properties of the mat fibers.

6. Ethnobotanical aspects should know of this plant by rural people.

CONCLUSION

More steps should be taken for the upliftment of the socio-economic conditions of the artisans. Scientific projects should be made for the development of the mat industry itself. Floods of Keleghai and Kapaleswari must be controlled, with a view to protect the field of mat grass, as well as, the dwelling places of the artisans. Research schemes are welcomed for the improvement of mat sticks and also of the weaving implements. An auxiliary jute industry is required near Sabong to produce more fine threads from jute, so it is essential for the mat industry.

It is expected that the Central government of India and the State government of West Bengal will be sympathetic enough to do the needful for the protection and improvement of the mat industry, so that it may have a glorious place in the world. Madur grass is a most promising economically important plant for the rural people. A variety of products developed from madur stick with varying designs and colours such as floor cushions boxes, bed covers, foldable floor mats, sofa covers and bags purses. The madur stick is attractive durable and is eco-friendly substances which easily decomposed by organisms. So it does not create any environmental hazards. By using science and technology processing of madur sticks, and production of different substances also be possible to improved the quality and attraction for the people who uses these goods. Market demand of these product will be increased and the economic stability will also be increase for the rural people who are involved in this work.

Abbreviations

Cp. : *Cyperus pangorei*

M.S. : Mat Sticks

S.M. : Single Mat

D.M. : Double Mat

M.S. : Masland Mat

F.M. : Folding Mat

Conflict of Interest

We have no conflict of interest.

ACKNOWLEDGEMENTS

We would like to express thanks to our parents for their encouragements, helped for the success and support from time to time. We would like to express sincere thanks to the local people who helped in research and survey. We express special thanks to Mr. Ayan Kumar Naskar research scholar of the Plant Taxonomy, Biosystematics and Molecular Taxonomy Laboratory for help and encouragements all the time. Grateful thanks are due to rest of the research scholars of the Plant Taxonomy, Biosystematics and Molecular Taxonomy Laboratory for their valuable guidance and help throughout the study.

REFERENCES CITED

Chatterjee RK,Choulia N, Chatterjee A, Arora P. 2011. Anthelmintic Evaluation of the Rhizomes of the *Cyperus tegetum* Roxb. *Research Article*. 8.

Ghosh SK. 1993. *Gramin Kuther Shilpa*. Deys Publication. 104-107.

Maity M. 1996. Economic of Mat Industry: A Study of P. S. Sabong, District Midnapore, West Bengal, Finance India10(3): 709-716.

Samat N. 2008. *Sabong Darpan*. Published by: Narayan Samat on behalf on Alore Mela Weekly News Paper, Sabong .Paschim Medinipore. 1-30.

Singha A. 1958. *Kuther Shilp and Parikalpana*. Bengal Publishers Pvt. Ltd..

19

Biosynthesis of Silver Nanoparticles from Leaf Extracts of Three Medicinally Important Plants and its Effects on Seed Germination and Seedling Growth on Mungbean

Sk. Md. Abu Imam Saadi and Amal Kumar Mondal*

Plant Taxonomy, Biosystematics and Molecular Taxonomy Laboratory
UGC-DRS-SAP and DBT-BOOST-WB Funded Department, Department of Botany
and Forestry, Vidyasagar University, Midnapore-721102, West Bengal, INDIA
Email: saadivu@gmail.com, amalcaebotvu@gmail.com;
*akmondal@mail.vidyasagar.ac.in

ABSTRACT

The use of engineered nano-materials has been increased as a result of their positive impact on many sectors of the economy including agriculture. Silver nanoparticles (AgNPs) are now used to enhance seed germination, plant growth and as anti-microbial agents to control plant diseases. This study investigated the synthesis of silver nanoparticles using the aqueous solution of three medicinally important plants (such as *Spondius mombin* L., *Stachytarpheta jamaicensis* (L.) Vahl and *Syzygium samarangense* (Blume) Merr & Perry leaf extract in room temperature (35^0C). AgNPs present in the leaf confirmed by colour change, UV-VIS spectrum analysis, average size of AgNPs confirmed by PAS. FTIR measurement were carried out to identify the other possible biomolecule groups such as alkenes, esters, phenols, alcohols and aromatic. Leaf extract AgNPs were shown significantly higher antimicrobial activities against four species of bacteria. This findings suggest that the seed germination percentage, relative seed germination rate, relative shoot and root growth and germination index of the tested plant depends upon concentrate gradient of AgNPs.

Keywords: Silver nano-particles, anti-microbial activity, Seed Germination

INTRODUCTION

Nanoparticles are special group of materials with unique features and extensive applications in diverse fields (Matei et al. 2008). Nano-particles exhibit completely new or improved properties based on specific characteristics such as size, distribution and morphology. Among the various inorganic metal nanoparticles, silver nanoparticles has received substantial attention for various reasons – silver is an effective antimicrobial agent and exhibits low toxicity (Jain et al. 2009 and Sondi 2004), silver nanoparticles have diverse in vitro and in vivo applications (Haes 2002 and McFarland 2003). Silver nanoparticles varying between 1- 100 nm in size. Use of plant extract for the synthesis of nanoparticles could be advantageous over other environmental biological processes by eliminating the elaborate process of maintaining cell cultures. Very recently green silver nanoparticles have been synthesized using various natural products like green tea (*Camellia sinensis* (L.) Kuntze) (Vilchis-Nestor 2008), neem (*Azadirachta indica* A.Juss, leaf broth (Shiv Shankar 2004), starch (Vigneshwaran 2006), aloe vera plant extract (Chandran 2006), lemon grass leaves extract (Shankar 2004 and 2005), leguminous shrub (*Sesbania drummondii* (Rydb.) Cory (Sharma 2007). Green synthesis provides advancement over chemical and physical method as it is cost effective, environment friendly, easily scaled up for large scale synthesis and in this method there is no need to use high pressure, energy, temperature and toxic chemicals.

Silver is known for its anti-microbial properties and has been used for years in medical field for anti-microbial applications and even has shown to prevent HIV binding to host cells (Elechiguerra 2005). The Ag-NPs are reported to be nontoxic to human and effective against bacteria, virus and other eukaryotic micro-organisms at very low concentration and without any side effects (Jeong 2005). Because of their large specific surface area, silver nanoparticles are highly active and can play a crucial role in inhibiting bacterial growth in aqueous and solid media. The antimicrobial activity of colloidal silver is influenced by the size of the particles. Smaller the particle size more is its anti-microbial effect (Shahverdi 2007).The anti-microbial capability of SNPs allows them to be suitably employed in numerous household products such as textiles, food storage containers, home appliances and in medical devices (Marambio-Jones 2010).

The impact of AgNPs on higher plants appears to depend on species and age of plants size and concentration of the nanoparticles; experimental conditions such as temperature duration and method of exposure. Seed germination of *Boswellia ovalifoliolata* N.P. Balakr & A.N.Henry (Savithramma 2012) of *Pennisetum glaucum* (L) R.Br (Parveen 2014) has been shown to be positively affected by treatment with AgNPs. Nanotechnology application to the agriculture and food sectors is relatively recent compared with its use in drug delivery and pharmaceuticals.

Hence the aim of present study is to develop a novel approach for the green synthesis of silver nanoparticles using Indian herbal plants extracts as a reducing and stabilizing agent. We have carried out a unique protocol for synthesizing of Ag nanoparticles (temperature, time, extract preparation method and storage). The present study highlights (i) the methods employed in the synthesis of Ag nanoparticles, characterized through UV-VIS absorption, FTIR, PSA analysis, (ii) the role of Ag nanoparticles in antibacterial experiments and (iii) the effects of silver nanoparticals on seed germination. We investigated the impact of AgNP application on the seed germination and seedling growth of mung bean plants.

Collection of Plant Materials

Medicinally important taxa - *Spondius mombin* L. (Anacardiaceae), *Stachytarpheta Jamaicensis* (L.) Vahl (Verbenaceae) and *Syzygium samarangense* (Blume) Merr & Perry (Myrtaceae) leaves were collected from Vidyasagar University campus, Midnapore, India and washed thoroughly with distilled water. Leaves were then put into hot air oven with temperature maintain 50-60°C for 72h and dried leaves were ground by morter and pestle. The powder stored in an air tight container for further use.

Preparation of Plant Extracts

The 1 gm leaf powder was weighed and mixed with 10 ml distilled water. The mixture was poured in a test tube and boiled in water bath for 10 min at 60°C. Then the mixture was filtered and filtrate used as a plant extract.

Preparation of Silver Nitrate solution

1 mM silver Nitrate solution was prepared (by mixing 1.8gm $AgNO_3$ into1000 ml water)

Synthesis of Silver nanoparticles from plant extracts

90ml of 1mM AgNO3 was taken in conical flask and 1ml of plant extract was added to it. Then the mixture was stirred by glass rod for 1h and then incubated in normal temperature for 3h.

Characterization of silver nanoparticles by following instruments

UV-VIS Spectroscopy: After 3h incubation small amount of aliquot was diluted using distilled water (1:9).Then biosynthesized Ag+ nanoparticles were measured by the UV-VIS spectrum analysis (UV-3600) in 400-500 nm.

PSA Analysis: PSA analysis was carried out for the sample for determination of nanoparticles size [Malvern (Nano-ZS90)].

FTIR Analysis: The dried sample was mixed with potassium bromide crystals and the sample was then characterized by FTIR (Model no - Spectrum 2) for size conformation. The FTIR spectrum was obtained in the mid IR region of 400-4000cm-1. It is an important technique for identification and characterization of a substance.

Materials and methods for Anti-microbial Activity

Pure cultures of four bacterial stains -*Vibrio parahaemolyticus*, *Staphylococcus aureus* and *Bacillus cereus* were used for the study, supplied from the laboratory of microbiology, Department of Botany and Forestry, Vidyasagar University.

Preparation of Nutrient Agar Media

For 400 ml culture media: Nutrient broth-3 gm; Agar powder-8gm; Distilled water-400ml

Experimental Procedure for Anti-microbial Activity

For antimicrobial activity, nutrient broth bacterial culture mediums were used. After solidification of bacterial culture sterile discs were dipped in $AgNO_3$ solution and placed in nutrient agar plate and kept for incubation at room temperature for 24h. Then zones of inhibition for control SNPs were measured.

Materials and Methods for Seed Germination Test

The mungbean seeds were purchased from market; these seeds were kept in a dry place in the dark under the room temperature before using.

Preparation of Silver Nano-particle solution

1mM $AgNO_3$ solution was prepared. Powder of each selected plant's leaves was weighted at different gradients (0.2gm, 0.6gm & 1.0gm). After, the entire gradient mixed with 80 ml prepared $AgNO_3$ solution and stirred for 10 minute using glass rod.

Seedling Exposure

The seeds were checked for their viability by suspending them in double distilled water. The seeds which are settled to the bottom were selected for further study. Then surface sterilizations of seeds were done by immersing them into 5% sodium hypochloride solutions for 10 minute. After this seeds were rinsed in double distilled water thrice and then surface sterilization of seeds was done completely. The germinations process was done by added 5 ml of nanoparticle solution into Petridis. 10 sterilized seeds were transferred into each Petridis and kept it for 3hr. In this experiment control seeds (treated with distilled water) were also taken for comparison with the treated ones. After 3hr the soaked seeds were put in prepared pots and placed under shadow condition.

Seed Germination Test Equation

This test was conducted on mungbean seeds. Germination percentage = $(G_s/T_s) \times 100$; Relative seed germination Rate = $(Sc/Ss) \times 100$; Relative root growth = $(R_s/R_c) \times 100$; Relative shoot growth = $(S_s/S_c) \times 100$; Germination index = $(RSG/RRG) \times 100$ Where, G_s = Number of germinated seed; T_s = Total number of sowing seed; S_s = The number of seed germinated in sample; S_c = The number seed germinated in control; R_s = The average root length in sample; R_c = The average root length in control; S_s = The average shoot length in sample; S_c = The average shoot length in control.

Root and Shoot Length

Root length was taken from the point below the hypocotyls to the end of the tip of the root. Shoot length was measured from the base of the root-hypocotyl transition zone up to the base of the cotyledons. The root and shoot length was measured with the help of a thread and scale.

DISCUSSION

The colour change was observed after 3h incubation which revealed that a clear indication of the formation of silver nanoparticles (SNPs) in reaction mixture.

UV-VIS Spectroscopy

The sample was observed under UV-VIS spectrophotometer for its maximum absorbance and wavelength to confirm the reduction of Silver nitrate. AgNO3 particles are known to exhibit a UV-Visible absorption maximum in the range of 400-500nm due to this property (Sastry et al. 1997). All of the plant materials showed that the maximum peak was found between 400-450nm (Figure 1, 2nd 3).

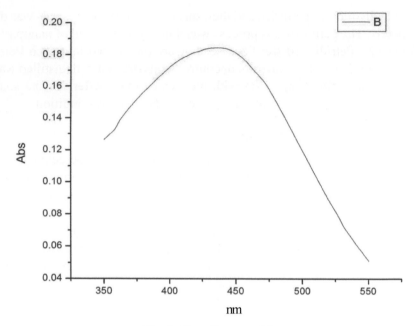

Fig. 1: *Spondius mombin*

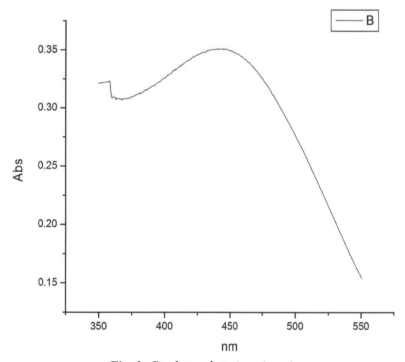

Fig. 2: *Stachytarpheta jamaicensis*

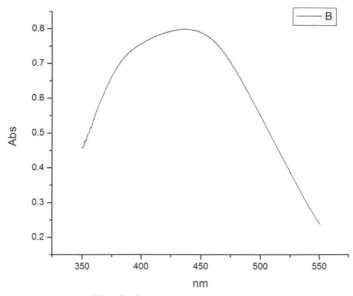

Fig. 3: *Syzygium samarangense*

PSA Analysis

Particle Size Analyser shows that the size range of the synthesized nanoparticles are 79.23 nm in *Spondius mombin*, (Figure 4), 47.33 nm in *Stachytarpheta jamaicensis*, (Figure 5) and 26.67 in *Syzygium samarangense* (Figure 6).

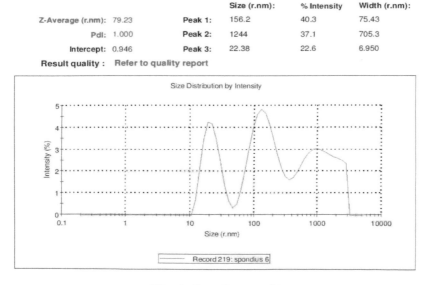

Results

			Size (r.nm):	% Intensity	Width (r.nm):
Z-Average (r.nm):	79.23	Peak 1:	156.2	40.3	75.43
PdI:	1.000	Peak 2:	1244	37.1	705.3
Intercept:	0.946	Peak 3:	22.38	22.6	6.950
Result quality :	Refer to quality report				

Fig. 4: *Spondius mombin*

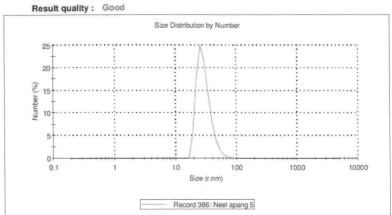

Results

		Size (r.nm):	% Number	Width (r.nm):
Z-Average (r.nm): 47.33	Peak 1:	29.74	100.0	8.689
PdI: 0.168	Peak 2:	0.000	0.0	0.000
Intercept: 0.940	Peak 3:	0.000	0.0	0.000
Result quality : Good				

Fig. 5: *Stachytarpheta jamaicensis*

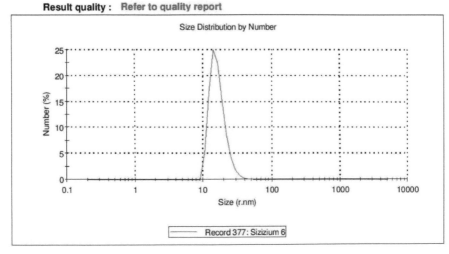

Results

		Size (r.nm):	% Number	Width (r.nm):
Z-Average (r.nm): 26.67	Peak 1:	16.48	100.0	4.577
PdI: 0.214	Peak 2:	0.000	0.0	0.000
Intercept: 0.942	Peak 3:	0.000	0.0	0.000
Result quality : Refer to quality report				

Fig. 6: *Syzygium samarangense*

FTIR Analysis

FTIR spectroscopy measurements are carried out to identify the biomolecules that present in leaf extract of three species. The characterization of bio molecules depends on the absorption bands which were observed at different spectra like-in case of *Spondius mombin*, band 3455.4, 2912.6, 2127, 1638 and 1078.33cm^{-1} contain N- H stretching; C-H stretching (alkane); C=C stretching (alkyne); C=C stretching (alkene), N-H bending; S=O stretching, C=S stretching, C-N vibrations, O-H stretching (alcohols) group respectively (Figure 7). In case of *Stachytarpheta jamaicensis*, band 2928.1& 2858, 1743.2, 1577.6, 1361, 1201.7, 1080.6 and 8443cm^{-1} contain H stretching (alkane); C=O stretching (ester); C=C stretching (aromatic), N=N stretching, N-H bending; S=O stretching, C-N vibrations, C-O stretching (phenols); C-N vibrations, S=O stretching; C=S stretching, S=O stretching, C-N vibrations, O-H stretching (alcohols); C-H bending (aromatic) group respectively (Figure 8). In case of *Syzygium samarangense*, band 2925, 2151, 1603, 1488.4, 1281.3, 1319.5 and 1117 & 10873cm^{-1} contain C-H stretching (alkane); C=C stretching (alkyne); N=N stretching; C=C stretching (aromatic), S=O stretching, C-N vibrations, C-O stretching (phenols); S=O stretching, C-N vibrations, O-H bending (alcohols); S=O stretching, C=S stretching, C-N vibrations, O-H stretching (alcohols) group respectively (Figure 9).

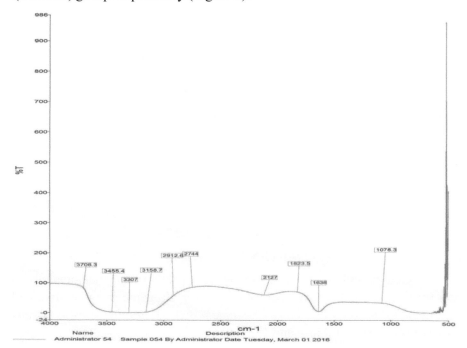

Fig. 7: *Spondius mombin*

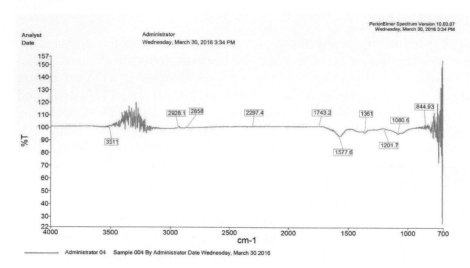

Fig. 8: *Stachytarpheta jamaicensis*

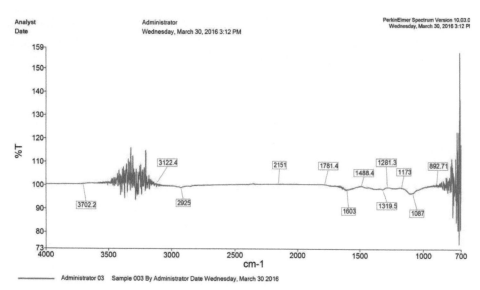

Fig. 9: *Syzygium samarangense*

Anti-microbial Activity of Silver Nano-particles

The biologically synthesized silver nanoparticles (SNPs) using medicinal plants were found to be highly toxic against different pathogenic gram positive and gram negative bacteria like *Vibrio parahaemolyticus*, *Staphylococcus aureus* and *Bacillus cereus* (Table 1 and Fig. 10)

Table 1: Measurement of zone of inhibition by different species of Bacteria

Plants Name	*Vibrio parahaemolyticus*	*Staphylococcus aureous*	*Bacillus cereus*
Spondius mombin	18.5	15	10.5
Stachytarpheta jamaicensis	25	16	13
Syzygium samarangense	21.5	16.5	11
Con	10.5	9.5	10

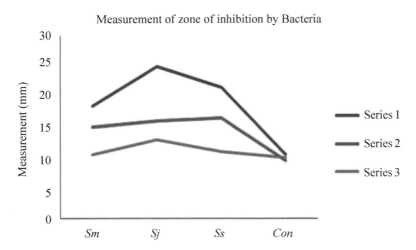

Fig. 10: Measurement of zone of inhibition by different species of Bacteria

Application of Silver Nano-particles on Seed Germination and Seedling Growth

The seed germination test was conducted on mungbean seeds by the use of silver nanoparticles. We used three different concentrations of silver nanoparticles like 0.2, 0.6 and 1.0 gm/ml to know the impact of seed germination percentage, relative seed germination rate, relative shoot & root growth and germination index. This study carried out 7 days after then we take result. Our finding reveals that seed germination percentage of *Stachytarpheta jamaicensis* and *Syzygium samarangense* is better than *Spondius mombin* in all the concentration gradients of AgNPs. The relative seed germination rate comparatively better in 0.2 and 0.6 gm/ml concentratios and the relative shoot & root growth comparatively better in 1.0 gm/ml concentrations. *Spondius mombin* shows very slow shoot growth in 0.2 and 0.6 gm/ml concentrations.

CONCLUSION

The present work focused on the development of a green method for the synthesis of AgNPs using Indian medicinal plants. Silver nanoparticles prepared in this process are fast, suitable, eco-friendly, and can be potentially applied in variety of extracts for preparing different metal nanoparticles. Some biomolecules characterized by FTIR spectroscopy on the basis of bands or peaks such as alkane, aromatic, phenols, alcohols, polysaccharides, carboxylic, aldehyde, ester, ketone, acid and amino group. From these molecules we can prepare different kind of drugs against different disease. The present study also included the bio-reduction of silver ions through medicinal plants extracts and testing for their antimicrobial activity. Hence has a great potential in the preparation of drugs used against bacterial diseases. Seed germination and root elongation is a rapid and widely used acute phytotoxicity test owing to sensitivity, simplicity, low cost and suitability for unstable chemicals. Seed coats, which can have selective permeability, play a very important role in protecting the embryo from harmful external factors. Pollutants as nano-metals could penetrate root system causing obviously root growth inhibition, may not affect seed germination if they cannot pass through seed coats. This may explain that seed germination is not affected by exposure to AgNPs suspension. Exposure to nanomaterials can encourage earlier plant germination and improve plant production. The outcomes of this study can be useful for determining the biocompatibility of AgNPs and for identifying potential agricultural applications for nanoparticles in crop improvement and food production.

Abbreviations

AgNPs/SNP :   Silver nanoparticles

UV-VIS   :   Ultraviolet–visible

PSA   :   Particle size analysis

FTIR   :   Fourier Transform Infrared

Conflict of Interest

There is no conflict of interest.

REFERENCES CITED

Chandran SP, Chaudhary M, Pasricha R, Ahmad A, Sastry M. 2006. Synthesis of gold nanotriangles and silver nanoparticles using *Aloe vera* plant extract. *Biotechnology Progress* 22:577–583.

Elechiguerra JL, Burt JR. 2005. Morones, Interaction of silver nanoparticles with HIV- 1. *Journal of Nanobiotechnology* 3(6): 34.

Jain J. 2009. Silver Nanoparticles in Therapeutics: Development of an Antimicrobial Gel Formulation for Topical Use. Molecular Pharmaceutics.

Jeong S, Yeo S, Yi S. 2005. Antibacterial characterization of silver nanoparticles against *E. coli* ATCC-15224. *Journal of Material Science* 40:5407.

Marambio-Jones C, Hoek EMV. 2010. A review of the antibacterial effects of silver nano materials and potential implications for human health and the environment. *J Nanopart Res* 12:1531-1551.

Matei A, Cernica I, Cadar O, Roman C, Schiopu V. 2008. Synthesis and characterization of ZnO-polymer nanocomposites. *The International Journal of Material Forming* 1: 767-770.

McFarland A, Van Duyne R. 2003. Single silver nanoparticles as real-time opticalsensors with zeptomole sensitivity. *Nano letters* 3(8): 1057-1062.

Parveen A, Rao S. 2014. Effect of nanosilveron seed germination and seedling growth in *Pennisetum glaucum. J. Clust. Sci* 21(1):13-17.

Sastry M, Mayya KS and Bandyopadhyay K. 1997. pH Dependent changes in the opticalproperties of carboxylic acid derivatized silver colloidal particles. *Colloids Surf.* A 127:221-228.

Savithramma N, Ankanna S, Bhumi G. 2012. Effect of nanoparticles on seed germination and seedling growth of *Boswellia ovalifoliolata*– anendemic and endangered medicinal tree taxon. *Nano Vision* 2(1-3):61-68.

Shahverdi AR, Fakhimi A, Shahverdi A, Minaian MS. 2007. Synthesis and effect of silver nanoparticles on the antibacterial activity of different antibiotics against Staphylococcus aureus and Escherichia coli. *J. Nanomedicine* 3(2):168-171 .

Shiv Shankar S, Rai A, Ahmad A, Sastry M. 2004. Rapid synthesis of Au, Ag, and bimetallic Au core–Ag shell nanoparticles using neem (*Azadirachta indica*) leaf broth. *J. Colloid Interf. Sci.* 275: 496–502.

Shankar SS, Rai A, Ankamwar B, Singh A, Ahmad A, Sastry M. 2004. Biological synthesis of triangular gold nanoprisms. *Nat. Mater* 3: 482–488.

Shankar SS, Rai A, Ahmad A, Sastry M. 2005. Controlling the optical properties oflemongrass extract synthesized gold nanotriangles and potential application ininfrared-absorbing optical coatings. *Chem. Mater* 17: 566–572.

Sharma NC, Sahi SV, Nath S, Parsons JG, Gardea-Torresdey JL, Pal T. 2007. Synthesis of plant-mediated gold nanoparticles and catalytic role of biomatrix embedded nanomaterials, Environ. *Sci. Technol* 41: 5137–5142.

Sondi I, Salopek-Sondi B. 2004. Silver nanoparticles as antimicrobial agent: A case studyon E. coli as a model for Gram-negative bacteria. *J Colloid Interface Sci* 275: 177 - 182.

Vilchis-Nestor AR, Sanchez-Mendieta V, Camacho-Lopez MA, Gomez- Espinosa, RM, Camacho-Lopez MA, Arenas-Alatorre JA. 2008. Solventless synthesis and optical properties of Au and Ag nanoparticles using *Camellia sinensis* extract *Mater. Lett* 62: 3103–3105.

Vigneshwaran N, Nachane RP, Balasubramanya RH, Varadarajan PV. 2006. A novelone pot 'green' synthesis of stable silver nanoparticles using soluble starch, *Carbohydr. Res.* 341: 2012–2018.

20

Genetic Improvements of Medicinal Plants through Tissue Culture Techniques

Gitasree Borah, Manabi Paw and Mohan Lal*

Medicinal Aromatic and Economic Plants Group
Biological Science and Technology Division
CSIR-North East Institute of Science and Technology
Jorhat-785006, Assam, INDIA
*Email: drmohanlal80@gmail.com

ABSTRACT

Plant tissue culture is an important and efficient technique for the production of desired plant products with different and improved variety in aseptic conditions. Plant tissue culture methods have a wide scope for the creation, conservation, and utilization of genetic variability for the genetic improvement of field, fruit, vegetable, forest crops, medicinal and aromatic plants. The main objective of this communication were to collects the information and techniques of tissue culture to genetic improvement of the medicinal plants. Tissue culture technique in combination with molecular techniques has been successively used to incorporate specific traits through gene transfer. Modified varieties of medicinal plants is mostly required, as some compounds like alkaloids, terpenoids, steroids, saponins, phenolics, flavanoids, and amino acids etc are extracted from these plants. From the ancient time up to 20th century, most of the human-being are dependent on the medicinal plant for basic healthcare needs and improving such a crop is the most important for this billionaire world. As far conventional methods of genetic improvements the tissue culture techniques takes cheap and lesser time required for traits specific improvements in the MAPs.

Keywords: Tissue culture, Medicinal plants, Plant breeding, *In-vitro* propagation, Protoplast culture, Genetic improvements.

INTRODUCTION

Tissue culture is an important and world-wide modern technique that gives an efficient result for the improvement of the medicinal plants, especially for the endangered species. Tissue culture has been exploited to create genetic variability from which crop plants can be improved, to improve the state of health of the planted material and to increase the number of desirable germplasms available to the plant breeder. Combination of molecular methods and tissue culture techniques has been giving a successful result for the development of the medicinal as well as other plant species *In-vitro* techniques for the culture of protoplasts, anthers, microspores, ovules and embryos have been used to create new genetic variation in the breeding lines (Brown 1995).

The World Health Organization has estimated that more than 80% of the world's population in developing countries depends primarily on herbal medicine for basic healthcare needs (Vieira et al. 1993). To fulfil the demand of the entire human population, good variety of the medicinal plants must be used in industrial level for optimum yield and higher useful alkaloid.

Historical background of tissue culture

The science of plant tissue culture takes its roots from the discovery of cell followed by propounding of cell theory. In the year 1838, Schleiden and Schwann proposed that cell is the basic structural unit of all living organisms. They visualized that cell is capable of autonomy and therefore it should be possible for each cell if given an environment to regenerate into whole plant. Based on this premise, in 1902, a German Physiologist, Gottlieb Haberlandt for the first time attempted to culture isolated single palisade cells from leaves in Knop's salt solution enriched with sucrose. The cells remained alive for up to one month, increased in size, accumulated starch but failed to divide. Though he was unsuccessful but laid down the foundation of tissue culture technology for which he is regarded as the father of plant tissue culture (Hussain et al. 2012)

From the time of the discovery of plant tissue culture up-to these days, tissue culture is tried to be improved for the fulfilment of the demand of this billioner earth. To improve the quality of important plants, different tissue culture methods have been applied in medicinal plants. Some of those important techniques are-described below.

Plant Breeding

Plant breeding can be broadly defined as alterations caused in plants as a result of their use by humans, ranging from unintentional changes resulting from the advent of agriculture to the application of molecular tools for precision breeding.

The vast diversity of breeding methods can be simplified into three categories: (i) plant breeding based on observed variation by selection of plants based on natural variants appearing in nature or within traditional varieties; (ii) plant breeding based on controlled mating by selection of plants presenting recombination of desirable genes from different parents and (iii) plant breeding based on monitored recombination by selection of specific genes or marker profiles, using molecular tools for tracking within-genome variation. The continuous application of traditional breeding methods in a given species could lead to the narrowing of the gene pool from which cultivars are drawn, rendering crops vulnerable to biotic and abiotic stresses and hampering future progress. Several methods have been devised for introducing exotic variation into elite germplasm without undesirable effects. Cases in rice are given to illustrate the potential and limitations of different breeding approaches (Breseghello 2013).

Protoplast Culture

The ultimate objective in protoplast culture is the reconstruction of plant from the single protoplast. The strategy for plant regener-ation has been to recover rapidly growing cal-lus from protoplasts and to transfer the callus to a species specific regeneration medium. It is generally noted that plant regener-ation occurs very easily in some plant species while others are recalcitrant.

Genetic Transformation

Transformation is the process by which genetic makeup of an organism is altered by the insertion of new gene (or exogenous DNA) into its genome. This is usually done using vectors such as plasmids. The main objective of using transformation technique in plant tissue culture is to produce a better crop yields to improve the varietal trait and protection against their pests, parasite and harsh weather conditions. The combination of biotechnology with natural tissue culture process have been resulting a successful improvement for the medicinal plant as well as for other crop also.

Pollen Culture

Pollen culture is the in vitro technique by which the pollen grains (preferably at the microscope stages) are squeezed from the intact anther and then cultured on nutrient medium where the microspores without producing male gametes (Rani et al. 2017).

Anther or Microspore Cultures

Anther culture is the *in-vitro* development of haploid plants originating from potent pollen grains through a series of cell division and differentiation.

Androgenesis has become a powerful tool for the rapid production of haploid and inbred lines used for obtaining hybrid cultivars (Sopory et al. 1996).

EXPERIMENT DESIGN

The main objective of this study is to review the genetic improvement of medicinal plant through tissue culture and to study some of the improved technique of plant tissue culture applied in the medicinal plants.

TECHNIQUES

Plant tissue culture is a collection of techniques used to maintain or grow plant cells, tissues or organs under aseptic conditions. It has some unique methods which are performed with different materials. To get a better result from the plant tissue culture, one should always strictly maintain the environment of the tissue culture laboratory and its basic requirements. The basic requirements of plant tissue culture laboratory are-

- Healthy explants
- Aseptic conditions
- Control of temperature
- Proper culture media with proper plant growth regulator
- Sub-culturing

Healthy Explants

The plant parts which are used for regeneration in plant tissue culture are called explants that can be leaf, pieces of stem, a protoplast, a tissue, or an organ. Only healthy explants can give a better product.

Aseptic Condition

Successful tissue culture depends on a clean and microbes free environment, because the explants which are taken from the outer environment of the laboratory may contain many micro-organisms like bacteria, fungi or viruses. Besides these, they may arise from the non-sterile supplies, media, reagents, airborne particles laden with microorganisms, unclean incubators, and dirty work surfaces also. The condition which is maintained to provide a barrier between the microorganisms and the tissue culture techniques to make a microbes free environment is called aseptic conditions. These include surface sterilization of the explants with different reagents like tween 20 mixed with liquid soap, bevistin, 70% ethanol and mercuric chloride. Distilled water; proper autoclaved media and instruments at 120°C for 15 minutes, inoculation at laminar airflow sterilized with UV rays and good personal and laboratory hygiene.

Control of Temperature and Light

The temperature and the light period of tissue culture incubation room should be maintained properly. Temperature should be at 25 ±2°C and 50 to 60% relative humidity in the presence of fluorescent light of 1000-2000 lux. Photoperiod of the incubation room is 16hr/8hr-light/dark. Light and temperature vary from species to species and sometimes during the various stages of developments.

Proper Culture Media and Plant Growth Regulator

Plant tissue culture medium contains all the nutrients required for the normal and growth and developments of the plants. These nutrients mainly composed of macronutrients, micronutrients, vitamins, other organic compounds, plant growth regulators, carbon source, and some gelling agents in case of solid medium Hussain et al. (2012). Usually pH of the media for most of the medicinal plants remains in between 5.6-5.8. Murashige and Skoog medium (MS Medium) is the most acceptable media used for the vegetative propagation of many medicinal plants. Composition of MS media is shown below in the Table 1.

Table 1: Composition of Murashige and Skoog Medium (1962)

Macronutrients (10x)		
mM (1x MS)	Salt	10x stock (g/L)
41.2	NH_4NO_3	16.5
18.8	KNO_3	19
3.0	$CaCl_2$ anhydrous	3.3
1.5	$MgSO_4.7H_2O$	3.7
1.25	KH_2PO_4	1.7
Micronutrients (100x)		
μM (1x MS)	Salt	100x stock (g/L)
100.0	H_3BO_3	0.62
100.0	$MnSO_4 H_2O$	1.69
30.0	$ZnSO_4.7H_2O$	0.86
5.0	KI	0.083
1.0	$Na_2MoO_4.2H_2O$	0.025
0.1	$CuSO_4.5H_2O$	0.0025
0.1	$CoCl_2.6H_2O$	0.0025
Fe-EDTA (100x)		
mM (1x MS)	Salt	100x stock (g/L)
Na 0.2	$Na_2.EDTA$	3.73
Fe 0.1	$FeSO_4.7H_2O$	2.78

Sub-Culturing

Transfer of tissue or callus from old culture media to fresh culture media in aseptic conditions is called sub-culturing. It has a great value in maintenance of the amounts of nutrients as they depleted according to the growth of the newly formed plant

Techniques Applied for Improvement of Medicinal Plants

To improve the medicinal plants and their varities, different plant tissue culture methods have to use as mentioned above. The methodology of these techniques are given below-

POLLEN CULTURE

This technique can be considered as the basic protocol for pollen culture and involves the following procedure.

1. Selection of suitable unopened flower bud, sterilization, excision of anther without filaments.

2. About 50 anthers are placed in small sterile beaker containing 20 ml of liquid basal medium (MS or White or Nitsch and Nitsch).

3. Anthers are then pressed against the side of beaker with a sterile glass piston of a syringes to squeeze out the pollens.

4. The homogenized anthers are then filtered through a nylon sieve to remove anther tissue debris.

5. The filtrate or pollen suspension is then centrifuged at low speed (500-800 rpm/min) for five minutes. The supernatant containing fine debris is discarded and pillet of pollen is suspended in fresh liquid medium and washed twice by repeated centrifugation and resuspension in fresh liquid medium.

6. Pollens are mixed finally with measured volume of liquid basal medium so that it makes the density of 10 3-10 4 pollen/ml.

7. A 2.5 ml of pollen suspension is pipetted off and is spread in 5 cm petridish. Pollens are best grown in liquid medium but, if necessary, they can be grown by plating very soft agar added medium. Each dish is sealed with cello tape to avoid dehydration.

8. Petridishes are incubated at 27-30°C under low intensity of white cool light (500 lux, 16 hrs).

9. Young embryoids can be observed after 30 days. The embryoids ultimately give rise to haploid plantlets.

10. Haploid plantlets are then incubated at 27-30°C in a 16 hrs day light regime at about 2000 lux. Plantlets at maturity are transferred to soil as described in anther culture.

ANTHER CULTURE

Tobacco is the ideal material for anther culture. So the basic protocol described below should be applicable to anther culture in general with modifications. The immature anthers containing uninucleate pollens at the time of first mitosis are the most suitable material for the induction of haploids (Fig -1).

1. Flower buds are collected at the onset of flowering. Flower buds of 17-22 mm are selected in length when the length of the sepals equals that of the petals. Flower buds which are beginning to open are rejected.

2. Selected flower buds are transferred to the laminar airflow. Each flower bud contains five anther and these are normally surface sterile in closed buds. The flower buds are surface sterilized by immersion in 70% ethanol for 10 seconds followed immediately by 10 minutes in 20% sodium hypochlorite. They are washed three times with sterile distilled water. Finally the buds are transferred to sterile petridish.

3. To remove the anthers, the side of the bud are slited with a sharp scalpel and removed them, with a pair of forceps; the five anthers are placed with the filaments to another petridish. The filaments are cut gently. Damaged anthers should be discarded as they often form callus tissue the damaged parts other than the pollens.

4. Anthers are placed on agar solidified basal MS or White or Nitsch and Nitsch medium.

5. The culture is kept initially in dark. After 3-4 weeks, the anthers normally undergo pollen embryogenesis and haploid plantlets appear from the cultured anther. In some cases, anther may undergo proliferation to form callus tissue which can be induced to differentiate into haploid plants.

6. At this stage the cultures are incubated at 24-28 °C in a 14 hrs day light regime at about 2000 lux.

7. Approximately 50mm tall plantlets are freed from agar by gently washing with running tap water and then transferred to small pots containing autoclaved potting compost. Cover each plant with glass beaker to prevent desiccation and maintain in a well-lit-humid green house. After some week, remove the glass beaker and transfer the plant to larger pots when the plants will mature and finally flower.

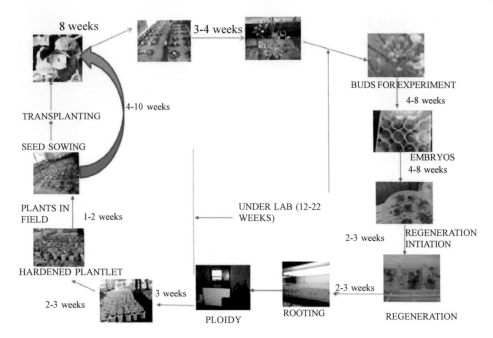

Fig. 1: Procedure of microspore culture in Cabbage

PROTOPLAST CULTURE

Protoplast culture is done by two parts. First is isolation of protoplast, then it is followed by culture of protoplast. These two methods are given below-

Isolation of Protoplast

Protoplasts can be prepared from a variety of tissue but among them leaf mesophyll tissue from a wide range of plants has been proved to be the most ideal source of plant material for protoplast isolation.

1. Young fully expanded leaves from the up-per part of 7-8 weeks old plants growing in a greenhouse are detached and leaves are washed thoroughly with tap water.

2. Surface sterilization used to be carried out by first immersing in 70% ethanol for 60 seconds then were dipping into 0.4-0.5% sodium hypochlorite solution for 30 minutes. For sterilization of the leaves, sterilized container (s) should be used. Sterilization is done in front of laminar air flow.

3. After 30 minutes, the sterilant is poured off and leaves are washed aseptically 3-4 times with autoclaved distilled water to remove every trace of hypochlorite.

4. With the help of long sterilized forceps (8 inches), one leaf is transferred on a steril-ized floor tile. The impermeable lower epi-dermis of the surface sterilized leaves is peeled off as completely as possible. During this process, a sterilized fine jeweler's force is inserted into a junction of the midrib and a lateral vein and the epidermis, is care-fully peeled away at an angle to the main axis of the leaf.

   Where peeling of the leaf is not possible, slicing of the leaf into thin strips may be suffi-cient to allow entry of the enzymes through the cut edges of the strip.

5. Peeled leaf pieces are placed lower surface down onto 30 ml sterilized CPW 13M solution in a 14 cm petridish. When the liquid surface is completely covered with peeled leaf pieces then the Cell and protoplast washing medium (CPW) 13M solution will be pipetted off from the beneath of leaf pieces. The CPW 13M solution is replaced by bac-terial filter sterilized solution of enzyme containing 2% cellulase (Onozuka R1O), 0.5% macerozyme in 13% manitol added CPW (pH 5.5).

6. Leaf pieces in enzyme solution are incubat-ed in the dark at 24-26°C for 16-18 hrs.

7. Without disturbing the digested leaf pieces the enzyme solution is gently replaced by CPW 13M. Then digested leaf pieces are gently agitated and squeezed with sterile fine forceps to facilitate the release of the protoplast. The protoplast suspension is then allowed to pass through a 60 μ-80μ nylon mesh to remove the larger pieces of undigested tissue.

8. The filtrate is transferred to screw-capped centrifuge tube and is spinned for 5 minutes at 100g.

9. The protoplasts form the pellet. The super-natant is pipetted off and the pellet is re- suspended in CPW 21S solution. It is again centrifuged for 5-7 minutes at 200 g. The viable protoplasts will float at the top surface of CPW 21S in the form of dark green band while the remaining cells and debris will sink at the bottom of the tube.

10. The viable protoplasts are collected from the surface and are re-suspended again in CPW 13M to remove the sucrose. Centrifugations are repeated two-three times for washing.

11. Finally the protoplasts are suspended in measured volume of liquid Nagata and Tabeke medium (1971) supplemented with NAA (3 mg/L), 6-BAP (1 mg/L) and 13% mannitol.

Table 2: Compositions of Cell and Protoplast Washing medium (CPW media)

Compositions	Concentrations (mg/L)
KH_2PO_4	27.2 mg/L
KNO_3	101 mg/L
$CaCl_2 2H_2O$	1480 mg/L
$MgSO_4.7H_2O$	246 mg/L
KI	0.16 mg/L
$CuSO_4.5H_2O$	0.025 mg/L
Conditions:	
P^h= 5.8	
CPW 13M = CPW + 13% mannitol (13M)	
CPW 21S = CPW + 21% sucrose (21S)	

Isolation of Protoplast from Cell Suspension Culture

Rapidly growing cell suspension cultures are the most suitable material for protoplast isola-tion. A new cell suspension does not yield many protoplast until it has been sub-cultured at least twice. For protoplast isolation, suspension cul-tures are generally harvested at its early expo-nential growth phase or log phase.

Older suspension cultures have a tendency to form elongated giant cells with thick wall. So it is very difficult to isolate the protoplast from such culture. Again, the presence of large num-ber of cell aggregates in suspension culture is not desirable for the isolation of protoplast.

Addition of colchicine and some chelating chemicals in suspension culture generally prevents the for-mation of cell aggregates. Sometimes very low concentration of cellulase (0.1%) is added in cell suspension culture two days before use to dis-courage the formation of thick wall.

Step 1: Filtering of Cell Suspension

The harvested cell suspension is passed through a coarse nylon sieve so that filtrate contains single cells as well as very small cell clumps.

Step 2: Preparation of Liquid Medium Free Cells for Plasmolysis

The filtrate is allowed to settle out of the medium. Most of the medium is decanted off and the cells are transferred (by pouring) to a flask. Using the Pasteur pipette, all of the cul-ture medium is removed. This is best achieved by drawing off medium from the base of the cell layer.

Step 3: Pre-plasmolysis

The cells are suspended in CPW13M solution for 1 hr. After 1 hr., the plasmolyticum is pipetted off.

Step 4: Enzyme Incubation

The enzyme solution is added to the cells. The flask containing cells and enzyme are placed on the platform of a slow-ly rotating gyratory shaker (ca 30-40 rpm) for standardized period (4-6 hrs.).

Step 5: Washing and Purification

Protoplast sus-pension is filtered through 60/i-100/i stainless steel sieve to remove the larger debris. The filtrate is transferred into centrifuge tubes and is spinned at 80 g for 5-10 minutes so as to sediment the protoplasts and then supernatant is pipetted off. The pellet is re-suspended in CPW 21S solution.

The protoplast suspension is again spinned at 100 g for 5-7 minutes. Viable and de-bris free protoplasts are collected at the surface. The protoplasts are washed with CPW 13M by repeated centrifugation and finally protoplasts are re-suspended in measured volume of liquid culture medium. Finally, the yield of protoplast is measured by counting in haemocytometer.

Protoplast Culture

Isolated protoplast can be cultured in several ways of which agar embedding technique in small petridish is commonly followed. In this technique, protoplast suspension is mixed with equal volume of melted 1.6% 'Difco' agarified medium (37 °C) and the protoplast-agar mixture are poured into small petridish.

In petridishes, embedding of protoplasts in solid agar medium is known as plating of protoplasts. The plated pro-toplast can be handled very easily and the agar medium provides a good support to the proto-plast. In situ developmental stages of embed-ded protoplast can be studied under compound microscope. Besides this, separated clones de-rived from individual protoplast can be moni-tored. The method is described below-

1. The protoplasts in liquid Nagata and Takebe (NT) medium 1971 (components of NT medium are shown in table 3) are counted with the help of haemocytometer. The protoplast deficit is adjusted to 1×10^5 to 2×10^5 protoplast/ml.

2. Agar solidified (1.6% 'Difco' agar) NT medium is melted.

3. The tight lid of Falcon plastic petridish (35 mm diameter 5 mm thickness) is opened and 1.5 ml of protoplast suspension is taken. To this equal aliquot of melted agar medium is added when it cools down at 37°C to 40°C.

4. The lid is quickly replaced tightly, and whole dish is swirled gently to disperse the protoplast-agar medium mixture uniformly throughout the dish.

5. The medium is allowed to solidify. The petridish is then inverted.

6. The culture is incubated at 25° C with 500 lux illumination (16 hrs light) initially.

7. The cultures are sub-cultured periodically in the same solid medium (0.8% agar) with gradually reducing mannitol.

Several other methods are also found for the culture of protoplasts such as droplet culture, co-culture, feeder layer, hanging droplets and immobilized/bead culture etc.

Table 3: Compositions of NT medium

Constituents	Amounts in mg/L
Macronutrients:	
NH_4NO_3	825
KNO_3	950
$CaCl_2,2H_2O$	220
$MgSO_4,7H_2O$	1233
KH_2PO_4	680
Micronutrients:	
KI	0.83
H_3BO_3	6.2
$MnSO_4,4H_2O$	22.3
$ZnSO_4,4H_2O$	8.6
$Na_2NoO_4,5H_2O$	0.25
$CuSO_4,5H_2O$	0.025
Iron source:	
$FeSO_4,7H_2O$	27.8
$Na_2EDTA, 2H_2O$	37.3
Vitamin:	
Meso-inositol	100
Thiamine HCL	1
Carbohydrate source:	
Sucrose	1%
Growth Substances:	
α napthalene-acid (NAA)	3
6- Benzylaminopurine (6-BAP)	1
Plasmolyticum:	
Mannitol	13%-0%
pH	5.8
Agar (For solid medium)	1.6% or 0.8%

Plant Breeding

Plant breeding can be considered as a co-evolutionary process between humans and plants. Breeders try to change in the plants that were used for agriculture and in turn, those has highly importance in medicinal as well as other commercial value. Civilization could not exist without agriculture, and agriculture could not sustain the civilized world without modern crop varieties. From this point of view, it becomes clear that plant breeding is one of the main foundations of civilization.

Genetic Transformation in plant tissue culture

Genetic transformation can be done by two methods. First one is indirect method, example, *Agrobacterium* mediated gene transfer and second one is direct method, example, Particle bombardment, Chemical methods, Microinjection, Macroinjection, Electroporation, Pollen transformation etc. *Agrobacterium* mediated gene transfer can be done by two ways- (i) Co-culture with tissue explants, and (ii) Through plant transformation. Here some of the methods of direct transformation in plants are discussed below-

Particle Bombardment

Prof Sanford and colleagues at Cornell University (USA) developed the original bombardment concept in 1987 and coined the term 'biolistics' for both the process and the device. Kikkert et al. (2005). This process is done by following method.

The biolistic gun employs the principle of conservation of momentum and uses the passage of helium gas through the cylinder with arrange of velocities required for optimal transformation of various cell types. It consists of a bombardment chamber which is connected to an outlet for vaccum creation. The bombardment chamber consists of a plastic rupture disk below which macro carrier is loaded with micro carriers. These micro carriers consist of gold tungsten micro pellets coated with DNA for transformation.

The apparatus is placed in laminar air flow while working to maintain sterile conditions. The target cells/ tissue are placed in the apparatus and a stopping screen is placed between the target cells and micro carrier assembly. The passage of high pressure helium ruptures the plastic rupture disk propelling the macro carrier and micro carriers. The stopping screen prevents the passage of macro projectiles but allows the DNA coated micro pellets to pass through it thereby, delivering DNA into the target cells.

Micro-injection

DNA Microinjection was first proposed by Marshall in the early of nineteenth century. It involves delivery of foreign DNA into a living cell through a fine glass micropipette. The introduced DNA May lead to the over or under expression of certain genes. Microinjection process is done by following methods.

Delivery of foreign DNA is done under a powerful microscope using a glass micropipette tip of 0.5 mm diameter.

Cells to be microinjected are placed in a container. A holding pipette is placed in the field of view of the microscope that sucks and holds a target cell at the tip. The tip of micropipette is injected through the membrane of the cell to deliver the contents of the needle into the cytoplasm and then empty needle is taken out (NPTEL 2018)

DISCUSSION

Plant tissue culture technology is being widely used for large scale plant multiplication. It has become major industrial importance in the area of plant propagation, disease elimination, plant improvement and production of secondary metabolites. Endangered, threatened and rare species have successively been grown and conserved by Micro propagation because of high coefficient of multiplication and small demands on number of initial plants and space.

Plant tissue culture methods have a wide scope for the creation, conservation, and utilization of genetic variability for the improvement of field, fruit, vegetable, and forest crops and medicinal/aromatic plants. (Satbir 2012). The past decades of plant cell biotechnology has evolved as a new era in the field of biotechnology, focusing on the production of a large number of secondary plant products. Now a days, one of the most promising methods of producing proteins and other medicinal substances, such as antibodies and vaccines, is the use of transgenic plants. Transgenic plants represent an economical alternative to fermentation based production systems. A number of medicinally important alkaloids, anticancer drugs, recombinant proteins and food additives are produced in various cultures of plant cell and tissues. Advances in the area of tissue culture leads to the production of a wide variety of pharmaceuticals like alkaloids, terpenoids, steroids, saponins, phenolics, flavanoids and amino acids (Tabata 1976). Some of such outcomes of improvement of medicinal plants are discussed below-

Cell Suspension Culture

Successful attempts to produce well-known medicinal compounds in relatively large quantities by cell-cultures are illustrated by the following examples:

Kaul et al. (1969) have reported that diosgenin, a major raw material in the commercial production of corticosteroids and steroid contraceptives can be produced by suspension cultures of *Dioacorea deltoidea* with a 1.5 & dry weight content. The cell cultures of the valuable Chinese drug *Panax ginseng* have been found to produce ginsenosides in large amounts (21 & of dry weight as crude saponins) by Furuya and Ishii (1973).

Genetic Transformation

Legumes are a large, diverse family ranging from herbaceous annuals to woody perennials that, because of their capacity to fix nitrogen, are essential components in natural and managed terrestrial ecosystems that can produce food, feed, forage, fiber, industrial and medicinal compounds, flowers, and other end uses. Understanding the molecular basis of nitrogen fixation and the unique metabolic pathways that result in the myriad of end users of legumes is both a matter of scientific curiosity and of economic necessity because of their importance in the biosphere and to the sustainability of the human race Somers (2003).

Advancement of molecular genetics in legumes, e.g. gene over expression, gene suppression, promoter analysis, T-DNA tagging, and expression of genes for crop improvement, requires efficient transformation systems that produce low frequencies of tissue culture-induced phenotypic abnormalities in the transgenic plants. There is a current trend toward increasing the use of *A. tumefaciens* for DNA delivery in crop improvement programs compared with microprojectile bombardment. This is driven by recent development of highly virulent strains and binary vectors that are useful for legume transformation and its ease of use and researcher familiarity. The development of the plant transformation system for Arabidopsis radically accelerated research in basic plant molecular biology (Clough and Bent 1998).

CONCLUSION

Plants are not only the most important producers of natural products including foods, wood, fibres and oils, but also the richest sources of medicinal substances. Now a days it is difficult to secure an ample supply of medicinal plants because of a drastic decrease in plants resources due to human disturbance of the natural environment, ruthless exploitation, increasing labour cost, and technical difficulties in cultivating wild plants. Tissue culture is a bypass method to reduce these difficulties to conserve and cultivate the important medicinal plants and trying to improve their qualities by introducing various methods. It will help to accelerate the conventional multiplication rate which can be a great benefit to many countries where a disease or some climatic disaster wipes out crops. As such, improvement of the medicinal plants through tissue culture is very much important and it has a great value for human kind.

Abbreviations

MS: Murashige and Skoog, CPW: Cell and protoplast washing medium NT Media: Nagata and Takebe Media, ml: millilitre

ACKNOWLEDGEMENT

Authors are thankful to the Director, CSIR-North East Institute of Science and Technology, Jorhat for kind supports and facility.

REFERENCES CITED

Arora S. 2018. Project report on plant tissue culture, www.biologydiscussion.com, retrieved on 19.3.2018.

Boutilier K, Offringa R, Sharma VK, Kieft H, Ouellet T, Zhang L, Hattori J, Liu C-M, van Lammeren AAM, Miki BLA, et al. 2002. Ectopic expression of BABY BOOM triggers a conversion from vegetative to embryonic growth. *Plant Cell* 14:1737–1749.

Brown DCW, Thorpe TA.1995. Crop improvement through tissue culture. *World Journal of Microbiology & Biotechnology* 11: 409-415.

Clough S J, Bent AF.1998. Floral dip: a simplified method for Agrobacterium mediated transformation of Arabidopsis thaliana. *Plant Journal* 16:735-743.

David AS, Deborah AS, Paula MO. 2003. Recent Advances in Legume Transformation .*Plant Physiology* 131(3):892-9.doi.org/10.1104/pp.102.017681.

Desfeux C, Clough SJ, Bent AF .2000. Female reproductive tissues are the primary target of *Agrobacterium*-mediated transformation by the *Arabidopsis* floral-dip method. *Plant Physiology* 123:895–904.

Flavio B, Alexandre S, Guedes C .2013. Traditional and modern plant breeding methods with examples in Rice (*Oryza sativa* L.). *Journal of Agriculture and Food Chemistry* 61 (35): 8277–8286.

Gosal SS, Kang MS. 2012 . Plant tissue culture and genetic transformation for crop improvement Book edited by Satbir Singh Gosal and Shabir Hussain Wani. *Cell and Tissue Culture Approaches in Relation to Crop Improvement* 1-55.

Hinchee MAW, Connor W DV, Newell CA, Mc Donnell RE, Sato SJ, Gasser CS, Fischh off DA, Re DB,Fraley RT, Horsch RB .1988. Production of transgenic soybean plants using *Agrobacterium*-mediated DNA transfer. *Bio Technology* 6:915–922.

Hussain A, Qarshi AI, Hummera N, Ikram U. 2012. Plant Tissue Culture: Current status and Opportunities. *In book: Recent Advances in Plant in vitro Culture* doi.org/10.5772/50568

Kaul B, Stohs S, Staba EJ. 1969. Dioscorea tissue culture influence of various factors on diosgenin production by Dioscorea deltoidea cellus and suspension culture. *Lloydia* 32:347-359.

Kikkert JR, Vidal JR, Reisch BI. 2005. Stable transformation of plant cells by particle bombardment/ biolistics. *Methods Molecular Biology* 286:61-78.

McCabe DE, Swain WF, Martinell BJ, Christou P .1988. Stable transformation of soybean (*Glycine max*) by particle acceleration. *Bio Technology* 6:923–926.

NPTEL.2018. Gene transfer techniques: Physical or mechanical methods. www.nptel.ac.in, retrieved on 21-3-18.

Rani A, Kumar S. 2017. Tissue culture as a plant production technique for medicinal plants: a review. Paper presented at *International conference on innovative research in Science, Technology and Management*. From 22-23 January 2017 at Modi Institute of Management &Technology Dadabari Kota Rajasthan.

Sopory S K, Munshi M. 1996.Anther Culture in vitro haploid production in higher Plants, eds. SM Jain, Sopory SK, Veilleus RE) Kluwer Academic Publishers, Netherlands.145-176.

Tsutomu F, Hisashi K, Kunihiko S, Takafumi I, Kazumichi U, Motohiro N. 1973. Isolation of saponins and sapogenins from callus tissue of Panax ginseng. *Chemical and Pharmaceutical Bulletin* 21:98-101.

Vieira R F, and Skorupa LA.1993. Brazilian medicinal plants gene bank. *Acta Horticulture* 330: 51-58.

21

Aromatic Wealth of Himalaya: Value Addition and Product Development from Essential Oils

Bikarma Singh[1,2*], Sneha[3] and Rajneesh Anand[4]

[1]*Plant Sciences (Biodiversity and Applied Botany), [4]Instrumentation, CSIR-Indian Institute of Integrative Medicine, Jammu 180001, Jammu and Kashmir, INDIA*
[2]*Academy of Scientific and Innovative Research, New Delhi-110001 INDIA*
[3]*CSIR-Central Institute of Medicinal and Aromatic Plant, P.O. CIMAP Lucknow-226015, INDIA.*
*Corresponding Email: drbikarma@iiim.res.in; bikarma81@yahoo.co.in.

ABSTRACT

Plants are the basic requirement upon which all other living organisms depend. Plant resources have been used over the decades for human welfare in promotion of health as medicine, flavour and fragrance. Natural essential oils derived from plants are used as cosmetics, perfumes and flavors in foods and beverages. Development of processsing industries are directly linked to specific needs, available resources and intervention of technological capabilities. Resourgence of public interest in herbal products has created a huge market for plant based products which not only satisfies human needs, but also provides quality and safety insurance. Himalaya range in Southeast Asia is a hub of medicinal and aroma bearing plants globally recognized for their chemical constituents as well as endemic and threatened nature. These plant resources has high demand in international market for development of value added products, for isolation of commercial important compounds used in drug discovery programmes, preparation and formulation for development of new nutraceutical products and semi-synthetic derivatives for medicines. Aromatic plants are valued for their aromas, tastes and their applications in treatment of various illness, and mostly prefered in cosmatics, perfumes, confectionery food items and medicines. More than 3000 aroma bearing plants have been characterized world wide, of which, 116 aromatic plants are recorded from Indian Himalayas. *Lavender, Geranium, Lemongrass* and other aroma bearing plants are the main primary ingredient for many new and existing herbal products. An attempt has been made in this communication to present an account of different aromatic wealth of Himalaya, extraction techniques and their major chemical constituents, which will be helpful to pharmaceutical industries in development of value added products in time to come.

Keywords: Aromatic Wealth, Indian Himalaya, Extraction Techniques, Value Addition, Product development, Aroma Sector

INTRODUCTION

New value added products prepared from natural herbal botanicals are replacing the existing market products on a regular basis due to their best quality improvement from safety point of view. Aromatic plants have been used for centuries in cosmetic, medicine perfumery, and aroma therapy. These plants have been added as a source of value ingredient component to culinary food as part of culinary herbs or spices. As many as 3000 species of essential oil bearing plants have been discovered (Jumaat et al. 2017), of which more than 300 species has applications in pharmaceutical industries (Teixeira et al. 2013). Essential oils are aromatic and volatile liquids extracted from plant materials such as from whole plant samples, woods, barks, leaves, flowers, roots, fruits and seeds. These volatile liquids are considered as secondary metabolites such as alcohols, hydrocarbons, phenols, aldehydes, esters, ketones, and constituents of hundreds of organic components including hormones, vitamins and other natural elements. These secondary metabolites produced by plant is important for plant defence as they possess anti-microbial properties and act as a defence mechanism against insects, infections, as well as attract pollinators.

Recently, more attention has been given to essential oils as a significant natural resource for phytochemicals. The reason is due to their amazing and wide range of potential applications, which is applicable to different fields such as pharmaceutical, food, fragrance and cosmetic industries. In addition, such phytochemicals has non-phytotoxic properties, compared to the synthetic compounds and their application is generally safe, easy decomposition and environmentally friendly.

Aroma bearing crops have added lots of value to the growth of flavor and fragrance industries as well as boasted agriculture. Essential oils are backbone to the process of globalization which started about a decade before, and is now growing rapidly. The usage of aroma products has been a trend among consumers and mostly preferred in cosmatics, perfumes, confectionery food and medicines (Dutra et al. 2016, Jumaat et al. 2017). Industrial data indicates that essential oils of tulsi, citronella, lemongrass, lavender, palmarosa, patchouli, sandalwood, geranium and varieties of mints are finding increasing use in formulations of aroma value added products, and also act as a roadway to therapeutical use in aromatherapy. It is believed that the herbal aromatherapy can provide good effect to body without causing any side effect to human's life. Besides, the usage of aromatic plants in various sector have supported the economy of the country. Perfumes, essential oils and aroma extracts are some of the products which indicate religious values, living standards, personality development for personal use and adornment from years back.

Himalayan Biodiversity and Aromatic Wealth

As per Barthlott et al. (2005) and Jumaat et al. (2017), Himalaya is one of the 'Center of Plant Diversity' recognized world-wide. Composition of biodiversity depends on monsoon rains, which goes upto 10,000 mm rainfall annually, altitudinal zonation consisting of tropical lowland rainforests (100–1200 m above sea level) to alpine meadows (4800–5500 m above sea level). There is a report that Indian Himalaya is home to more than 8000 species of vascular plants, of which 4000 species are endemic and 1748 species are known for their medicinal properties (Hara et al. 1978, Singh and Hajra 1996, Samant et al. 1998).

Total 116 wild aromatic plants species reported from Himalaya is known for their ethno-pharmacology, volatile phytochemistry and biological activities (Joshi et al. 2016). Out of these, 75% (87) species localised their distribution in Jammu and Kashmir State. This include 8 species of gymnosperms and 79 species of angiosperms (Table 1). Pinaceae and Cupressaceae are aroma bearing gymnospermic families, where as Asteraceae, Apiaceae, Araceae, Araliaceae, Aristolochiaceae, Cannabaceae, Caprifoliaceae, Convolvulaceae, Cyperaceae, Fabaceae, Juglandaceae, Lamiaceae, Lauraceae, Myrtaceae, Oleaceae, Piperaceae, Poaceae, Rosaceae, Rutaceae, Solanaceae, Verbenaceae and Zingiberaceae are angiospermic group of families. Member species of Lamiaceae, Asteraceae, Lauraceae, Fabaceae, Rosaceae, Apiaceae and Cupressaceae are dominant in Himalayas.

Table 1: Major essential oil bearing aromatic plant species of Indian Himalayas (Modification after Joshi et al. 2016)

Scientific Name/Family	Major Oil Components
Abies pindrow (Royle ex D.Don) Royle/ Pinaceae	Limonene (21.0%), camphene (19.9%), alpha-pinene (16.8 %), beta-pinene (6.5%), myrcene (6.7%)
Achillea millefolium L. /Asteraceae	beta-Caryophyllene (16.2%), 1,8-cineole (15.1%), beta-pinene (10.6%), borneol (0.2%)
Acorus calamus L./Araceae	Asarone (78.1 %), (E)-asarone (1.9%-4.0%)
Aegle marmelos (L.) Correa/Rutaceae	Limonene (64.1%), (E)-beta ocimene (9.7%), germacrene B (4.7%)

Contd.

Ageratum conyzoides L./Asteraceae	Ageratochromene (42.5%), demothoxya-geratochromene (22.5%), beta-caryophyllene (20.7%)
Ajuga parviflora Benth./ Lamiaceae	Beta-caryophyllene (22.4%), gama-muurolene (12.7%), gama-terpinene (6.3%), caryophyllene oxide (6.2%)
Amomum subulatum Roxb./ Zingiberaceae	1,8-Cineole (60.8%), alpha-terpineol (9.8%), alpha-pinene (6.4%), beta-pinene (8.3%)
Anisomeles indica (L.) Kuntze/Lamiaceae	Isobornyl acetate (64.6%), isothujone (6.0%)
Aralia cachemirica Decne./Araliaceae	Alpha-pinene (41.0%), beta-pinene (35.1%)
Aristolochia indica L./Aristolochiaceae	Trans-pinocarveol (24.4%), alpha-pinene (16.4%), pinocarvone (14.2%)
Artemisia dracunculus L./Asteraceae	Capillene (58.4%), (Z)-beta-ocimene (8.6%), beta-phellandrene (7.0%)
Artemisia dubia Wall. ex Besser/Asteraceae	Chrysanthenone (29.0%), coumarin (18.3%), camphor (16.4%)
Artemisia gmelinii Weber ex Stechm./ Asteraceae	Artemisia ketone (28.2 %), 1,8-cineole (13.0%), sabinene (6.6%)
Artemisia indica Willd./Asteraceae	Ascaridole (15.4%), isoascaridole (9.9%), trans-p-mentha-2,8-dien-1-ol (9.7%), trans-verbenol (8.4%)
Artemisia japonica Thunb./Asteraceae	Linalool (27.5%), germacrene D (11.2%), (E)-beta-ocimene (6.5%), 1,8-cineole (5.5%)
Artemisia maritima L./Asteraceae	Alpha-thujone (63.3%), sabinene (7.8%), 1,8-cineole (6.5%)
Artemisia nilagirica (C.B.Clarke) Pamp./Asteraceae	Camphor (12.6%), artemisia ketone (10.2%), caryophyllene oxide (7,4%), borneol (5.3%)
Artemisia parviflora Buch.-Ham. ex D.Don /Asteraceae	Germacrene D (41.01%), beta-caryophyllene (10.58%), alpha-humulene (7.86%)
Artemisia roxburghiana Besser/Asteraceae	beta-Thujone (65.3%)
Artemisia scoparia Waldst. & Kit./ Asteraceae	Capillene (42.1%), beta-caryophyllene (12.5%), myrcene (9.2%), beta-pinene (8.6%)
Artemisia vulgaris L./Asteraceae	alpha-Thujone (30.5%), 1,8-cineole (12.4%), camphor (10.3%)
Blumea lacera (Burm.f.) DC./Asteraceae	(Z)-Lachnophyllum ester (25.5%), (Z)-lachnophyllic acid (17.0%), germacrene D (11.0%)
Boenninghausenia albiflora (Hook.) Rchb. ex Meisn./Rutaceae	Germacrene D (4.2%-18.2%), beta-caryophyllene (4.6%-13.1%)
Callistemon citrinus (Curtis) Skeels / Myrtaceae	1,8-Cineole (52.1%), alpha-terpineol (14.7%), eugenol (14.2%)
Cannabis sativa L. /Cannabinaceae	beta-Caryophyllene (20.4%), alpha-humulene (7.0%), alpha-bisabolol (5.8%)
Carum carvi L. /Apiaceae	Carvone (65.8%-78.8%), limonene (19.4%-31.6%)
Cassia fistula L. /Fabaceae	Eugenol (25.0%), (E)-phytol (21.5%0, camphor (13.5%), limonene (11.0%), salicyl alcohol (10.4%)
Cassia tora L./ Fabaceae	Elemol (26.9%), linalool (19.6%), palmitic acid (15.3%)
Cedrus deodara (Roxb. ex D.Don) G. Don /Pinaceae	α-Himachalene (38.3%), β-himachalene (17.1%), γ-himachalene (12.6%)

Contd.

Centella asiatica (L.) Urb./Apiaceae — Isocaryophyllene(9.2%–32.3%), β-caryophyllene (7.5%–24.5%), α-humulene (0.1%–17.1%),(E)-β-farnesene (1.7%–18.9%)

Chaerophyllum villosum Wall. ex DC. /Apiaceae — γ-Terpinene (74.9%), p-cymene (10.0%), carvacrol methyl ether (31.1%), myristicin (19.1%), thymol methyl ether (18.6%), γ-terpinene (11.7%)

Chenopodium ambrosioides L. / Amaranthaceae — α-Terpinene (8.3-44.7%), p-cymene (21.3%–27.1%), ascaridole (17.9%–45.0%)

Chrysanthemum cinerariifolium (Trevir.) Vis./Asteraceae — Camphor (11.0%), chrysanthenone (7.6%), α-cadinol (4.8%),γ-muurolene (4.6%) and cis-chrysanthenol (4.4%)

Cinnamomum glanduliferum (Wall.) Meisn. /Lauraceae — 1,8-Cineole (41.4%), α-pinene (20.3%),α terpineol (9.4%), germacrene D-4-ol (6.1%) and α-thujene (5.10%)

Cinnamomum camphora (L.) J. Presl./ Lauraceae — Camphor (82.4%), camphor (81.5%), camphor (36.5%), camphene (11.7%), limonene (9.0%), sabinene (6.3%), β-pinene (6.3%)

Cinnamomum glaucescens Hand.-Mazz. /Lauraceae — Methyl (E)-cinnamate (40.5%) 1,8-cineole (24.8%), α-terpineol (7.4%), elemicin (92.9%)

Cinnamomum tamala (Buch.-Ham.) T. Nees and Nees / Lauraceae — Camphor (35.0%), linalool (10.6%), p-cymene (8.5%), o-cymene (6.8%), and 1,8-cineole (6.1%), (E)-cinnamaldehyde (79.4%), (E)-cinnamyl acetate (3.7%), linalool (5.4%), linalool (52.5%), (E)-cinnamaldehyde (26.4%), 1,8-cineol (4.2%)

Curcuma angustifolia Roxb./ Zingiberaceae — Xanthorrhizol isomer (12.7%), methyl eugenol (10.5%), camphor (21.3%), germacrone (12.8%)

Curcuma longa L./ Zingiberaceae — Rhizome: alpha-Turmerone (30–32%), β-turmerone (15%–18%). Leaf: α-phellandrene (18.2%), 1,8-cineole (14.6%), p-cymene, α-turmerone (44.1%)

Cuscuta reflexa Roxb./ Convolvulaceae — cis-3-Butyl-4-vinylcyclopentane (26.4%), limonene (5.1%), (E)-nerolidol (9.5%)

Cymbopogon distans (Nees ex Steud.) Watson /Poaceae — Leaves: Alpha-oxobisabolene (68%), Geraniol (9.5%), geranyl acetate (15.0%), α-terpinene (24.9%), piperitone (45.3%), cis-p-menth-2-en-1-ol (22.7%), trans-p-menth-2-en-1-ol (10.8%), cis-piperitol (13.0%), trans-piperitol (5.6%)

Dodecadenia grandiflora Nees / Lauraceae — Germacrene D (26.0%), furanodiene (13.7%)

Elsholtzia flava (Benth.) Benth. / Lamiaceae — Piperitenone (30.8%), carvacrol (4.8%), (Z)-anethole (4.4%), γ-elemene (4.8%)

Eryngium foetidum L. Apiaceae — (E)-2-Dodecenal (58.1%), dodecanal(10.7%), 2,3,6-trimethylbenzaldehyde (7.4%), (E)-2-tridecenal (6.7%)

Eupatorium adenophorum Spreng. / Asteraceae — p-Cymene (16.6%), bornyl acetate (15.6%), amorph-4-en-7-ol (9.6%), camphene (8.9%), 1-naphthalenol (17.5%), α-bisabolol (9.5%), bornyl acetate(9.0%)

Ficus religiosa L./ Moraceae — Eugenol (27.0%), itaconic anhydride (15.4%),3-

Contd.

methyl-cyclopenetane-1,2-dione (10.8%), 2-phenylethyl alcohol (8.0%)

Filipendula vestita (Wall. ex G.Don)
Maxim./ Rosaceae

Methyl salicylate (56.0%), salicaldehyde (15.6%), santene (9.4%)

Gualtheria fragrantissima Wall. / Ericaceae

Methyl salicylate (94.2%)

Hedychium spicatum Buch.-Ham.
ex Sm. /Zingiberaceae

1,8-Cineole (15.5%–58.2%), linalool (0.8%–10.6%), terpinen-4-ol (0.7%–15.2%), elemol (0.7%–16.6%),10-epi—eudesmol (0.2%–13.9%), _α-cadinol (4.5%–11.2%), α-pinene (9.6%), _-pinene (40.9%), 1,8-cineole(11.9%). Root: β-pinene (8.9%), 1,8-cineole (48.7%), α-terpineol 11.8%)

Inula cappa (Buch.-Ham. ex D.Don)
DC./ Asteraceae

β-Caryophyllene (27.5%), cis-dihydro-mayurone (6.7%), _-bisabolene (6.5%), (E)-farnesene (5.6%)

Jasminum mesnyi Hance / Oleaceae

Coumarin (48.9%), linalool (14.8%)

Juglans regia L. /Juglandaceae

α-Pinene (15.1%), β-pinene (30.5%), β-caryophyllene (15.5%) germacrene D (14.4%), limonene (3.6%), eugenol (27.5%), methyl salicylate (16.2%), germacrene D (21.4%), (E)- β-farnesene (8.2%), germacrene D (16.1%–22.1%), β-caryophyllene (10.4%–13.5%), -copaene (6.5%–10.1%)

Juniperus communis L./ Cupressaceae

α-pinene (35.4%), limonene (23.8%), α-pinene (10.8%), limonene (15.1%), terpinene-4-ol (8.8%)

Juniperus indica Bertol./ Cupressaceae

Sabinene (19.4%–31.3%), terpinen-4-ol (3.7%–13.0%), β-thujone (4.5%–25.8%), trans-sabinyl acetate (7.6%–24.3%)

Juniperus macropoda Boiss. /
Cupressaceae

Sabinene (27.5%), terpinen-4-ol (9.4%), cedrol (14.1%), α-elemene (42.5%) trans-sabinene hydrate (8.8%), α-cubebene (7.9%)

Juniperus recurva Buch.-Ham. ex
D. Don /Cupressaceae

3-Carene (13.6%), γ-cadinene (10.2%), γ -cadinol (5.5%), γ-muurolol (5.5%), γ-cadinol (13.1%)

Kyllinga brevifolia Rottb./Cyperaceae

α-Cadinol (40.3%), γ-muurolol (19.5%), germadrene D-4-ol (12.5%)

Lantana camara L. /Verbenaceae

Ermacrene D (27.9%), germacrene B (16.3%), β-caryophyllene (9.6%) 3,7,11-trimethyl-1,6,10-dodecatriene (28.9%), β-caryophyllene (12.3%), zingiberene (7.6%), γ-curcumene (7.5%)

Lawsonia inermis L. /Lythraceae

Limonene (20.0%), (E)-phytol (27.5%), linalool (7.0%),1,8-cineole (6.9%), β-pinene (18.1%), p-cymene (14.7%), 1,8-cineole (58.6%)

Leucas aspera (Willd.) Link /Lamiaceae

1-Octen-3-ol (30.6%), β-caryophyllene (23.4%), caryophyllene oxide (24.4%)

Lindera neesiana (Wall. ex Nees)
Kurz /Lauraceae

Geranial (15.1%), neral (11.9%), citronellal (6.7%), 1,8-Cineole (8.8%), α-pinene (6.6%), β-pinene (5.6%)

Lindera pulcherrima (Nees)
Hook.f. /Lauraceae

Curzerenone (17.6%), furanodienone (46.6%)

Matricaria recutita L./Asteraceae

(E) β-Farnesene (44.2%), α-bisabolol oxide A

Contd.

	(22.3%), (E,E)-α-farnesene (8.3%)
Mentha piperita L. /Lamiaceae	Menthol (22.6%–42.8%), menthone (0.8%–33.8%), menthyl acetate (0.6%–32.8%), myrcene (0.0%–15.5%), 1,8-cineole (2.3%–13.9%), menthofuran (0.0%–17.9%)
Mentha arvensis L. /Lamiaceae	Menthol (61.9%–82.2%), menthone (3.6%–19.3%)
Mentha longifolia (L.) Huds. /Lamiaceae	Carvone (61.1%–78.7%), dihydrocarveol (0.4%–9.5%), cis-carvyl acetate (0.2%–6.4%), germacrene D (1.3%–5.7%), piperitenone oxide (54.2%), trans-piperitone oxide (24.1%), cis-piperitone oxide (7.0%)
Mentha spicata L. /Lamiaceae	Carvone (76.7%), limonene (9.6%)
Morina longifolia Wall. ex DC. / Caprifoliaceae	Germacrene D (10.8%), α-pinene (4.8%), bicyclogermacrene (4.3%), β-cadinol (4.3%), (E)-citronellyl tiglate (4.2%) β-phellandrene (3.2%)
Murraya koenigii (L.) Spreng. /Rutaceae	α-Pinene (51.7%), sabinene (10.5%), β-pinene (9.8%)
Nardostachys grandiflora DC. / Caprifoliaceae	β-Gurjunene (9.4%), valerena-4,7(11)-diene (7.1%), nardol A (6.0%), 1(10)-aristolen-9 β-ol (11.6%), jatamansone (7.9%)
Neolitsea pallens (D.Don) Momiy. & Hara /Lauraceae	Furanogermenone (59.5%), β-caryophyllene (6.6%)
Nepeta clarkei Hook.f. /Lamiaceae	Iridodial β-monoenol acetate (25.3%),-sesquiphellandrene (22.0%), germacrene D (13.0%), α-guaiene (10.0%), kaur-16-ene (36.6%), pimara-7,15-dien-3-one (19.7%), caryophyllene oxide (14.1%)
Nepeta discolor Royle ex Benth. / Lamiaceae	1,8-Cineole (25.5%), β-caryophyllene (18.6%), p-cymene (9.8%)
Nepeta elliptica Royle ex Benth. / Lamiaceae	(7R)-Trans-nepetalactone (83.4%), β-elemene (23.4%), α-humulene (11.8%), bicyclogermacrene (13.1%)
Nepeta govaniana (Wall. ex Benth.) Benth. / Lamiaceae	Isoiridomymecin (35.2%), pregeijerene (20.7%), prejeijerene (38%), geijerene (6.8%), pregeijerene (56.9%), germacrene D (9.4%), β-caryophyllene (6.1%), torreyol (5.1%)
Nepeta juncea Benth./ Lamiaceae	Nepetalactone (71.8%)
Nepeta laevigata (D.Don) Hand.-Mazz. / Lamiaceae	Citronellol (16.5%), β-caryophyllene (10.8%), germacrene D (19.4%), α-bisabolol oxide B (12.4%), 1,8-cineole (11.1%), β-caryophyllene (5.7%), caryophyllene oxide (15.2%), manool (7.9%)
Nepeta leucophylla Benth./ Lamiaceae	Iridodial β-monoenol acetate (25.4%), dihydroiridodial diacetate (18.2%), iridodial dienol diacetate (7.8%)
Nepeta raphanorhiza Benth./ Lamiaceae	(Z)- α-Farnesene (49.2%), β-3-carene (12.3%), γ-bisabolene (9.4%), germacrene D-4-ol (5.8%)
Nepeta royleana Stewart /Lamiaceae	1,8-Cineole (75%)

Contd.

Nepeta spicata Wall. ex Benth. /Lamiaceae β-Caryophyllene (27.0%), linalool (25.1%) germacrene D (20.1%), caryophyllene oxide (10.6%)

Nyctanthes arbor-tristis L. /Oleaceae Linalool (11.3%), (3Z)-hexenyl benzoate (11.0%) palmitic acid (26.4%), (E)-phytol (13.6%) β-eudesmol (17.1%), á-eudesmol (8.7%), palmitic acid (34.3%)

Ocimum basilicum L. / Lamiaceae Linalool (50.8%–58.3%), geraniol (5.2%–13.7%) eugenol (0.0%–19.1%), γ-cadinol (5.1%–5.9%) 1,8-cineole (0.8%–7.3%)

Origanum vulgare L. / Lamiaceae Thymol (53.2%), p-cymene (10.3%), carvacrol (3.9%), thymol (82.0%), myrcene (14.2%) α-humulene (9.2%), γ-terpinene (21.8%), linalool (13.1%)

Perovskia abrotanoides Kar./ Lamiaceae α-Pinene(18.2%–23.2%), 1,8-cineole (24.4% – 27.1%), borneol (7.9%–10.4%), β-caryophyllene (5.7%–12.3%), γ-3-carene (4.7%–9.3%)

Persea duthiei (King) Kosterm. / Lauraceae α-Pinene (10.0%), β-pinene (10.0%), limonene (10.1%), (E)-nerolidol (13.2%)

Persea duthiei (King)Kosterm./ Lauraceae α-Pinene (10.0%), _-pinene (10.0%), limonene (10.1%), (E)-nerolidol (13.2%)

Persea gamblei (King ex Hook.f.) Kosterm. /Lauraceae β-Caryophyllene (22.1%), γ –gurjunene (16.8%)

Persea odoratissima (Nees) Kosterm Lauraceae α-Pinene (16.6%), sabinene (13.1%), β-caryophyllene (10.4%), (E)-nerolidol (13.2%)

Phoebe lanceolata (Nees) Nees / Lauraceae 1,8-Cineole (18.2%), β-caryophyllene (27.4%)

Pinus roxburghii Sarg. / Pinaceae beta-Caryophyllene (31.7%), terpinen-4-ol (30.1%), α-humulene (7.3%), eugenol (11.4%) linalool (6.4%), terpinen-4-ol (16.2%), γ-3-carene (6.8%)

Piper betle L. / Piperaceae Chavibetol (80.5%), chavibetol acetate (11.7%) allylpyrocatechol diacetate (6.2%)

Pleurospermum angelicoides (Wall. ex DC.) Benth. ex C.B. Clarke / Apiaceae Nothoapiole (87.3%), limonene (48.4%), α-asarone (23.2%), γ -terpinene (11.0%), α-pinene (22.3%), α-asarone (20.7%), perilla aldehyde (16.8%), limonene (14.8%) [

Rhododendron anthopogon D.Don / Ericaceae α-Pinene (37.4%), β-pinene (16.0%), limonene (13.3%), γ-cadinene (9.1%)

Selinum tenuifolium Salisb. / Apiaceae Nona-3,5-diyne (85.6%), α-bisabolol (71.8%)

Senecio nudicaulis Buch.-Ham. ex D.Don / Asteraceae Caryophyllene oxide (25.0%), humulene epoxide-II (21.3%), α-humulene (18.8%), β-caryophyllene (9.7%)

Senecio rufinervis DC. / Asteraceae Germacrene D (33.7%), γ-cadinene (5.5%) γ-cadinene (5.5%), germacrene D-4-ol (5.4%) germacrene D (32.9%), germacrene A (19.5%) γ-elemene (7.6%)

Skimmia anquetilia Tayl. & Airy Shaw / Germacrene B (11.6%), linalool (9.5%), linalyl

Contd.

Rutaceae	acetate (7.3%), α-bisabolol (7.2%), β-gurjunene (6.6%), β-phellandrene (18.6%), geijerene (15.1%) linalyl acetate (11.2%), linalool (9.4%)
Skimmia laureola (DC.) Decne. / Rutaceae	Linalyl acetate (33.0%), linalool (25.0%), limonene (8.1%), α-terpineol (5.9%), geranyl acetate (5.9%) linalool (34.9%), linalyl acetate (26.7%) α-terpineol (12.8%), geranyl acetate (6.6%)
Solanum xanthocarpum Schrad. & J.C. Wendl. / Solanaceae	Benzyl benzoate (21.7%), (E,E)-geranyllinalool (12.6%), heptacosane (20.0%), (E)-phytol (8.4%) solavetivone (22.9%)
Stachys sericea Wall. ex Benth. / Lamiaceae	Germacrene D (37.7%), β-caryophyllene (17.4%) γ-cadinene (6.0%)
Tanacetum gracile Hook.f. & Thomson / Asteraceae	Lavendulol (21.5%), 1,8-cineole (15.2%), (Z)-β-ocimene (6.4%), α-bisabolol (28.0%) chamazulene (8.4%), α-phellandrene (6.9%)
Tanacetum longifolium Wall. ex DC. / Asteraceae	trans-Sabinyl acetate (43.2%) and trans-sabinol (12.7%)
Tanacetum nubigenum Wall. ex DC. / Asteraceae	cis-Chrysanthenol (37.0%), sabinene (10.7%), cis-chrysanthenyl acetate (5.8%), cis-chrysanthenyl isobutyrate (5.7%), bornyl acetate (39.7%), borneol (10.6%), (E)-β-farnesene (6.6%), 1,8-cineole (5.8%), 1,8-cineole (30.0%), sabinene (15.6%), eudesmol (11.2%), camphor (8.0%)
Thuja orientalis L./ Cupressaceae	α-Pinene (29.2%), γ-3-carene (20.1%), α-cedrol (9.8%), β-caryophyllene (7.5%), α-humulene (5.6%)
Thymus linearis Benth. / Lamiaceae	Thymol (38.4%), carvacrol (30.7%), terpinene (10.1%), geraniol (67.8%), geranyl acetate,
Thymus serpyllum L./ Lamiaceae	Thymol (16.5-18.8%), 1,8-cineole (14.0%–18.0%), thymol (19.4%–60.1%), -terpinene (0.3%–13.8%) and p-cymene (3.5%–10.4%)
Valeriana hardwickii Wall. / Caprifoliaceae	Bornyl acetate (11.2%), cuparene (7.1%), valeracetate (11.6%), methyl linoleate (21.1%), methyl linoleate (11.7%), cuparene (10.4%), α-cedrene (6.2%), kessanyl acetate (22.2%), maaliol (13.4%), bornyl acetate (7.4%), β-gurjunene (5.4%)
Valeriana jatamansi Jones /Caprifoliaceae	Maaliol (64.3%), viridiflorol (7.2%), β-gurjunene (7.2%), seychellene (17.6%), α-santalene (8.7%), patchouli alcohol (40.2%), α-bulnesene (10.7%), seychellene (8.2%), viridiflorol (5.2%)
Vitex negundo L./ Verbenaceae	Ethyl 9-hexadecenoate (28.5%), α-bulnesene (18.0%), caryophyllene oxide (10.2%), β- Caryophyllene (5.0%)
Zanthoxylum armatum DC./ Rutaceae	Linalool (55.3%), limonene (22.5%), methyl cinnamate (8.8%), 2-undecanone (55.7%), linalool (11.5%), β-caryophyllene (4.6%), 1,8-cineole (4.3%), 2-decanone (20.9%), 2-tridecanone (8.9%), β-fenchol (9.4%), β-phellandrene (6.0%)

Aroma Bearing Parts of Plants

In general, essential oils presents in minute quantities in plant material is generally less than 5% of dry matter. Essential oils are produced in tiny secretary structures found in various parts of plant such as leaves (e.g. *Eucalyptus*, *Ocimums*, Mints, *Tagetes*), berries (e.g. *Juniperes*), needles (Pines, *Cedrus*, Junipers), grasses (e.g. Lemongrass, Rosagrass, Tawirosa, Palmarosa), flowering twigs (e.g. *Lavender*, *Jammu Monarda*, *Cannabis*), petals (e.g. Rose), roots (e.g. *Angelica*, *Vetivers*, *Acorus*), zest of fruit (e.g. Orange), resins (e.g. Boswellia, Frankincense), Wood (e.g. *Cedar*, Pine).

Aroma Colours

Red (Clove, Cinnamon, Rose, Cedar, Musk patchouli, Sandalwood), Orange (Bergamot, Geranium, Jasmine, Neroli, Sweet Orange, Bitter Orange and Patchouli), Yellow (Lemon verbena, Lemon grass, Vetiver, Angelica, Jasmine), Light blue (Clarysage, Eucalyptus).

TECHNIQUES FOR EXTRACTION OF ESSENTIAL OILS

Quantities, chemical composition and aroma-active ingredients in aromatic plants depend on the method of extraction and solvent used (Cowan 1999). Natural essential oils typically obtained by distillation, and having the characteristic odour of the plants, or sources from which they are extracted. EOs is concentrated hydrophobic liquid containing volatile aroma compounds from plants and therefore, steam can helps a lot in separation.

The most common conventional methods that are employed for essential oils extraction are hydro-distillation and steam distillation. The general concept in using these methods is to heat up the plant materials using the boiling water or hot water steam to allow the aromatic volatile compounds, mainly essential oils, to escape from the plant matrix; then, collecting the evaporated compounds by the condensation process. Although these methods are simple and cost effective, however, the heating process could modify and break up the chemical structure of such mixture of valuable compounds, which negatively affect the quality of the extracts. The needed amount from plant material is usually high comparing to the obtained essential oils. For example, at laboratory scale, there is need of around 1 kg of dry or fresh plant matter to produce, may be one drop of essential oils or 1 or 2 ml volatile oil . In addition, this process consumes a lot of time and includes many of impurities, which make such conventional ways inefficient in terms of extracts quality, operating time, efforts, and resources conservation (Al-Marzouqi et al. 2007, Pourmortazavi and Hajimirsadegh 2007).

Fig. 1: Extraction of essential oil in laboratory through hydro-distillation technique

Recently, different extraction methods intensively explored to extract the best quality production from essential oils. Supercritical fluid CO_2 is an innovative, simple, clean, fast, selective and environmental friendly technology to extract essential oils. Many studies reveals that this new technology is capable to extract essential oils with superior qualities comparing to other conventional techniques such as hydro-distillation and steam distillation. It can be expected that there would be flourish future for the application of such technology and essential oils for new product developments.

Percentage of essential oil (%) = (Weight of Oil (g)/Weight of plant sample Used)*100

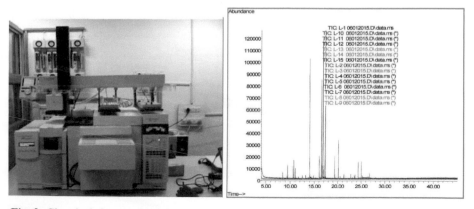

Fig. 2: Chemical characterization and analysis of essential oils through FID and MS Detectors

Essential Oil Markets and Future Perspectives

Essential oils (EOs) and their components extracted from plants accomplishing increasing interest in the food, cosmetic and pharmaceutical industries. The growing interest of consumers in substances of natural origin has resulted in the use of aromatic plants, their extracts and essential oils, as functional ingredients in the pharmaceutical, food and botanical industries. More than 3000 plants have been identified as potential oil bearing plants distributed in different agroclimatic zones, and out of which about 400 species have high demand in global market. Less than 50% of the aromas bearing plants are cultivated, and remaining are harvested from wild plants. China, France, Germany, Italy, Japan, Spain, United Kingdom and United States of America are major consumers of essentials oils. In USA, the market of essential oil is 4 Billion US $ or 16,000 Crores INR, where as Indian Fragrance and Flavour Market is only 225 Million US $. India are major exporter of spice oleoresins.

Major Essential Oil Bearing Plants Cultivated in CSIR-IIIM Jammu

Currently, CSIR-Indian Institute of Integrative Medicine Jammu has developed agrotechnology of high yielding varieties of aroma bearing crops, and involved in captive cultivation and extension of these crops in J&K state and elsewhere in India. However, the mandate of this institute is to discover new drugs, therapeutic approaches from natural products, both of plant and microbial origin,

Lemongtass CKP-25   Lemongrass CPK-F2-38   Lemongrass -Krishna

Rosagrass RRL(J) CN-5   Rosagrass IIIM(J)CK-10, Himrosa   Palmarosa-Harsh

Photoplate 1: Aroma bearing plants in CSIR-IIIM Jammu experimental farm enabled by biotechnology, to develop technologies, drugs and products of high value for the national and international markets

Besides, CSIR-IIIM Jammu has four farms located at different regions of J&K where germplasm of large numbers of medicinal and aromatic plants are being maintained and cultivated for commercial exploitation. These farms are situated at Jammu (Chattha farms ~ 50 acres), Pulwama (Bonera farms ~ 150 acres), Verinag (Verinag farms ~ 5 acres) and Gulmarg (Yarikha farms ~ 25 acres). Several new varieties of high valued aromatic plants have been developed by CSIR-IIIM. Quality planting material of these varieties is provided to the farmers throughout the country for cultivation and for their socio-economic upliftment (Photoplate 1). Table 2 provides various essential oil yielding plants available at IIIM as cultivated or wild species in major and minor scale having potential for future deveopment of value added products.

Table 2: Essential oil bearing plants in CSIR-IIIM Jammu

Scientific Name	Family	Habit	Major oil Components
Acorus calamus L.	Araceae	Herb	Asarone (78.1 %), (E)-asarone (1.9%-4.0%)
Aegle marmelos (L.) Correa	Rutaceae	Tree	Limonene (64.1%), (E)-beta-ocimene (9.7%), germacrene B (4.7%)
Ammomum subulatum Roxb.	Zingiberaceae	Herb	1,8-Cineol, limonene, terpinene, terpineol, terpinyl acetate, sabinene
Artemisia dracunculus (L.) L.	Asteraceae	Herb	Estragole (Methyl chavicol)
Artemisia maritima L.	Asteraceae	Herb	1,8-Cineole, chrysanthenone
Artemisia vestita Wall.	Asteraceae	Herb	Grandisol, camphor
Artemisia annua	Asteraceae	Herb	Artimisinin
Artimisia vulgaris L.	Asteraceae	Herb	Thujone
Cinnamomum camphora Nees.	Lauraceae	Shrub	Camphoric acid, cineole, cymene, dipentine, eugenol, phellandrene, pinene, safrole, sesquiterpenes, terpineole
Coleus ambonicus Lour.	Lamiaceae	Herb	Carvachrol, caryophylene, patchoulane
Cymbopogon commulatus (Steud.) Stapf (Tawi Rosa RRL-CC-1	Poaceae	Herb	Geraniol (78-85%)
Cymbopogon flexuosus (Nees ex Steud) W.Watson	Poaceae	Herb	Citral
Cymbopogon jawarncusa (Jones) Schult × *C. nardus* var. conferiflorus	Poaceae	Herb	Geraniol
Cymbopogon khasianus (Hack) Stapf. ex Bor (Rosa grass "IIIM(J)CK-10")	Poaceae	Herb	Geraniol (71.88%), cis-Ocimene (10.15%), geranyl acetate (71.88%)

Contd.

Cymbopogon khasianus (Hack.) Stapf. ex Bar × *C. pendulus* (Nees ex Steudel) (Lemon grass "CKP-25")	Poaceae	Herb	Beta-Citral (80-85%), Cis-Ocimene (5%), Geraniol butyrate
Cymbopogon nardus (L.) Rendle (Rosagrass "RRL(J) CN-5")	Poaceae	Herb	Geraniol (45-60%), Geranyl acetate (15-25%), Cis-Ocimene (12-13%)
Cymbopogon pendulus (Hack.) Stapf. ex Bar × *C. khasianus* (Nees ex Steudel) (Lemon grass "CPK-F2-38")	Poaceae	Herb	Citral (75-80%)
Eucalyptus citridora Hook.	Myrtaceae	Tree	Citronellal
Geranium dissectum L.	Geraniaceae	Herb	α-Pinene, Myrcine, Limonene, Menthone, Linalool, Citranellol, Geranial, Geranyl butyrate
Lavendula angustifolia Mill. (Kashmir lavender RRL-12)	Lamiaceae	Herb	Linalool (36.24%), Linalyl acetate (32.26%), others includes 1,8-Cineole , Borneol, Caryophyllene, Terpineol, Ocimenes and Lavandulyl acetate
Mentha × *citrata* Enrich.	Lamiaceae	Herb	Linalool (42.91%), Linalyl acetate (25.15%)
Mentha longifolia (L) Hudson var. *incana* (Willd) Dinson (Anant carvomint RRL(J)ML-4)	Lamiaceae	Herb	L-Carvone, Limonene
Mentha piperita L.	Lamiaceae	Herb	Menthol (44.45%), Menthone (27.46%), Carvone (3.95%)
Mentha spicata L.	Lamiaceae	Herb	Limonene (25.96%), Carvone (49.45%), beta-Bourbonene (3.32%)
Monarda citriodora Cerv. ex Lag. (Jammu Monarda IIIM(J) MC-02)	Lamiaceae	Herb	Thymol (65.39%), para-Cymene (9.17%), Carvacrol (4.2%)
Ocimum americanum L.	Lamiaceae	Shrub	Citral, Linalool, Geraniol, Citronellol
Ocimum basilicum var. *glabratum* L. (Tulsi RRL-OB-15)	Lamiaceae	Shrub	Methyl chavicol (85-90%)
Ocimum carnosum (Spreng) Link & Otto ex Benth.	Lamiaceae	Shrub	Elemicin
Ocimum gratissimum L. var. *clocimum* (Tulsi RRL-OB-14)	Lamiaceae	Shrub	Eugenol (80-85%)
Pelargonium graveolens (Rose scented Geranium PG-IIIM-101)	Gerinacieae	Herb	Citronellol (25.77%), Geraniol (20.81%), Linalool (11.94%), Citronellyl formate (8.04%) and isomenthone (7.58%).
Psidium guajava L.	Myrtaceae	Tree	1-8-Cineole, 2-3-4-6 tetra-0-galloyl glucose
Rosa damascena L. (Kashmir Rose)	Rosaceae	Shrub	Citronellol (54.92%), Geraniol (12.72%)

Contd.

Salvia officinalis L.	Lamiaceae	Herb	Cineole, tannic acid, oleic acid, ursonic acid, cornsolic acid, fumaric acid, chlorogenic acid, caffeic acid, niacin, nicotinamide
Salvia sclarea L.	Lamiaceae	Herb	Linalool, linalyl acetate, caryophyllene, α- Terpineol, geraniol, neryl acetate, sclareol, germacrene
Tanacetum vulgare L.	Asteraceae	Herb	Thujone , isopinocamphone, camphor, Borneol, camphene, artemisone
Thymus vulgare L.	Lamiaceae	Herb	a-thujone, a-pinene, camphene, α-terpinene, linalool, borneol, thymol, carvacrol.
Valeriana Jatamansi Jones	Caprifoliaceae	Herb	Maalioxide, 2-Acetyl pyrrol, 8-Epikessanol, pyrrol-Ketone, Caffeic acid, Capronic acid, linolenic acid, valerenolic acid.

Value Added Product from Essential Oils of IIIM Jammu

Keeping the pre-historic importance of aroma bearing crops, Scientists of IIIM Jammu has developed a aroma value added product **"Nature Fresh...Aroma Value Kit....Sense of Living"**. This aroma kits are of two kinds: 6 ml capacity and 3 ml capacity. The 6 ml capacity aroma kit contains five types of essential oils, viz., mint oil, lavender oil, lemongrass oil, himros oil and jammu monarda oil. The 3 ml capacity aroma kit has six types of essential oils, such as rosemary oil, mint oil, lavender oil, lemongrass oil, himrosa oil and jammu monarda oil. The essential oils bears the name of the plant from which it was extracted. The crop lavender and rosemary are high altitude high value crops, where as a mint, lemongrass, himrosa and jammu monarda crops are usually cultivated in tropical and sub-tropical belts only (Photoplate 2). The detail of aromatic species used in the preparation of aroma value kit is given below:

1. Jammu Monarda (*Monarda citriodora* Cerv. ex Lag. [IIIM(J)MC-02]: Developed and released for commercial cultivation on the occasion of Foundation day on 1st December 2010. *Characteristics:* A rich source of thymol (about 60-75%) and better economic returns (100-125 kg/ha). Essential oil active against cancerous cell line (HL60), acceptability by flavour and pharmaceutical industries. It is also used as an anti-septic, expectorant and cough medication, to treat nail fungus infection.

2. Himrosa (*Cymbopogon khasianus* Bor) [IIIM (J) CK-10]: Developed and released by CSIR-IIIM Jammu on 1st December, 2012 on the occasion of IIIM(J) Foundation day 2012. *Characteristics*: Having high drought and salt tolerance ability, and is a rich source of Geraniol (75-85 %). The

oil is extensively used as perfumery raw material in soaps, "oral rose-like perfumes, cosmetics preparations, and also used in manufacture of mosquito repellent products. The essential oil has a scent similar to that of rose oil, and named Himrosa.

| Lavender (RRL-12) | Mint | Jammu Monarda |
| Vetiver | Java citronella | Patcholi |

Photoplate 2: Aroma bearing plants growing in CSIR-IIIM campus as potential plants for value addition.

3. Anant carvomint *(Mentha longifolia* (L.) Hudson var. *incana* (Willd) Dinson) [RRL(J)ML-4]: Released for commercial cultivation on 30th January, 2007. *Characteristics:* Hyper productive strain developed through clonel multiplication, wider adaptability (Kashmir to Kanya Kumari). It is in rich in L-carvone 67±5 %). It has essential oil content (0.5 to 0.9%w/w FWB) depending on season to season. It has wider acceptability shown by perfume, flavour and pharmaceutical industries.

4. Lemon grass (*Cymbopogon khasianus* x *C. pendulus*) [CKP-25]: Released in 2002. *Characteristics*: It is an interspecific hybrid and the oil content is 0.5%. Its main constituents are citral (80-85%), Cis-ocimet and geraniol butyrates. It is very useful in perfumery, flavoring and pharmaceutical industry.

5. Lavender (*Lavandula officinales*) [RRL 12]: Released in 1972. *Characteristics*: Lavender is a high altitude high-valued crop, and is an

incredible and much sought aromatic plant having significant position in trade all over the world due to its essential oil which has multifarious uses and market outlets. Main constituents are linalool, linalyl acetate, 1,8 cineole, borneol, caryophyllene, terpineol, ocimenes, lavandulyl acetate. It is useful in perfumery, flavor and cosmetic industry.

6. Rose Mary (*Rosmarinus officinalis* L.), *Characteristics*: Rose Mary is yet another high altitude, high valued crops and its main constituents are p-cymene (40-44.02%), linalool (18-20.5%), gamma-terpinene (14-16.62%), thymol (1-1.81%), beta-pinene (2-3.61%), alpha-pinene (1-2.83%) and eucalyptol (1-2.64%), which is very useful in perfumery, flavor and cosmetic industry.

CONCLUSION

Research contribution reveals the value of use of aromatic plants, and thereby integrating essential oils and justifying their therapeutic applications, agricultural scientists and botanist have been motivated to expand cultivations. Now this has created a huge competitive global market. Essential oils are expensive phytochemicals produced and extracted from specific species belonging to particular families in the plant kingdom. Volatile compounds are characterized by a strong aroma, produced by plant materials as secondary metabolites, and acting as defense chemicals. This aroma is the result of complex interactions between hundreds of compounds which consist mainly of terpenes, oxygenated terpenes, sesquiterpenes and oxygenated sesquiterpenes (Reverchon 1997, Burt 2004, Wink 2003, Rios et al. 1988). High altitude plants has played key roles in the lives of tribal peoples living in the Himalaya by providing forest products for both food and medicine. Traditional herbal medicine continues to play a role in many tribal areas, and numerous medicinal plants and their essential oils have shown remarkable biological activities. The Ministry of Environment and Forests, Govt. of India has identified and documented over 9,500 species of medicinal and aromatic plants that are significant for the pharmaceutical industries. Future of the essential oils seems to be bright, in terms of discovering new raw resources, and products however, there are many critical challenges related to this field that must be defeated in near future. Finally, this is very important to best explore, manage and conserve the available natural resources, at national and international level, in a sustainable way, which guarantee safe and healthy availability for the next generations. There is a need of proper conservation of economic plants for protection human of generation to come.

ACKNOWLEDGEMENT

Authors would like to thanks Director IIIM Jammu, Dr. Ram A Vishwakarma for facilities and encouragement. Also like to thanks instrumentation division for working on analysis of essential oils and labour force working in different forms for conservation and maintainance of aromatic and medicinal plants. This article represent institutional publication number IIIM/2232/2018

REFERENCES CITED

Al-Marzouqi AH, Rao MV, Jobe B. 2007. Comparative evaluation of SFE and steam distillation methods on the yield and composition of essential oil extracted from spearmint (Mentha spicata). *Journal of Liquid Chromatography & related Technologies* 30 (4): 463- 475.

Barthlott W, Mutke J, Rafiqpoor D, Kier G, Kreft H. 2005. Global centers of vascular plant diversity. *Nova Acta Leopoldina* 92: 61–83.

Burt S. 2004. Essential oils: their antibacterial properties and potential applications in foods—a review. *International Journal of Food Microbiology* 94(3): 223-253.

Cowan MM. 1995. Plant products as antimicrobial agents. *Clinical Microbiological Review* 12(4): 564-582.

Dutra RC, Compos MM, Santos AR, Calixto JB. 2016. Medicinal plants in Brazil: Pharmacological studies, drug discovery, challenges and perspective. *Pharmacological Research* 112: 4-29.

Hara H, Stearn WT, Williams HJ. 1978. An Enumeration of the Flowering Plants of Nepal. British Museum of Natural History, London, UK.

Joshi RK, Satyal P, Setzer WW. 2016. Himalayan aromatic medicinal plants: a review of their ethnopharmacology, volatice phytochemistry and biological activities. *Medicine* 3(1):6 doi: 10.3390/dedicines 3010006.

Jumaat SR, Tajuddin SN, Sudamoon R, Chaneerach A, Abdullah UH, Mohamed R. 2017. Chemical constituents and toxicity screening of three aromatic plant species from peninsular Malayasia. *Bioresources* 12 (3): 5878-5895.

Pourmortazavi SM, Hajimirsadeghi SS. 2007. Supercritical fluid extraction in plant essential and volatile oil analysis. *Journal of Chromatography* 1163(1): 2-24.

Reverchon E. 1997. Supercritical fluid extraction and fractionation of essential oils and related products. *The Journal of Supercritical Fluids* 10(1): 1-37.

Rios J, Recio M, Villar A. 1988. Screening methods for natural products with antimicrobial activity: a review of the literature. *Journal of Ethnopharmacology* 23(2): 127-149.

Samant SS, Dhar U, Palni LMS. 1998. Medicinal Plants of Indian Himalaya: Diversity Distribution Potential Values; G.B. Pant Institute of Himalayan Environment and Development: Almora, India.

Singh DK, Hajra PK. 1996. Floristic diversity. In: Changing Perspective of Biodiversity Status in the Himalaya; Gujral GS, Sharma V, Eds.: British Council Division, British High Commission Publication, Wildlife Youth Services: New Delhi, India.

Teixeira B, Maques A, Ramos C, Neing NR, Nogueira JM, Saraiva JA, Nunes MZ. 2013. Chemical composition and antibacterial and antioxidant properties of commercial essential oils. *Industrial Crops and Products* 43(1): 567-595.

Wink M. 2003. Evolution of secondary metabolites from an ecological and molecular phylogenetic perspective. *Phytochemistry* 64(1): 3-19.

Glossary of Terms

Abscess: Localised collection of pus caused by suppuration in a tissue.

Acne: An inflammatory disease occurring in or around the sebaceous glands.

Aflatoxin: Poisonous carcinogens that are produced by certain molds which grow in soil, decaying vegetation, hay and grains.

Agronomy: The science of crop production and soil management.

Allosteric Modulator: Drug that binds to a receptor at a site distinct from the active site. Induces a conformational change in the receptor, which alters the affinity of the receptor for the endogenous ligand. Positive allosteric modulators increase the affinity, whilst *negative allosteric modulators* decrease the affinity.

Amino acid: Amino acids are organic compounds containing amine and carboxyl functional groups, along with a side chain specific to each amino acid.

Amorphous: without a clearly defined shape or form.

Anaemia: Lack of enough blood in the body causing paleness.

Anaesthetic: Inducing loss of feeling or consciousness.

Analgesic: Relieving pain.

Analytic studies: Studies with control groups, namely case-control studies, cohort studies, and randomized clinical trials.

Annual: A type of flower or plant that lives for only one year.

Antagonist: Drug that attenuates the effect of an agonist. Can be *competitive* or *non-competitive*, each of which can be *reversible* or *irreversible*. A *competitive antagonist* binds to the same site as the agonist but does not activate it, thus blocks the agonist's action. A *non-competitive antagonist* binds to an allosteric (non-agonist) site on the receptor to prevent activation of the receptor. A *reversible antagonist* binds non-covalently to the receptor, therefore can be "washed out". An irreversible antagonist binds covalently to the receptor and cannot be displaced by either competing ligands or washing.

Antidiarrheal: Preventing or controlling diarrhea.

Antidote: An agent which neutralizes or opposes the action of a poison.

Apoptosis: Death of cells which occurs as a normal and controlled part of an organism's growth or development.

Ascocarp: An ascocarp is the fruiting body of an ascomycete phylum fungus. It consists of very tightly interwoven hyphae.

Autophagy: Autophagy (or autophagocytosis) is the natural, regulated mechanism of the cell that disassembles unnecessary or dysfunctional components.

Bioactivity: Biological activity describes the beneficial or adverse effects of a drug on living matter.

Biodiversity hotspot: Biogeographic region that is both a significant reservoir of biodiversity and is threatened with destruction. The term biodiversity hotspot specifically refers to 25 biologically rich areas around the world that have lost at least 70 percent of their original habitat.

Biodiversity: reflects the number, variety and variability of living organisms.

Biological resources: Those components of biodiversity of direct, indirect, or potential use to humanity.

Biome: A major portion of the living environment of a particular region characterized by its distinctive vegetation and maintained by local climatic conditions.

Bovine: cattle.

Calibration: the action or process of calibrating something.

Carbohydrate: Biomolecule consisting of carbon, hydrogen and oxygen atoms, usually with hydrogen-oxygen atom ratio of 2:1.

Chromatin: Mass of genetic material composed of DNA and proteins that condense to form chromosomes during eukaryotic cell division.

Chromatography: Laboratory technique for the separation of mixture.

Clinical pharmacology: The study of the effects of drugs in humans.

Coniferous: The conifers are a division of vascular land plants containing gymnosperms, cone-bearing seed plants.

Crystallization: Chemical solid–liquid separation technique, in which mass transfer of a solute from the liquid solution to a pure solid crystalline phase occurs.

Cytotoxicity: Quality of being toxic to cells. Examples of toxic agents are an immune cell or some types of venom.

Deciduous: Shedding the leaves annually, as certain trees and shrubs.

Deciduous: A type of tree that sheds its leaves before the colder months.

Decoction: Concentrated liquor resulting from heating or boiling a substance, especially a medicinal preparation made from a plant.

Distillation: A process of evaporation and recondensation used for purifying liquids.

Diuretic: Diuretics, also called water pills, are medications designed to increase the amount of water and salt expelled from the body as urine.

Dose-response relationship: A relationship in which a change in amount, intensity, or duration of exposure is associated with a change in risk of a specified outcome.

Drip irrigation: The use of pipes to bring water into contact with the roots of plants.

Drug product: A finished dosage form, for eg., a tablet, capsule or solution that contains a drug substance

Drug substance: An active ingredient that is intended to furnish pharmacological activity or other direct effect in diagnosis, cure, mitigation, treatment or prevention of diseases or to effect the structure or any function of the human body

Drug: An agent that is used therapeutically to treat diseases. It may also be defined as any chemical agent and/or biological product or natural product that affects living processes

EC_{50}: Molar concentration of an agonist that produces 50% of the maximum possible response for that agonist.

Ecology: the study of the environment and how living things interact with it.

Ecosystem: a community of living and non-living things that interact by exchanging matter and energy.

Environment: physical surroundings; all that is around you.

Enzymes: proteins that start a chemical reaction.

Ecosystem: A dynamic complex of plant, animal and micro-organism communities and their non-living environment interacting as a functional unit.

ED_{50}: *In vitro* or *in vivo* dose of drug that produces 50% of its maximum response or effect.

Efficacy: Describes the way that agonists vary in the response they produce when they occupy the same number of receptors. High efficacy agonists produce their maximal response while occupying a relatively low proportion of the total receptor population. Lower efficacy agonists do not activate receptors to the same degree and may not be able to produce the maximal response.

Endemic: Distribution restricted to a particular area: used to describe a species or organism that is confined to a particular geographical region, for example, an island or river basin.

Essential Oil: Volatile perfumery material derived from a single source of vegetable or animal origin by a process, such as hydrodistillation, steam distillation, dry distillation or expression.

Ethanol: Form of natural gas that can be produced from corn.

Experimental studies: Studies in which the investigator controls the therapy that is received by each participant, generally using that control to randomly allocate patients among study groups.

Extract: A concentrate of dried, less volatile aromatic plant part obtained by solvent extraction with a polar solvent.

Extraction: The process of isolating essential oil with the help of a volatile solvent.

Fatty acid: A carboxylic acid consisting of a hydrocarbon chain and a terminal carboxyl group, especially any of those occurring as esters in fats and oils.

Flavonoid: Class of plant and fungus secondary metabolites.

Flavour: Refers to that characteristic quality of a material as affects the taste or perception.

Gastronomist: A connoisseur of good food; a gourmet.

Glutamic acid: An á-amino acid that is used by almost all living beings in the biosynthesis of proteins. It is non-essential in humans, meaning the body can synthesize it.

Half-life: Half-life ($t\frac{1}{2}$) is an important pharmacokinetic measurement. The metabolic half-life of a drug *in vivo* is the time taken for its concentration in plasma to decline to half its original level. Half-life refers to the duration of action of a drug and depends upon how quickly the drug is eliminated from the plasma. The clearance and distribution of a drug from the plasma are therefore important parameters for the determination of its half-life.

Heart palpitations: Abnormally rapid and irregular beating of the heart

Heterothallism: The term is applied particularly to distinguish heterothallic fungi, which requires two compatible partners to produce sexual spores from homothallic ones.

Humus: The organic component of soil, formed by the decomposition of leaves and other plant material by soil microorganisms.

Hydro-Distillation: Distillation of a substance carried out in direct contact with boiling water.

IC$_{50}$: In a functional assay, the molar concentration of an agonist or antagonist which produces 50% of its maximum possible inhibition. In a radioligand binding assay, the molar concentration of competing ligand which reduces the specific binding of a radioligand by 50%.

In vitro: Taking place in a test-tube, culture dish or elsewhere outside a living organism.

In vivo: Taking place in a living organism.

Incidence rate: Measure of the frequency of the disease or outcome. The number of new cases which develop over a defined time period in a defined population at risk, divided by the number of people in that population at risk.

Infusion: A process of treating a substance with water or organic solvent, with or without heating.

Insecticide: A type of chemical used to kill insects.

Liana: Any of various long-stemmed, woody vines that are rooted in the soil at ground level and use trees, as well as other means of vertical support, to climb up to the canopy to get access to well-lit areas of the forest.

Microorganisms: Tiny living things that can only be seen with a microscope.

Migraine: A periodic condition with localised headaches, frequently associated with vomiting and sensory disturbances

Monitoring: The performance and analysis of routine measurements aimed at detecting changes in the environment or health status of populations.

Morel: Name given to genus Morchella, an edible fungi closely related to the anatomically simpler cup fungi. These distinctive fungi have a honeycomb appearance due to the network of ridges with pits composing their cap.

Mushroom: A mushroom, or toadstool, is the fleshy, spore-bearing fruiting body of a fungus, typically produced above ground on soil or on its food source.

Mycelia: The vegetative part of a fungus, consisting of a network of fine white filaments (Hyphae).

Mycologist: The person deals with the study of fungi, including their genetic and biochemical properties, their taxonomy and their use to humans as a source for medicine, food, as well as their dangers, such as toxicity or infections.

Nutraceutical: A food stuff that is held to provide health or medicinal benefits in addition to its basic nutritional value also called functional food.

Odour: hat property of a substance which stimulates and is perceived by the olfactory sense.

Organic acid: An organic compound with acidic properties.

Organic farming: Producing foods without the use of laboratory made fertilizers, growth subtances, or pesticides.

Organic matter: Dead plants, animals and manure converted by earthworms and bacteria into humus.

Osteoporotic: A disease where increased bone weakness increases the risk of a broken bone. It is the most common reason for a broken bone among the elderly.

Perfume: A suitably blended composition of various materials of synthetic and/ or natural origin to give a desired odour effect. It is carried in a suitable medium to the extent of not more than 20 percent.

pH: A scale of measurement by which the acidity or alkalinity of soil or water is rated. A pH of 6 to 7.5 is considered "ideal" for most agricultural crops. Each plant (specie-type), however, has its own "ideal" pH range.

Pharmacology: The study of the effects of drugs. The branch of biology concerned with the study of drug action, where a drug can be broadly defined as any man-made, natural, or endogenous (from within the body) molecule which exerts a biochemical or physiological effect on the cell, tissue, organ, or organism (sometimes the word pharmacon is used as a term to encompass these endogenous and exogenous bioactive species).

Phytochemistry: The study of phytochemicals, which are chemicals derived from plants.

Plant: Multicellular predominantly photosynthetic eukaryotes of the kingdom Plantae. Historically, plants were treated as one of two kingdoms including all living things that were not animals, and all algae and fungi were treated as plants.

Polymorphic: Occurrence of two or more clearly different morphs or forms, also referred to as alternative phenotypes, in the population of a species.

Polysaccharides: A carbohydrate (e.g. starch, cellulose, or glycogen) whose molecules consist of a number of sugar molecules bonded together.

Population: The number of living things that live together in the same place. In biology, a population is all the organisms of the same group or species, which live in a particular geographical area, and have the capability of interbreeding.

Potency: A measure of the concentrations of a drug at which it is effective.

Saprophyte: A plant, fungus, or microorganism that lives on dead or decaying organic matter.

Sclerotia: Compact mass of hardened fungal mycelium containing food reserves.

Screening: The presumptive identification of unrecognized disease or defect by the application of tests, examinations or other procedures which can be applied rapidly. Screening is an initial examination only and positive responders require a second diagnostic examination.

Side effect: Any unintended effect of a pharmaceutical product occurring at doses normally used in humans which is related to the pharmacological properties of the drug.

Signal: Reported information on a possible causal relationship between an adverse event and a drug, the relationship being unknown or incompletely documented previously. Usually more than a single report is required to generate a signal, depending upon the seriousness of the event and the quality of the information.

Solitary: Existing alone.

Species: A group of living organisms consisting of similar individuals capable of exchanging genes or interbreeding. The species is the principal natural taxonomic unit, ranking below a genus and denoted by a Latin binomial, e.g. *Homo sapiens*.

Specificity: The ability of a method, system or tool to correctly classify the proportion of persons who truly do not have a characteristic, as not having it.

Spectroscopy: The study of the interaction between matter and electromagnetic radiation.

Steam Distillation: Distillation of a substance by bubbling steam through it.

Stimulant: Making a body organ active.

Succulent: A plant (especially a xerophyte) having thick fleshy leaves or stems adapted to storing water.

Surveillance: Ongoing scrutiny, generally using methods distinguished by their practicability, uniformity, and rapidity, rather than by complete accuracy. Its main purpose is to detect changes in trends or distribution in order to initiate investigative or control measures.

Tannin: A yellowish or brown bitter tasting organic substance present in some galls , barks and other plant tissues.

Taxon (plural taxa): In biology, a taxon is a group of one or more populations of an organism or organisms seen by taxonomists to form a unit. Although neither is required, a taxon is usually known by a particular name and given a particular ranking, especially if and when it is accepted or becomes established.

Technology: Instruments, tools or inventions developed through research to increase efficiency.

Tincture: A cold alcoholic extract of natural fragrant material of vegetable or animal origin, the solvent being left in the extract as a diluent.

Topographic: Relating to the arrangement of the physical features of an area.

Traditional knowledge: The term traditional knowledge generally refers to knowledge systems embedded in the cultural traditions of regional, indigenous, or local community.

Urbanization: the growth of the city into rural areas.

Validation: Establishing documented evidence which provides a high degree of assurance that a specific process will consistently produce a product meeting its pre-determinant specifications and quality attributes.

Volatile: A material is said to be volatile when it has the property of evaporating at room temperature when exposed to atmosphere.

Yield: the amount of a crop produced in a given time or from a given place.

spectacle: The ability of a method, when used to... correctly classify true properties of persons who truly do not have a characteristic... not having it.

spectroscopy: The study of the interaction between matter and electromagnetic radiation.

Thaumaturgy: The control of a substance by enabling team thought...

Index to Scientific Names

A

Abies pindrow 363
Acanthus ilicifolius 195
Achillea millefolium 363
Achromobacte 204, 209
Achromobacter piechaudii 209
Aconitum 5
Acorus 363, 370, 373
Acorus calamus 363, 373
Aegle marmelos 363, 373
Ageratum conyzoides 364
Agrobacterium 357
Ajuga parviflora 364
Amomum subulatum 364
Amoora rohituka 98, 106
Anacardiaceae 333
Angelica 370
Angelica glauca 5, 90
Anisomeles indica 364
Apiaceae 89, 363, 364, 365, 368
Arabidopsis 236, 359
Araceae 363, 373
Aralia cachemirica 364
Araliaceae 363, 364
Aristolochiaceae 363, 364
Aristolochia indica 363, 364
Arnebia benthamii 5
Artemisia 167, 168, 169, 170, 172, 176
Artemisia annua 167, 168, 169, 171, 173, 176, 177, 373
Artemisia capillaries 170
Artemisia cashemirica 170
Artemisia desertorum 170
Artemisia dracunculus 168, 170, 171, 364, 373
Artemisia dubia 170, 364
Artemisia gmelinii 364
Artemisia indica 170, 364

Artemisia japonica 170, 171, 364
Artemisia laciniata 170
Artemisia macrocephala 170
Artemisia maratima 170, 171, 364, 373
Artemisia moorcroftiana 170
Artemisia myriantha var. *pleocephala* 170
Artemisia nilagarica 170, 171, 364
Artemisia parviflora 171, 364
Artemisia roxburghiana 171, 364
Artemisia scoparia 170, 171, 364
Artemisia sieversiana 171
Artemisia vestita 373
Artemisia vulgaris 168, 171, 364, 373
Aspergillus niger 138
Asteraceae 363, 364, 365, 366, 368, 369, 373, 375
Azadirachta indica 40, 332
Azospirillum 204, 205
Azospirillum brasilense 203
Azotobacter 205

B

Bacillus 204, 206
Bacillus amylolequifacians 202
Bacillus cereus 334, 340, 341
Bacillus edaphicus 207
Bacillus firmus 202
Bacillus licheniformis 202
Bacillus megaterium 202, 203
Bacillus mucilaginosus 207
Bacillus polymyxa 203
Bacillus pumilis 203
Bacillus subtilis 203
Betula utilis 5
Bipolaris sorokiniana 245
Blumea lacera 364
Boenninghausenia albiflora 364